Environmental Policy
and Public Health

Environmental Policy and Public Health

Principal Health Hazards and Mitigation, Volume 1

Third Edition

Barry L. Johnson, MSc, PhD, FCR

RADM (Ret.), U.S. Public Health Service
Adjunct Professor
Rollins School of Public Health
Emory University

Maureen Y. Lichtveld, MD, MPH

Dean, Graduate School of Public Health
Professor, Environmental and Occupational Health
Jonas Salk Chair in Population Health
University of Pittsburgh

CRC Press
Taylor & Francis Group
Boca Raton London New York

CRC Press is an imprint of the
Taylor & Francis Group, an **informa** business

Third edition published 2022
by CRC Press
6000 Broken Sound Parkway NW, Suite 300, Boca Raton, FL 33487-2742

and by CRC Press
2 Park Square, Milton Park, Abingdon, Oxon, OX14 4RN

First edition published by CRC Press 2006

CRC Press is an imprint of Taylor & Francis Group, LLC

ISBN: 978-1-032-12174-1 (hbk)
ISBN: 978-1-032-18194-3 (pbk)
ISBN: 978-1-003-25335-8 (ebk)

DOI: 10.1201/9781003253358

Typeset in Times
by codeMantra

Contents

Preface

Environmental Policy and Public Health is a two-volume textbook addressing key threats in the environment that impact public health. Volume 1, *Principal Health Hazards and Mitigation*, is complemented by Volume 2, *Emerging Health Hazards and Mitigation*. Principal health hazards are those with the greatest background of published data on the specifics of their adverse health effects. Mitigations discussed in this volume are twofold: policies that are intended for control of specific hazards and suggested hazard interventions.

The health of the environment is inextricably linked to that of people. Regarding principal hazards, fundamental to environmental health is the quality of water—surface, ground, and drinking water. Likewise, air pollution, out- and indoor air, has been linked to many adverse health conditions impacting respiratory and cardiovascular health. Respiratory diseases, such as asthma and chronic obstructive pulmonary disease, are exacerbated by smoking tobacco and other products. Food safety and security represent global health threats. The quality of our air, water, and food is threatened by the presence of toxic substances. Countering climate change is a global health priority already impacting our health and environment. The role of policy in addressing each of these key environmental health areas is extensively discussed in this volume of *Environmental Policy and Public Health* and is accompanied by practice questions to facilitate interactive classroom and group learning.

We have also endeavored in these two volumes to objectively describe the major environmental health policy redirections that have occurred in the U.S. as a consequence of sociopolitical changes in the U.S. federal government's environmental policies. In particular, we record here the major differences in environmental policies associated with the Obama, Trump, and Biden presidential administrations. And because we consider climate change as the most threatening global environmental hazard to human and ecological health, we have labored to include climate change impacts wherever possible throughout several chapters. Additionally, the year 2020 was subject to two significant social and public health challenges that merited special attention and description: social justice and the COVID-19 pandemic. Both subjects are described in those chapters' whose contents were influenced by these two societal impacts.

Authors

Barry L. Johnson, MSc, PhD, FCR, is adjunct professor at the Rollins School of Public Health, Emory University. Dr. Johnson conducted environmental research at the USPHS, USEPA, and CDC. An elected Fellow, Collegium Ramazani, he retired from the USPHS with the rank of Rear Admiral. Dr. Johnson is author of *Legacy of Hope* (2016), *Environmental Policy and Public Health* (2007, 2018), *Impact of Hazardous Waste on Human Health* (1999), and senior editor of *Hazardous Waste: Impacts on Human and Ecological Health* (1997); *Hazardous Waste and Public Health* (1996); *Advances in Neurobehavioral Toxicology* (1990); and *Prevention of Neurotoxic Illness in Working Populations* (1987).

Maureen Y. Lichtveld MD, MPH, is dean, professor, Environmental and Occupational Health and Jonas Salk Chair in Population Health, University of Pittsburgh, Graduate School of Public Health. Previous roles: professor, Freeport McMoRan Chair of Environmental Policy, Tulane University's School of Public Health and Tropical Medicine. She is a member of the National Academy of Medicine with decades of environmental public health experience. Research: environmentally induced disease, health disparities, environmental health policy, disaster preparedness, climate and health, and community resilience. Dr. Lichtveld is a board member of the Consortium of Universities for Global Health. Honors: Johns Hopkins' Society of Scholars; CDC's Environmental Health Scientist of the Year.

1 Climate Change Crisis

1.1 INTRODUCTION

The bodies of climate science together with observable changes already occurring in planet Earth signal the greatest current threat to much of life on our globe. The threat is climate change. For these reasons, the authors of this chapter chose to affix the word *crisis* to the chapter's title. Many scientists and policymakers consider climate change as the current single most consequential threat to planet Earth and its living inhabitants, one of which is illustrated in Figure 1.1, struggling to survive as polar ice melts due to temperature increase. While this strong assertion might appear to some as hyperbole, the science of climate change supports this assertion, as do signs of changes already evident in the planet's atmosphere, oceans, migratory patterns of pests, and food security. For instance, the melting of polar ice and shrinkage of glaciers, along with increased numbers and severity of episodes of extreme weather, are evidence of a change in our climate. As a reminder, climate is defined as the weather conditions prevailing in an area in general or over a long period.

While some naysayers might argue that these kinds of changes are a normal cycling of Earth's climate, the science of climate change disagrees. As an observation, the history of environmental health is replete with examples of denial of an existing body of science for personal or commercial reasons. These examples of denied health science include tobacco smoking, lead additives in gasoline, and workplace exposure to asbestos. Moreover, taking

> Climate is defined as the weather conditions prevailing in an area in general or over a long period.

FIGURE 1.1 Melting of polar ice due to climate change. (U.S. House of Representatives, Select Committee on Energy Independence and Global Warming, 2016.)

DOI: 10.1201/9781003253358-1

into consideration the body of climate change science and the observed signs of change, a precautionary stance in policymaking seems prudent. Putting it in different words, can humankind afford to be wrong about climate change? Can we afford to ignore climate change and its consequences to future generations of living creatures, including humans?

1.2 OVERVIEW OF CLIMATE CHANGE

Before proceeding, as a matter of environmental history and subsequent policymaking, it is important to acknowledge the work and leadership of one individual for making climate change a concern of policymakers and the general public alike. Dr. James Edward Hansen is a native of Iowa, trained in physics and astronomy in the space science program of James Van Allen at the University of Iowa, where he was awarded a PhD in Physics in 1967 (Figure 1.2). He later developed and refined models to understand the Earth's atmosphere, and in particular, the effects that aerosols and trace gases have on the Earth's climate [1]. His development and use of global climate models contributed to the further understanding of the Earth's climate. From 1981 to 2013, he was the director of the NASA Goddard Institute for Space Studies in New York City, a part of the Goddard Space Flight Center. As of 2014, Dr. Hansen directs the Program on Climate Science, Awareness and Solutions at Columbia University's Earth Institute.

Dr. Hansen testified as a NASA scientist before the U.S. Senate Committee on Energy and Natural Resources on June 23, 1988. He testified that "Global warming has reached a level such that we can ascribe with a high degree of confidence a cause-and-effect relationship between the greenhouse effect and observed warming…It is already happening now" and "The greenhouse effect has been detected and it is

FIGURE 1.2   Dr. James Hansen, NASA climate scientist. (NASA, 2020.)

changing our climate now...We already reached the point where the greenhouse effect is important." Dr. Hansen said that NASA was 99% confident that the warming was caused by the accumulation of greenhouse gases in the atmosphere and not a random fluctuation. Dr. Hansen reported there was a clear cause-and-effect relationship with the greenhouse effect and lastly that due to global warming, the likelihood of freak weather was steadily increasing. In a policymaking sense, Dr. Hansen was a herald of the consequences of climate changes.

Climate change will affect every living creature on the planet if policies to mitigate its causes are unsuccessful. Some societies will be affected more than others, but all will in some way be affected through such impacts as reduced food supply, less healthy seas, increased incidence of environmental-related diseases, increased number and duration of droughts, more powerful hurricanes and tornados, and disturbances in ecological systems. As but one example of the latter, warmer areas of the globe will attract unwelcome carriers of disease, e.g., mosquitoes and ticks.

> Given the global nature and impact of climate change, any effort to reverse climate change must necessarily be a global endeavor.

For instance, cases of Lyme disease, a potentially debilitating condition primarily transmitted by black-legged ticks, have doubled over the past two decades to about 30,000 cases a year in the U.S. These ticks have spread into the upper reaches of New England and the mid-west, while other tick species normally found in warmer southern states, such as the long-horned tick and lone star tick, are now appearing in New York and New Jersey [2].

In addition to the adverse effects on human and ecosystem health, the global economy will be severely injured. One study estimates that global warming will cost the world economy more than **£1.5 trillion** a year (emphasis added) in lost productivity by 2030 as it becomes too hot to work in many jobs [3]. In the U.S., an official with the Federal Reserve Bank estimated in 2019 that climate- and weather-related events have directly cost the U.S. more than $500 billion over the preceding 5 years [4].

As background, technology in its various forms and climate are inextricably connected, where technology is used here in the context of "machinery and equipment developed from the application of scientific knowledge." In a historical sense, machinery and equipment came forth from the application of intuition and human experience, rather than the application of principles of science. For example, the invention of the plow for purpose of land cultivation likely came in the absence of any application of science. That having been said, in addition to technology, many forces can shape local, regional, and global patterns of climate. For example, forces of nature such as volcanic eruptions, bursts of radiant energy from the Sun, and the impact of orbiting celestial materials can affect changes in climate on local to global scales. These impacts of non-anthropogenic causes of climate change have—and will continue to be—part of Earth's climate-shaping factors. Notwithstanding the importance of nature's climate shapers, human activity has grown increasingly crucial for shaping climate change, with the impact of technology at the forefront of impact shapers.

One could assert that from our very appearance as a separate species, we have lived in ways that have affected the climate. For example, the use of fire for food preparation and protection from feral animals resulted in the release into ambient air of products of incomplete combustion, carbon monoxide, and carbonaceous particles. The impacts on the climate were insignificant due to the small amounts of pollutants released into ambient air and waste into soils and water. Changes in climate were insignificant and sustainable.

This benign relationship between human activity and climate began to change with the appearance of the Industrial Revolution. As described by one source,

"The Industrial Revolution, which took place from the 18th to 19th centuries, was a period during which predominantly agrarian, rural societies in Europe and America became industrial and urban. Prior to the Industrial Revolution, which began in Britain in the late 1700s, manufacturing was often done in people's homes, using hand tools or basic machines. Industrialization marked a shift to powered, special-purpose machinery, factories and mass production. The iron and textile industries, along with the development of the steam engine, played central roles in the Industrial Revolution, which also saw improved systems of transportation, communication and banking. While industrialization brought about an increased volume and variety of manufactured goods and an improved standard of living for some, it also resulted in often grim employment and living conditions for the poor and working classes" [5].

The engines of the industrial revolution were energy, raw materials, transportation, and labor. These engines produced goods, services, and waste. Products and goods became matters of commerce, with corresponding increases in personal, regional, and national wealth. Jobs were created across the span of the Industrial Revolution. Steam-powered machinery joined and subsequently supplanted the energy exerted by workers' heavy labor. Roads and waterways were constructed for use in transporting goods and products manufactured in factories. Trade became global as goods were shipped and exchanged across continents. The Agrarian Age came to a close in what became known as "industrialized countries," as was the case in Europe and much of North America. But with the successes of the Industrial Revolution came a silent threat to the public's health. The threat was waste released into the environment as air pollutants, water contaminants, and food impurities.

Coal was the dominant fuel for energy production during and following the industrialization of national economies. Coal was abundant in Europe, North America, China, and elsewhere. Although labor-intensive, coal was relatively easy to mine. Combustion of coal could heat water and thereby produce steam and steam engines delivered power to new forms of machinery that in turn manufactured goods and products for commerce. But burning coal was and

> Coal was the dominant fuel for energy production during and following the industrialization of national economies. Coal was abundant in Europe, North America, China, and elsewhere.

remains a relatively inefficient process of combustion, resulting in tall chimneys built for release into the atmosphere of incomplete products of coal combustion. Little or no thought was given to the environmental consequences of fouling the air with soot, carbon monoxide, and hydrocarbons released into the air that was essential for human life. These consequences became global in occurrence and impact.

1.3 SIGNS AND SCIENCE OF CLIMATE CHANGE

As preface, Earth's climate is always changing. In the past, Earth's climate has gone through warmer and cooler periods, each lasting thousands of years. Observations show that Earth's climate has been warming. Its average temperature has risen a little more than one degree Fahrenheit during the past 100 years or so. This amount may not seem like much. But small changes in Earth's average temperature can lead to big impacts. Some causes of climate change are natural. These include changes in Earth's orbit and in the amount of energy coming from the Sun. Ocean changes and volcanic eruptions are also natural causes of climate change [6].

Many scientists consider that most of the global warming occurring since the mid-1900s is due to the combustion of fossil fuels (coal, oil, and gas) [6]. Combusting these carbon-based fuels produces much of the energy used daily worldwide. Energy is produced, but so are products of incomplete combustion. The most consequential releases are gases that add to those already present in the atmosphere. Heat-trapping gases, such as CO_2, are emitted into the air. These gases are called greenhouse gases, as discussed in the following section.

1.3.1 SOURCES AND LEVELS OF ATMOSPHERIC GREENHOUSE GASES

As background, greenhouse gases in Earth's atmosphere absorb infrared radiation (IR) from the Sun and release it. Some of the Sun's released heat reaches Earth, along with heat from the Sun that has penetrated the atmosphere. Both the solar heat and the radiated heat are absorbed by the planet and released; some are reabsorbed by greenhouse gases in order to perpetuate the cycle. The more of these gases that exist, the more heat is prevented from escaping into outer space and, consequently, the more the Earth heats. This increase in heat is called the greenhouse effect; the increase in heat is measured as increased air temperature.

> A greenhouse gas is a gas that absorbs infrared radiation (IR) and radiates heat in all directions.

The greenhouse gases are relatively inefficient in heat transference properties, thereby contributing to heat buildup on Earth, in effect forming an umbrella over the planet. Greenhouse gases contribute to the greenhouse effect, which is the warming of Earth's land, water, and air because of a lessened ability of the Earth to radiate energy upward through its atmosphere. According to scientific consensus, global warming of 1°C–3.5°C will occur over the next 100 years unless greenhouse gas concentrations are decreased.

1.3.1.1 Sources of Greenhouse Gases

Some greenhouse gases occur naturally in the atmosphere, while others result from anthropogenic activities. Naturally occurring greenhouse gases include water vapor, carbon dioxide, methane, nitrous oxide, and ozone. Anthropogenic activities, however, have increased the atmospheric concentrations of some greenhouse gases. For example, carbon dioxide is formed when solid waste, fossil fuels, wood, and wood products are burned. Methane is emitted from the production and transport of coal, natural gas, and petroleum; the decomposition of organic waste in landfills; and bovine flatulence.

EPA data indicates that CO_2 represents about 82% of the greenhouse gases, with methane (CH_4) and nitrous oxide (N_2O) representing 10% and 6%, respectively. These percentages are important for policymaking purposes, indicating which greenhouse gases should receive primary attention in efforts to curb climate change. The EPA lists the anthropogenic sources of greenhouse gas emission as follows [7]:

- **Carbon dioxide:** CO_2 enters the atmosphere through burning fossil fuels (coal, natural gas, and oil), solid waste, trees, and other biological materials, and also as a result of certain chemical reactions (e.g., manufacture of cement). Carbon dioxide is removed from the atmosphere (or "sequestered") when it is absorbed by plants as part of the biological carbon cycle.
- **Methane (CH_4):** CH_4 is emitted during the production and transport of coal, natural gas, and oil. Methane also results from emissions from cattle, rice paddies, leaks from oil and gas wells, and the decay of organic waste in municipal solid waste landfills. Trees are also now known to be a source of methane emissions. Remarkably, it was reported in 2006 that terrestrial plants under aerobic conditions produce significant amounts of methane, estimated to be 10%–30% of annual levels of atmospheric methane. In 2019 a surprising source of methane generation was found by researchers who showed that trees, especially in tropical wetlands, are a major source of methane [8]. It was found that trees, especially in the extensive flooded forests, were stimulating methane production in the waterlogged soils and mainlining it into the atmosphere. It now seems that most of the world's estimated 3 trillion trees emit methane at least some of the time. The researchers caution that their findings should not imply that trees are therefore bad for climate and should be cut down. Indeed, in most cases, their carbon storage capability easily outweighs their methane emissions.
- **Nitrous oxide (N_2O):** Nitrous oxide is emitted during agricultural and industrial activities, combustion of fossil fuels and solid waste, as well as during treatment of wastewater. About 40% of N_2O comes from human activities including soil degradation, fertilizer use, and industrial operations. A 2020 study points to an alarming trend affecting climate change: N_2O has risen 20% from preindustrial levels—from 270 ppb in 1750 to 331 ppb in 2018—with the fastest growth observed in the last 50 years due to emissions from human activities. The growing use of nitrogen fertilizers in the production of food worldwide is increasing atmospheric concentrations

of N_2O, a greenhouse gas 300 times more potent than CO_2 that remains in the atmosphere for more than 100 years [9].

- **Fluorinated gases:** Hydrofluorocarbons, perfluorocarbons, sulfur hexafluoride, and nitrogen trifluoride are synthetic, powerful greenhouse gases that are emitted from a variety of industrial processes. Fluorinated gases are sometimes used as substitutes for stratospheric ozone-depleting substances (e.g., chlorofluorocarbons, hydrochlorofluorocarbons, and halons). These gases are typically emitted in smaller quantities, but because they are potent greenhouse gases, they are sometimes referred to as High Global Warming Potential gases ("High GWP gases").

1.3.1.2 Carbon Dioxide Principles

On a global scale, the principal greenhouse gas released into Earth's atmosphere is carbon dioxide (CO_2) [7]. Fossil fuel use is the primary source of CO_2. But the way in which people use land is also an important source of CO_2, especially when it involves deforestation, since trees absorb CO_2.

1.3.1.3 Historic Levels of Greenhouse Gases

The increase in CO_2 levels in the atmosphere is illustrated by NASA data shown in Figure 1.3. The historical data shown in the figure were derived from carbon measurements of fossils and polar ice. The data in Figure 1.3 reveal a clear and dramatic increase in CO_2 levels that began in the mid-20th century and show a dramatic continuous upward trend.

> On a global scale, the principal greenhouse gas released into Earth's atmosphere is carbon dioxide (CO_2). Fossil fuel use is the primary source of CO_2.

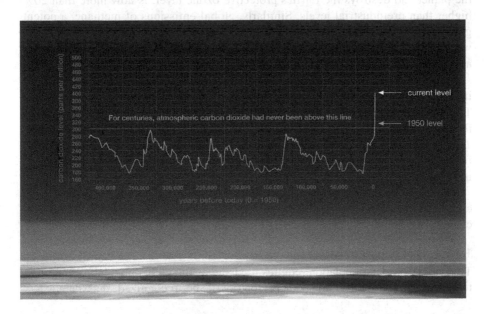

FIGURE 1.3 Historical global levels of CO_2 in Earth's atmosphere. (NASA [6].)

According to data from the National Oceanic and Atmospheric Administration (NOAA), the global average atmospheric CO_2 in 2017 was 405.0 parts per million (ppm), with a range of uncertainty of plus or minus 0.1 ppm [10]. Carbon dioxide levels today are higher than at any point in at least the past 800,000 years. The last time there was this much CO_2 in the Earth's atmosphere, modern humans didn't exist. The world's seas were up to 100 feet higher than today, and the global average surface temperature was up to 11°F warmer than presently. And there were trees growing at the South Pole [11]! Some progress in reducing CO_2 emissions is occurring, although not enough to meet global targets. For instance, global energy-related CO_2 emissions flattened in 2019 at around 33 gigatonnes (Gt), following 2 years of increases [12]. This resulted mainly from a sharp decline in CO_2 emissions from the power sector in advanced economies, due to the expanding role of renewable sources (mainly wind and solar PV), fuel switching from coal to natural gas, and higher nuclear power output.

> It is important to understand that the levels of greenhouse gases are at historic highest levels.

It is important to understand that the levels of these greenhouse gases are at historic highest levels. A study released in 2018 by the World Meteorological Organization discussed the levels of three primary greenhouse gases [13]. The investigators commented, "Carbon dioxide, methane and nitrous oxide are now far greater than preindustrial levels, with no sign of a reversal of the upward trend." Levels of CO_2 rose to a global average of 405.5 parts per million in the atmosphere in 2017—almost 50% higher than before the industrial revolution. Levels of methane, a potent greenhouse gas responsible for about 17% of global warming are now 2.5 times higher than in preindustrial times. Nitrous oxide, which also warms the planet and destroys the Earth's protective ozone layer, is now more than 20% higher than preindustrial levels. Similarly, global emissions of methane, a potent greenhouse gas, soared to a record high in 2017, the most recent year for which worldwide data are available, attributing the rise was driven by fossil fuel leaks and agriculture [14].

Of policy note, the Trump administration's Bureau of Land Management (BLM) in 2018 rolled back parts of the prior rule that limited the release of methane gas. The change was expected to allow for more methane leaks in a process called flaring and add to air pollution. On July 16, 2020, a federal court struck down the Trump administration's rule, stating the rule had failed to consider scientific findings and institutions relied upon by both prior Republican and Democratic administrations [15].

Having identified and discussed the role of greenhouse gases in contributing to climate change, it is useful at this point to discuss the signs and symptoms of climate change. In other words, what specifically is changing in the global environment?

1.3.1.4 Food System Issues

The world's human population requires food and as the global population continues to increase from 7.3 billion, more food will be recognized and demanded.

But lacking recognition is the relationship between food systems and climate change. That is because the global food system – everything from clearing land

> The global food system accounts for 30% of the world's GHG emissions.

and felling forests for cattle ranches to the arrival of meat and two vegetables on a suburban family dinner plate – accounts for 30% of the world's GHG emissions.

And to contain global heating later this century to no more than 1.5°C above the levels that existed before the Industrial Revolution, in addition to forgoing fossil fuels, revisions in the food system will be required [16]. Researchers found that the GHG emissions from food production alone would by 2050 take the world to the 1.5°C target, and to 2°C by the end of the century. Between 2010 and 2017, the global food system accounted for an average of 16 billion tons of CO_2 equivalent in emissions annually. In order to mitigate the contribution of food systems to emissions of GHGs, researchers recommend increased crop yields per hectare, reduced food waste, improved farm efficiency, and switching to **healthful calorie** supplies based increasingly on plant crops (emphasis added).

NASA: CURRENT SIGNS OF CLIMATE CHANGE

- Sea-level rise
- Global temperature rise
- Warming oceans
- Ocean acidification
- Shrinking ice sheets
- Declining Arctic sea ice
- Glacial retreat
- Extreme weather events
- Decreased snow cover

1.3.2 SIGNS AND SYMPTOMS OF CLIMATE CHANGE

There might be a year or years where Earth's average temperature is steady or even decreases, but the overall trend is expected to be temperature increase as a consequence of climate change. Earth's average temperature is expected to rise even if the amount of greenhouse gases in the atmosphere decreases. But such a rise would be less than if greenhouse gas amounts remain the same or increase. Some impacts of global warming are already evident, as enumerated by NASA, which provides the following pattern of signs of global climate change [6].

- **Sea-level rise:** Global sea level rose about 17 cm (6.7 inches) in the 20th century. The rate in the last decade, however, is nearly double that of the last century. The global average sea level has risen by 0.06 m (0.2 feet) since 2000. A rise of up to 0.8 m (2.8 feet) is expected by 2100 under moderate greenhouse emissions, 1.6 m (5.4 feet) by 2150, and 4.3 m (14 feet) by 2300, although one study speculates a rise of 50 feet by 2300 under a worst-case climate warming scenario [17]. As a consequence of sea-level rise, U.S. coastal communities should expect much more frequent flooding in coming decades as sea-levels rise, according to an NOAA report. Many places that are dry now could flood every day by the end of the century [17].

- **Global temperature rise:** Global surface temperature reconstructions show that the Earth has warmed since 1880. Earth's average temperature has risen by 1.5°F over the past century and is projected to rise another 0.5°F–8.6°F over the next 100 years. Most of this warming has occurred since the 1970s, with the 20 warmest years having occurred since 1981 and with all 10 of the warmest years occurring in the past 12 years. Earth's global average surface temperature in 2020 tied with 2016 as the warmest year on record, according to an analysis by NASA. Continuing the planet's long-term warming trend, 2020s globally averaged temperature was 1.84°F (1.02°C) warmer than the baseline 1951–1980 mean. A significant question facing climate scientists is how much would planet Earth warm should atmospheric CO_2 levels double those of preindustrial levels? In 2020, a team of researchers narrowed the range of increased temperatures to between 2.6°C and 3.9°C, a significant amount [18].
- **Warming oceans:** The oceans have absorbed much of this increased heat, with the top 700 m (about 2,300 feet) of ocean showing warming of 0.302°F since 1969.
- **Ocean acidification:** Since the beginning of the Industrial Revolution, the acidity of surface ocean waters has increased by about 30%. This increase is the result of humans emitting more carbon dioxide into the atmosphere and hence more being absorbed into the oceans. The amount of carbon dioxide absorbed by the upper layer of the oceans is increasing by about 2 billion tons per year.

> Earth's average temperature has risen by 1.5°F over the past century and is projected to rise another 0.5°F–8.6°F over the next 100 years.

- **Shrinking ice sheets:** The Greenland and Antarctic ice sheets have decreased in mass. Data show Greenland lost 150–250 cubic kilometers (36–60 cubic miles) of ice per year between 2002 and 2006, while Antarctica lost about 152 cubic kilometers (36 cubic miles) of ice between 2002 and 2005. The loss of Greenland's ice sheet reached a record loss in 2019, with an ice loss exceeding 530 billion metric tons. Melting of ice is occurring at both polar extremes of the planet Earth. A study found that Antarctica has lost about 3 trillion metric tons of ice since 1992 [19]. Melting ice contributes to rise in ocean levels.
- **Declining Arctic sea ice:** Both the extent and thickness of Arctic sea ice have declined rapidly over the last several decades. UK scientists in 2020 released the findings of a study of the amount of ice the Earth has lost [20]. Utilizing satellite surveys of the planet's poles, mountains, and glaciers, investigators estimated a total of 28 trillion tons of ice have disappeared from the surface of the Earth since 1994, attributing the loss to warmer climate temperatures. They further suggested that melting ice will raise sea levels by a meter by the end of the 21st century.

- **Glacial retreat:** Glaciers are retreating almost everywhere around the world—including in the Alps, Himalayas, Andes, Rockies, Alaska, and Africa. For example, glaciers in Alaska's Denali National Park are melting faster than at any time in the past 400 years because of rising summer temperatures [21]. A glacier in Patagonia may be the fastest-thinning glacier on the planet. The glacier, known as Hielo Patagónico Sur 12 (HPS-12), is perched in the Andes mountains in Chile. It has lost half its length in 30 years [22].

  > UN: From 2000 to 2019, there were 7,348 major recorded disasters claiming 1.23 million lives.

- **Extreme events:** The rising global temperatures have been accompanied by changes in weather and climate. Many places have seen changes in rainfall, resulting in more floods, droughts, or intense rain, as well as more frequent and severe heat waves. The number of record high-temperature events in the U.S. has been increasing, while the number of record low-temperature events has been decreasing, since 1950. The U.S. has also witnessed increasing numbers of intense rainfall events. Hurricanes are certainly an extreme event, given the widespread destruction of property, social cohesion, and adverse effects on human and ecological health. In October 2020, the UN Office for Disaster Risk Reduction released a report stating that climate change was a factor in most of the natural disasters over the last 20 years [23]. The UN analysis found there were 7,348 major recorded disasters from 2000 to 2019, claiming 1.23 million lives, affecting 4.2 billion people, and causing approximately $2.97 trillion in global economic losses. In a separate study, Japanese researchers showed that hurricanes that develop over warmer oceans carry more moisture and therefore stay stronger for longer after striking land. This means that as the world continues to warm, hurricanes are more likely to reach communities farther inland and be more destructive [24].

- **Decreased snow cover:** Satellite observations reveal that the amount of spring snow cover in the Northern Hemisphere has decreased over the past five decades and that the snow is melting earlier.

- **Forest deaths:** Researchers warn that climate change is accelerating the death of trees, stunting their growth, and making forests across the world younger and shorter [25].

- **Wildfires:** Wildfires occur more frequently and with greater severity in drier, hotter climates affected by global warming. Noteworthy were wildfires in 2020 that engulfed large areas of California, Oregon, and Washington. In California alone, more than two million acres have burned in 2020, fueled by high temperatures and high winds. Wildfires cause air pollution that in turn can result in adverse health effects, as discussed in Chapter 2.

These observed changes in Earth's climate are caused by the buildup of greenhouse gases (GHGs).

Perspective: The nature and sources of GHGs are important for environmental health policymaking. In particular, policies to reduce the emission of GHGs can be focused on eliminating or tempering their sources and dealing with the economic and social effects of source reductions, for example, replacing fossil fuels with renewable sources of energy. Moreover, climate change is a global condition and as such requires global policies of mitigation, as illustrated in the next section.

1.3.3 INTERNATIONAL PANEL ON CLIMATE CHANGE

On March 13, 2018, the Intergovernmental Panel on Climate Change (IPCC) celebrated its 30th anniversary. Established by the World Meteorological Society and the United Nations Environment Programme (UNEP) in 1988, the IPCC's mandate is "to provide the world with a clear scientific view on the current state of knowledge in climate change and its potential environmental and socio-economic impacts" [26]. Thousands of experts from around the world synthesize the latest scientific findings on the impacts of and the potential responses to climate change with the IPCC's comprehensive Assessment Reports released every 5–7 years. The importance of the IPCC and its work cannot be overstated. It represents humankind's best hope of surviving the dire consequences of long-term climate change.

1.4 IMPACT OF CLIMATE CHANGE ON HUMAN HEALTH

Climate change has been cast here as the current single greatest environmental hazard to the planet Earth and its inhabitants. This hazard has drawn considerable attention and action by national and international governments and private-sector organizations. Left unabated, climate change has been forecast as exerting a terrible toll on human life, especially a toll on children. Some naysayers of climate change might assert that the foregoing words are mere hyperbole and exaggerate the facts. But that assertion is disarmed by the collection of global environmental science and current ominous signs of adverse signs of the consequences of a hotter atmosphere, more acidic oceans, migration of pests into warmer, previously uninhabited geographic regions, and increased numbers and severity of droughts, storms, and other weather events.

The consequences of these projected changes in global and regional environments will cause serious impacts on public health, national economies, and environmental quality. Public health impacts have been categorized by the IPCC as follows [27]:

- Increased frequency of heat waves. Heat waves bring heat-related mortalities and morbidities, together with loss of economic productivity. A study by Rutgers University investigators estimated heat stress from extreme heat and humidity will annually affect areas now home to 1.2 billion people by 2100, assuming current greenhouse gas emissions [28].
- Regional variations in precipitation patterns would cause problems with freshwater supplies, producing an increase in waterborne diseases as human populations use impure water supplies in lieu of freshwater.

- Food production would be compromised in regions with lesser precipitation and increased ambient temperature. Research has shown that many food crops become less nutritious when grown under the CO_2 levels expected by 2050, with reductions of protein, iron, and zinc estimated at 3%–17%. Such changes could mean that by the middle of the century about 175 million more people develop a zinc deficiency, while 122 million people who are not currently protein deficient could become so. In addition, about 1.4 billion women of childbearing age and infants less than 5 years old will be living in regions where there will be the highest risk of iron deficiency [29].
- Coastal flooding due to rising ocean levels would place human populations at risk.
- Vector-borne diseases would increase as vectors such as mosquitoes expand their region of activity.

The IPCC advises that the severity of these health risks will depend on the ability of public health and safety systems to address or prepare for these changing threats, as well as factors such as an individual's behavior, age, and economic status. Impacts will vary based on where a person resides, their sensitivity to health threats, their level of exposure to climate change impacts, and how well they and their community are prepared to adapt to climate change.

> WHO: Increased frequency of heat waves. Heat waves bring heat-related mortalities and morbidities, together with loss of economic productivity.

The World Health Organization (WHO) forecast in 2018 an estimation of the human health impacts of climate change elaborated as follows [30]:

- It is estimated that approximately 150,000 deaths in the year 2000 were attributable to changes in global climate experienced over the preceding decade.
- Increased frequency of heat waves. Heat waves bring heat-related mortalities and morbidities, together with loss of economic productivity.
- In the long term, the primary prevention of global warming health effects must be through marked reductions in the generation and release into the atmosphere of greenhouse gases.
- A study of heat-related premature deaths quantified the global risk of heat-related mortality. The researchers estimate that about 30% of the world's population is currently exposed to climatic conditions exceeding a deadly mortality threshold for at least 20 days a year. By 2100, this percentage is projected to increase to approximately 48% under a scenario with drastic reductions of greenhouse gas emissions and about 74% under a scenario of growing emissions.
- Between 2030 and 2050, climate change is expected to cause approximately 250,000 additional premature deaths per year, from malnutrition, malaria, diarrhea, and heat stress.

> WHO: Between 2030 and 2050, climate change is expected to cause approximately 250,000 additional premature deaths per year.

- Areas with weak health infrastructure—mostly in developing countries—will be the least able to cope without assistance to prepare and respond.
- Reducing emissions of greenhouse gases through better transport, food, and energy-use choices can result in improved health, particularly through reduced air pollution.

Of note, because children are at elevated health risk associated with climate change, the American Academy of Pediatricians released a policy statement in 2015 as follows [31]:

- "There is wide consensus among scientific organizations and climatologists that the broad effects known commonly as 'climate change' are the result of contemporary human activities.
- According to WHO, more than 88% of the existing burden of disease attributable to climate change occurs in children younger than 5 years old.
- Climate change poses a threat to human health and safety, but children are uniquely vulnerable.
- Failure to take prompt, substantive action would be an act of injustice to all children" [31].

Several attempts have been made to estimate the global risk to human well-being of climate change. A study by the World Bank focused on the impact of disasters on human life and economic consequences [32]. The report notes that a rapid increase in climate change-related natural disasters occurring by 2050 will put 1.3 billion people at risk, and with total damages worth $158tn—double the total current annual output of the global. The World Bank further noted that the global community is badly prepared for a rapid increase in climate change-related natural disasters that by 2050 will put 1.3 billion people at risk. Densely populated coastal cities are at particularly high risk, according to the report.

In a companion report, the World Bank opines that without the right policies to keep the global poor safe from extreme weather and rising seas, climate change could drive more than 100 million more people into poverty by 2030. The bank's estimate of 100 million more poor by 2030 is on top of 900 million expected to be living in extreme poverty if development progresses slowly. In 2015, the bank put the number of poor at 702 million people [33].

> World Bank: A rapid increase in climate change-related natural disasters occurring by 2050 will put 1.3 billion people at risk.

Projections of the number of premature deaths in the U.S. associated with climate change were derived by a Duke University team in conjunction with NASA researchers [34]. The study estimated the U.S. could avoid 4.5 million premature deaths if it works to keep global temperatures from rising by more than 2°C. The same study found working to limit climate change could prevent about 3.5 million hospitalizations and emergency room visits and approximately 300 million lost workdays in America. The avoided deaths are valued at more than $37 trillion.

The impact of climate on mental health has been the subject of investigation. One set of researchers commented on published literature that found exposure to hurricanes and floods is associated with symptoms of acute depression as well as posttraumatic stress disorder, psychiatric hospital visits increase during hotter temperatures, and both heat and drought amplify the risk of suicide. Further, persons with preexisting mental health conditions and lower socioeconomic status are among the most vulnerable to these adverse environmental conditions, commented the researchers [35]. The researchers extended published work by coupling meteorological and climatic data with reported mental health difficulties drawn from nearly 2 million randomly sampled U.S. residents between 2002 and 2012. The analysis found that short-term exposure to more extreme weather, multiyear warming, and tropical cyclone exposure were each associated with worsened mental health.

Researchers coined the term "eco-anxiety" to describe chronic or severe anxiety related to humans' relationship with the environment [36]. In 2017, the American Psychiatric Association (APA) described eco-anxiety as "a chronic fear of environmental doom." Eco-anxiety is not currently listed in the *Diagnostic and Statistical Manual of Mental Disorders* (DSM-5), meaning that doctors do not officially consider it a diagnosable condition. However, mental health professionals do use the term eco-anxiety within the field of ecopsychology, a branch that deals with people's psychological relationships with the rest of nature and how this impacts their identity, well-being, and health. The immediate effects of climate change—such as damage to community groups, a loss of food, and reduced medical supply security—can cause acute harm to people's mental health.

Akin to the impact of climate change on mental health is the subject of migration of populations adversely affected by climate change. A study by the World Bank Group reported that the worsening impacts of climate change in three densely populated regions of the world could see more than 140 million people move within their countries' borders by 2050, creating a looming human crisis. Of this number, 63 million migrants in South Asia are forecast [37]. However, the analysts advised that with concerted action—including global efforts to cut greenhouse gas emissions and robust development planning at the country level—this worst-case scenario of more than 140 m could be dramatically reduced, by as much as 80%, or more than 100 million people.

> Eco-anxiety: chronic or severe anxiety related to humans' relationship with the environment.

Perspective: The estimates by the WHO and the World Bank of the potential toll on global human health are sobering. Particularly sobering are the predictions that vulnerable populations will suffer disproportionately due simply to geographic location, income level, age, and local sociopolitical status. Preparing for and responding to this human health portent will strain public health resources in affected areas when responding to these kinds of vulnerabilities.

1.5 IMPACT OF CLIMATE CHANGE ON ECOSYSTEM HEALTH

As a reminder, an ecosystem is an interdependent system of plants, animals, and microorganisms interacting with one another and with their physical environment.

An ecosystem can be as large as the Mojave Desert or as small as a local pond. Ecosystems provide people with food, goods, medicines, and many other products. They also play a vital role in nutrient cycling, water purification, and climate moderation [38]. Climate is an important environmental influence on ecosystems. Climate changes and the impacts of climate change affect ecosystems in a variety of ways. For instance, warming could force species to migrate to higher latitudes or higher elevations where temperatures are more conducive to their survival. The EPA has grouped some of the more important effects of climate change on ecosystems as being the following [38]:

Changes in the timing of seasonal life-cycle events: For many species, the climate where they live or spend part of the year influences key stages of their annual life cycle, such as migration, blooming, and breeding. Because the climate has warmed in recent decades, the timing of these events has changed in some parts of the U.S. Changes like these can lead to mismatches in the timing of migration, breeding, and food availability. Growth and survival are reduced when migrants arrive at a location before or after food sources are present.

> As temperatures rise, the habitat ranges of many North American species are moving northward in latitude and upward in elevation.

Range Shifts: As temperatures rise, the habitat ranges of many North American species are moving northward in latitude and upward in elevation. While this means a range expansion for some species, for others it means a range reduction or a movement into a less hospitable habitat or increased competition for food and shelter. Some species have nowhere to go because they are already at the northern or upper limit of their habitat. For example, some fish species are forecast to migrate hundreds of miles. A study forecasts that hundreds of species of fish and shellfish will be forced to migrate northwards to escape the effects of climate change, putting global fisheries at risk [39]. The researchers comment that sea creatures are highly sensitive to the temperature of water, and if it gets too warm, they will often shift to areas that suit them better. They predict that 700 species of fish and other creatures inhabiting the waters around North America will migrate northward to cooler waters.

Food web disruptions: Energy from the Sun and CO_2 are used for photosynthesis by phytoplankton which are consumed by zooplankton or create sedimentation. The sedimentation turns into organic deposits that are consumed by seafloor creatures. Fish eat the seafloor creatures and zooplankton and are subsequently consumed by larger animals like seals, which are then consumed by animals at the top of the food chain, like polar bears. Ultimately the energy from the Sun and CO_2 create the food source for all species within the food web, including humans.

Threshold effects: In some cases, ecosystem change occurs rapidly and irreversibly because a threshold, or "tipping point," is passed. One area of concern for thresholds is the Prairie Pothole Region in the north-central part of the U.S. This ecosystem is a vast area of small, shallow lakes, known as "prairie potholes" or "playa lakes." These wetlands provide essential breeding habitat for most North American waterfowl species. The pothole region has experienced temporary droughts in the past. Similarly, when coral reefs become stressed, they expel microorganisms that live within their tissues and are essential to their health. This is known as coral bleaching. As ocean temperatures warm and the acidity of the ocean increases, bleaching and coral die-offs are likely to become more frequent. Chronically stressed coral reefs are less likely to recover.

In September 2019, the IPCC released a special report concerning the impact of climate change on the world's oceans. They warned that without steep cuts to greenhouse gas emissions, fisheries will falter, the average strength of hurricanes will increase, and rising seas will increase the risk of flooding in low-lying areas

> EPA: Climate change and shifts in ecological conditions could support the spread of pathogens, parasites, and diseases, with potentially serious effects on human health, agriculture, and fisheries.

around the globe [40]. The report projects that sea levels could rise by up to 1.1 m by 2100 if greenhouse gas emissions continue to rise.

Pathogens, parasites, and disease: Climate change and shifts in ecological conditions could support the spread of pathogens, parasites, and diseases, with potentially serious effects on human health, agriculture, and fisheries (Chapter 1). For example, the oyster parasite, *Perkinsus marinus*, is capable of causing large oyster die-offs. This parasite has extended its range northward from Chesapeake Bay to Maine, a 310-mile expansion tied to above-average winter temperatures.

As another example, southern pine beetles are among the most destructive insects invading North America's pine forests. A study found the beetles are spreading farther north as global temperatures rise, putting entire ecosystems at risk and creating fuel for wildfires as they kill the trees they infest. The study warns that insects' range could reach Nova Scotia by 2020 and cover more than 270,000 square miles of forest from the upper Midwest to Maine and into Canada by 2080 [41].

A different type of insect invasion in 2020 is occurring in Africa. Swarms of desert locusts have invaded eastern Africa, ravaging crops, decimating pasture, and deepening a hunger crisis [42]. Studies have linked a hotter climate to more damaging locust swarms, leaving Africa disproportionately affected—20 of the fastest-warming countries globally are in Africa. Wet weather also favors the multiplication of locusts. Widespread, above-average rain that pounded the Horn of Africa from October to December 2019 were up to 400% above normal rainfall amount. As a consequence of locust infestation, the Food and Agriculture Organization estimated that 19 million people would be at risk of severe food insecurity and another 20 million on the brink of food insecurity. Pakistan is another area infected with swarms of desert locusts. An interesting response to the locust threat is the importation into eastern Pakistan of 100,000 Chinese ducks [43], rather than applying pesticides. Chinese authorities assert that each duck is capable of consuming 200 locusts daily.

Extinction risks: Climate change, along with habitat destruction and environmental pollution, is one of the important stressors that can contribute to species extinction. The Intergovernmental Panel on Climate Change estimates that 20%–30% of the plant and animal species evaluated so far in climate change studies are at risk of extinction if temperatures reach levels projected to occur by the end of this century [cited in 26]. Projected rates of species extinctions are ten times greater than recently observed global average rates and 10,000 times greater than rates observed in the distant past (as recorded in fossils).

> Climate change, along with habitat destruction and environmental pollution, is one of the important stressors that can contribute to species extinction.

Also noted as consequences of the effects of climate change on ecosystem health are reports of species in decline, ocean acidification, and spikes in air pollution.

- Frog populations are decreasing in several areas of the world due to global warming. Warmer temperatures have been associated with outbreaks of skin fungus that is fatal to frogs. The fungus proliferates with warmer temperatures.
- Coral reefs are at risk of dissolving as oceans become more acidic. A study by Australian scientists noted that coral reefs could start to dissolve before 2100 due to increased acidification of sea waters. Acidification will threaten sediments that are building blocks for reefs [42, 44].
- Acidification of oceans is occurring according to the British Royal Society, which observes that ocean water is now 8.1 pH, a decrease of 0.1 over that of 200 years ago, which translates to a 30% increase in hydrogen ions in the water. The Society predicts that the pH of ocean water near the surface will decrease to 7.7–7.9 by year 2100. The increased acidity could reduce populations of plankton, disrupting the ocean food chain and harming fisheries.
- Spikes in U.S. air pollution were linked by the American Lung Association (ALA) with warmer temperatures due to climate change. In particular, the ALA analysis suggests global warming is causing short-term spikes in air pollution. The spikes result from droughts and wildfires that temporarily increase particulate levels from dust and smoke [45].

Several investigations of the observed impact of climate change on ecosystem health are available. For example, according to NOAA, there is strong evidence that global sea level is now rising at an increased rate and will continue to rise during the 21st century. While studies show that sea levels changed little from CE 0 until 1900, sea levels began to climb in the 20th century.

> IPCC: 20%–30% of the plant and animal species evaluated so far in climate change studies are at risk of extinction.

The two major causes of global sea-level rise are thermal expansion caused by the warming of the oceans (since water expands as it warms) and the loss of land-based ice (such as glaciers) due to increased melting. Records and research show that sea level has been steadily rising at a rate of 0.04–0.1 inches per year since 1900.

Since 1992, new methods of satellite altimetry (the measurement of elevation or altitude) indicate a rate of rise of 0.12 inches per year. This is a significantly larger rate than the sea-level rise averaged over the last several thousand years [46].

In addition to rise in sea levels, the Earth's temperatures have continued to rise. NOAA data indicate that the combined average temperature over global land and ocean surfaces for July 2015 was the highest for July in the 136-year period of record, at 0.81°C (1.46°F) above the 20th century average of 15.8°C (60.4°F), surpassing the previous record set in 1998 by 0.08°C (0.14°F). As July is climatologically the warmest month of the year globally, this monthly global temperature of 16.61°C (61.86°F) was also the highest among all 1627 months in the record that began in January 1880. The July temperature is currently increasing at an average rate of 0.65°C (1.17°F) per century [47].

Researchers in China noted in 2020 that climate change has resulted in a long-term warming of the planet [48]. They noted that more than 90% of the excess heat is stored within the world's oceans, where it accumulates and causes increases in ocean temperature and currents. Because the oceans are the main repository of the Earth's energy imbalance, measuring ocean heat content (OHC) is one of the best ways to quantify the rate of global warming. The researchers noted that OHC data for the year 2019 revealed that the world's oceans (especially the upper 2000 m) in 2019 were the warmest in recorded human history.

The effect of climate change on ocean acidity has also been documented. According to one source, every day, 22 million tons of CO_2 are absorbed by the world's oceans [49]. This equates to approximately one ton of CO_2 per person on Earth annually. The oceans have become 30% more acidic because of the carbon pollution pumped into the atmosphere. This absorption makes seawater more acidic, spelling disaster for many marine animals, from plankton and coral up the food chain to sea stars, salmon, sea otters, whales—and ultimately people, who rely on oceans for food.

Perspective: For persons focused on human health in general and health effects of climate change specifically, these aforementioned effects of climate change on ecosystems may seem irrelevant or uninteresting. That would be an unfortunate myopic attitude. Humans occupy a niche in global and local ecosystems; they sustain us in many ways. Loss of healthy ecosystems will portend difficulties for human health and well-being. Warmer air and ocean temperatures, more acidic oceans, and rising sea levels are consequences of climate change and as such portend global challenges to human well-being.

1.6 IMPACT OF CLIMATE CHANGE ON AGRICULTURE AND SPECIES

Climate change will have significant global impacts on agriculture and the species that populate the planet. In recognition of these impacts, international, regional, and some national governments have developed statements and proffered advice specific to agriculture. For example, the UN's Food and Agriculture Organization (FAO) considers climate change to be a fundamental threat to global food security, sustainable development, and poverty eradication. Agriculture, including the forestry and fisheries sectors, must adapt to the impacts of climate change and improve the resilience of food production systems in order to feed a growing population [50]. While the FAO and some regional groups offer some technical and financial assistance, by and large, regional and individual national governments must assume the primary responsibility for implementing policies for application to agriculture operations.

Similar to the FAO's concerns about the impacts of climate change on agriculture, the EU has expressed both concern and advice on the subject [51]. In particular, they note that agriculture is highly sensitive to climate change, as farming activities directly depend on climatic conditions. Changing in rainfall will be a serious problem in many EU regions, as well as rising temperatures; variability and seasonality as well as extreme events, heat waves, droughts, storms, and floods across the EU. But agriculture also contributes to the release of greenhouse gases to the atmosphere. However, agriculture can also help to provide solutions to the overall climate change problem by reducing emissions and by sequestering carbon while not threatening viable food production.

> FAO: Agriculture is highly sensitive to climate change, as farming activities directly depend on climatic conditions.

Climate change across the EU will require adaptive measures (both at farm and sectorial level) in agriculture, ranging from technological solutions to adjustments in farm management or structures, and to political changes, such as adaptation plans. Concerning farm-level adaptation, possible short- to medium-term adaptive solutions might include:

- Adjusting the timing of farm operations, such as planting or sowing dates and treatments;
- Technical solutions, such as protecting orchards from frost damage or improving ventilation and cooling systems in animal shelters;
- Choosing crops and varieties better adapted to the expected length of the growing season and water availability, and more resistant to new conditions of temperature and humidity;
- Adapting crops with the help of existing genetic diversity and new possibilities offered by biotechnology;
- Improving the effectiveness of pest and disease control through for instance better monitoring, diversified crop rotations, or integrated pest management methods;
- Using water more efficiently by reducing water losses, improving irrigation practices, and recycling or storing water;
- Improving soil management by increasing water retention to conserve soil moisture, and landscape management, such as maintaining landscape features providing shelter to livestock;
- Introducing more heat-tolerant livestock [51].

Pertaining to the U.S., as characterized by the U.S. Department of Agriculture (USDA) in 2012, increases in atmospheric CO_2, rising temperatures, and altered precipitation patterns will affect agricultural productivity [52]. Increases in temperature coupled with more variable precipitation will reduce the productivity of U.S. crops. Effects will vary among annual and perennial crops, and regions of

the U.S.; however, all production systems will be affected to some degree by
climate change. Agricultural systems
depend upon reliable water sources,
and the pattern and potential magnitude of precipitation changes are not

USDA: Increases in temperature
coupled with more variable precipitation will reduce the productivity of
U.S. crops.

well understood, thus adding considerable uncertainty to assessment efforts. More
specifically:

- Livestock production systems are vulnerable to temperature stresses.
  An animal's ability to adjust its metabolic rate to cope with temperature
  extremes can lead to reduced productivity and, in extreme cases, death.
- Projections for crops and livestock production systems reveal that climate
  change effects over the next 25 years will be mixed. The continued degree
  of change in the climate by mid-century and beyond is expected to have
  overall detrimental effects on most crops and livestock.
- Climate change will exacerbate current biotic stresses on agricultural plants
  and animals. Changing pressures associated with weeds, diseases, and
  insect pests, together with potential changes in timing and coincidence of
  pollinator lifecycles, will affect growth and yields.
- Agriculture is dependent on a wide range of ecosystem processes that
  support productivity including maintenance of soil quality and regulation of water quality and quantity. Multiple stressors, including climate
  change, increasingly compromise the ability of ecosystems to provide these
  services [52].

The predicted higher incidence of extreme weather events will have an increasing
influence on agricultural productivity. Extremes matter because agricultural productivity is driven largely by environmental conditions during critical threshold periods of crop and livestock development. The vulnerability of agriculture to climatic
change is strongly dependent on the responses taken by humans to moderate the
effects of climate change [52]. Climate change will necessitate adaptation procedures
and access to contemporary, reliable sources of advice and data. Food security produced by agricultural operations will require no less.

The health of the world's soils hinges on the abundance and diversity of the
microbes and fungi they contain, and environmental changes, including from global
warming, will impair their ability to support humans and other species, according to
a study that examined microbial diversity in 78 drylands on all inhabited continents
and 179 sites in Scotland. The investigators found that the loss of varieties—such
as from climate change increasing arid zones—undermined the services the soils

provided. The authors commented, "As
the aridity of soils goes up, the microbial
diversity and abundance is reduced, As
the soils' multi-functions are reduced,
so there are social and economic consequences" [53].

The health of the world's soils hinges
on the abundance and diversity of
the microbes and fungi they contain.
Global warming will impair them.

A different study investigated the adaptability of soil microbes under changing climate conditions. A 17-year study into the effect of global warming on microbes—the tiny bacteria, fungi, and other microorganisms that determine soil health—reveals them to be far less adaptable to changing conditions than expected [54]. The study involved swapping soil samples between two sites on a mountainside in 1994—the higher location had a warmer, drier climate than the one 500 m below. Seventeen years later they went back to check on the microbes' activities, focusing on their rate of respiration—how quickly they convert carbon in the soil to carbon dioxide as they break down the organic matter—to get a broader sense of their ability to adapt to the changing conditions. But they found very little change. The microbes that had been native to the higher site naturally respired at a faster rate because they were used to greater levels of rainfall and vegetation, or carbon. They continued to respire at a faster rate at their lower elevation—even 17 years later. And the microbes taken from lower down the mountain demonstrated very little change when they were moved uphill, suggesting them to be far less adaptable to changing conditions than expected. The study's findings raise concerns the microbes will not be able to carry out essential functions, such as breaking down leaves and other organic matter in a process which converts them into nitrogen and other nutrients that plants need to grow.

In addition to concerns about how soil microbes are responding to climate change is a companion issue of how plants themselves are responding. A report by the UNEP says that crops such as wheat and maize are generating more potential toxins as a reaction to protect themselves from extreme weather [55]. But these chemical compounds are harmful to people and animals if consumed for a prolonged period of time. According to the UNEP report, under normal conditions, for instance, plants convert nitrates they absorb into nutritious amino acids and proteins. But prolonged drought slows or prevents this conversion, leading to more potentially problematic nitrate accumulating in the plant, the report said.

If people consume too much nitrate in their diets, it can interfere with the ability of red blood cells to transport oxygen in the body. According to UNEP, crops susceptible to accumulating too much nitrate in times of stress include maize, wheat, barley, soybeans, millet, and sorghum. Some drought-stressed crops, when then exposed to sudden large amounts of rain that lead to rapid growth, in turn accumulate hydrogen cyanide, more commonly known as prussic acid, UNEP reported. Prussic acid can interfere with oxygen flow in humans. Plants such as cassava, flax, maize, and sorghum are most vulnerable to dangerous prussic acid accumulation, the report said.

> UNEP: Crops such as wheat and maize are generating more potential toxins as a reaction to protect themselves from extreme weather.

Aflatoxins, molds that can affect plant crops and raise the risk of liver damage, cancer, and blindness, as well as stunting fetuses and infants, are also spreading to more areas as a result of shifting weather patterns as a result of climate change, scientists said. UNEP observed that about 4.5 billion people in developing countries are exposed to aflatoxins annually. Europe will be at growing risk from aflatoxins in locally grown crops if global temperatures rise by at least 2°C.

Related to agriculture and climate change is the idea of using crops to collect more atmospheric carbon and locking it into soil's organic matter to offset fossil fuel emissions. The idea was launched in Paris in 2015 at COP21, the 21st annual Conference of Parties to review the UN Framework Convention on Climate Change. However, a study of unique soils data from long-term experiments, stretching back to the middle of the 19th century, confirmed the practical implausibility of burying carbon in the ground to halt climate change, an option once heralded as a breakthrough [56]. This study illustrates that methods to mitigate climate change will neither be easy to achieve nor inexpensive.

Perspective: Global climate change portends significant impacts on agriculture and correspondingly food security. Droughts and other extreme weather events will severely impact agricultural methods and crop yields if adaptation practices are not implemented. Further, the impacts of climate change on soil and plant quality will exacerbate the overall impact on agriculture and food production.

The projected effects of climate change will also present serious consequences to species globally. In a study by the International Union for Conservation of Nature, which updated their list of endangered species, about 12% of the animals on the list are either endangered or critically endangered because of climate change. This equates to approximately 1,400 species. A sample of the threatened species includes: seahorses, the Kaputar pink slug, wombats and wallabies, whooping cranes and ibises, akikikas, sea otter, seals, and sea lions [57]. Rising sea temperatures is a significant factor in causing this species endangerment.

In a separate study, climate change could drive up to a sixth of animals and plants on Earth to extinction unless climate change mitigation occurs. Overall, it found that one in six species could be driven to extinction if greenhouse gas emissions are unchecked and temperatures rise by 4.3°C above preindustrial times by 2100, in line with one scenario from the Intergovernmental Panel on Climate Change (IPCC). The study averaged out 131 previous studies of climate change, whose projections of the number of species that could be lost to climate change ranged from zero to 54% of species. Species in South America, Australia, and New Zealand are most at risk since many live in small areas or cannot easily move away to adapt to heatwaves, droughts, floods, or rising seas [58].

According to one report, climate change has claimed its first mammalian species, according to a new report by Australian scientists. Researchers from Australia's University of Queensland and Queensland Government say a rodent, known as the Bramble Cay melomys, which lived on Bramble Cay, a small sandy island in the Great Barrier Reef, has died due to "rising sea levels and an increased incidence of extreme weather events," the first mammal on record to be declared extinct "due solely (or primarily) to anthropogenic climate change." While the researchers were certain the animals were washed away from their only known home, they did observe the "possibility that the species occurs elsewhere on islands in the Torres Strait," an area between Australia and Papua New Guinea comprised of more than 200 islands [59].

> One in six species could be driven to extinction if greenhouse gas emissions are unchecked and temperatures rise by 4.3°C greater than preindustrial times by 2100.

1.7   HISTORY OF KEY UN CLIMATE CHANGE POLICIES

The UN provides an illuminating history of climate change, the environmental concerns, and its global policymaking efforts.

"To fully understand the current debate, one must look at the rise in prominence of environmental issues on the global agenda and the evolution of climate change within that context. Environmental issues, much less climate change, were not a major concern of the UN in the period following the Organization's creation. During its first 23 years, action on these issues was limited to operational activities, mainly through the World Meteorological Organization (WMO), and when attention was paid to them, it was within the context of one of the major preoccupations of that time: the adequacy of known natural resources to provide for the economic development of a large number of UN members or the 'underdeveloped countries', as they were then termed" [60].

"In 1949, the UN Scientific Conference on the conservation and utilization of resources (Lake Success, New York, August 17–September 6) was the first UN body to address the depletion of those resources and their use. The focus, however, was mainly on how to manage them for economic and social development, and not from a conservation perspective. It was not until 1968 that environmental issues received serious attention by any major UN organs. The Economic and Social Council on May 29 was the first to include those issues in its agenda as a specific item and decided—later endorsed by the General Assembly—to hold the first United Nations Conference on the Human Environment.

Held in Stockholm, Sweden, from June 5 to 16, 1972, the UN Scientific Conference, known also as the First Earth Summit, adopted a declaration that set out principles for the preservation and enhancement of the human environment, and an action plan containing recommendations for international environmental action. In a section on the identification and control of pollutants of broad international significance, the Declaration raised the issue of climate change for the first time, warning Governments to be mindful of activities that could lead to climate change and evaluate the likelihood and magnitude of climatic effects" [60].

"The UN Scientific Conference also proposed the establishment of stations to monitor long-term trends in the atmospheric constituents and properties, which might cause meteorological properties, including climatic changes. Those programs were to be coordinated by WMO to help the world community to better understand the atmosphere and the causes of climatic changes, whether natural or the result of man's activities.

Over the next 20 years, as part of efforts to implement the 1972 decisions, concern for the atmosphere and global climate slowly gained international attention and action. In 1979, the UNEP Governing Council asked its Executive Director, under the Earth Watch programme, to monitor and evaluate the long-range transport of air pollutants, and the first international instrument on climate—the Convention on Long-Range Transboundary Air Pollution—was then adopted. UNEP took it to another level in 1980, when its Governing Council expressed concern at the damage to the ozone layer and recommended measures to limit the production and use of chlorofluorocarbons F-11and F-12. This led to the negotiation and adoption in 1985 of the

Vienna Convention for the Protection of the Ozone Layer and the conclusion of a Protocol to the 1979 Transboundary Air Pollution Convention, which aimed at reducing sulphur emissions by 30%" [60].

> In 1979, the Convention on Long-Range Transboundary Air Pollution was adopted, the first international climate instrument.

"In 1987 the UN General Assembly gave real impetus to environmental issues, when it adopted the Environmental Perspective to the Year 2000 and Beyond—a framework to guide national action and international cooperation on policies and programmes aimed at achieving environmentally sound development. The perspective underlined the relationship between environment and development and for the first time introduced the notion of sustainable development" [60].

As the urgency for a stronger international action on the environment, including climate change, gained momentum, the UN General Assembly decided to convene in 1992 in Rio de Janeiro, Brazil, the United Nations Conference on Environment and Development. The Earth Summit, as it is also known, set a new framework for seeking international agreements to protect the integrity of the global environment in its Rio Declaration and Agenda 21, which reflected a global consensus on development and environmental cooperation.

"Chapter 9 of Agenda 21 dealt with the protection of the atmosphere, establishing the link between science, sustainable development, energy development and consumption, transportation, industrial development, stratospheric ozone depletion, and transboundary atmospheric pollution. The most significant event during the Conference was the opening for signature of the United Nations Framework Convention on Climate Change (UNFCCC); by the end of 1992, 158 States had signed it. As the most important international action thus far on climate change, the Convention was to stabilize atmospheric concentrations of 'greenhouse gases' at a level that would prevent dangerous anthropogenic interference with the climate system. It entered into force in 1994, and in March 1995, the first Conference of the Parties to the Convention adopted the Berlin Mandate, launching talks on a protocol or other legal instrument containing stronger commitments for developed countries and those in transition" [60].

The cornerstone of the climate change action was the adoption in Japan in December 1997 of the Kyoto Protocol to the UNFCCC, the most influential climate change action so far taken. It aimed to reduce the industrialized countries' overall emissions of carbon dioxide and other greenhouse gases by at least 5% below the 1990 levels in the commitment period of 2008–2012. The Protocol, which opened for signature in March 1998, came into force on February 16, 2005, 7 years after it was negotiated by more than 160 nations.

> The cornerstone of the climate change action was the adoption in Japan in December 1997 of the Kyoto Protocol to the UNFCCC, the most influential climate change action so far taken.

As discussed in the following material, the UN has coordinated several seminal meetings on issues of climate change. Five of the meetings resonate in importance as global efforts to mitigate the consequences of climate change. First, the Montreal Convention in 1987 was convened due to the discovery and global concern that the Earth's ozone layer was at risk of lessening its protective shield of planet Earth. Second, in 1988, the Intergovernmental Panel on Climate Change (ICCC) was established under UN auspices. Third, in 1992, the United Nations Framework Convention on Climate Change was adopted by the UN's Member States. Fourth, the Kyoto Protocol of the UNFCCC was adopted in 1997. Fifth, the culmination of these several meetings and conventions occurred in the 2015 United Nations Climate Change Conference, COP 21, convened in Paris. The Paris Agreement was achieved, following years of global diplomacy and national policymaking on climate control. This accord was the first global agreement on the vital need to commit to mitigation of climate change. Specific climate targets were set in place and interim goals were developed as global targets on temperature rise and carbon emissions. The meetings and conventions that forged the actions resulting in the Paris Agreement are summarized in more detail in the succeeding sections.

Perspective: The foregoing history of global policymaking on climate change is important as a lesson of the struggle to gain international cooperation on a matter of vital importance to life on Earth. Leadership by the UN on the matter of environmental health policymaking was essential, given the global impact of changes beginning to occur in the global environment. Simply put, climate change is a global problem; therefore, a global policymaking political resource was required. An examination of the UN's numerous attempts at consensus building on mitigation of climate change shows gradual success in gaining global acceptance of responsibility by national governments. The key meetings in the journey to build global consensus and action are described herein.

1.8   U.S. CLIMATE CHANGE POLICIES

There are many policy issues presented by actions intended for the mitigation of climate change. The challenges to climate change policymaking are for the purpose of this book grouped into issues of science, technology, economics, legal, and public's perspective. The following narrative begins with a description of U.S. issues and experience in climate change policymaking.

1.8.1   OVERVIEW OF U.S. CLIMATE CHANGE POLICYMAKING

The science of climate assessment, its analysis, and propagation of results is at the core of any consideration of policymaking on climate change. This is because the economic, political, and social consequences are so great in import that the science that supports mitigation of climate change must be well grounded and adhere to the standards expected of scientific investigation and reporting. Mere suspicion of climate change's portent cannot sustain and justify the extraordinary global revisions of national practices of energy development and use, food security, and sociopolitical stability. The science of climate change must be the foundation for policymaking on climate change. The efforts in the U.S. to develop environmental health policies for mitigation of climate change have involved both legal issues and executive branch factors and efforts, primarily by the U.S. President Barack Obama (D-IL) administration. Both factors are described in the subsequent sections.

1.8.2 GLOBAL CHANGE RESEARCH ACT, 1990

The U.S. Global Change Research Program (USGCRP) was established by U.S. presidential (George H. W. Bush (R-TX)) initiative in 1989 and mandated by Congress in the Global Change Research Act (GCRAct) of 1990. Its mandate is to develop and coordinate "a comprehensive and integrated U.S. research program which will assist the nation and the world to understand, assess, predict, and respond to human-induced and natural processes of global change [61]."

> The GCRAct requires a quadrennial assessment of the U.S. climate.

The USGCRP comprises 13 U.S. federal government agencies that conduct or use research on global change and its impacts on society. It functions under the direction of the Subcommittee on Global Change Research of the National Science and Technology Council's Committee on Environment, Natural Resources, and Sustainability. The USGCRP has three major sets of responsibilities: (1) coordinating global change research across the federal government, (2) developing and distributing mandated products, and (3) helping to inform decisions.

1.8.2.1 Fourth National Climate Assessment, 2018

One of the products mandated by the GCRA is a quadrennial assessment that USGCRP is to prepare and submit to the President and the Congress. This assessment, referred to as the National Climate Assessment (NCA), is directed by the GCRA to: Integrate, evaluate, and interpret the findings of the Program and discuss the scientific uncertainties associated with such findings; Analyze the effects of global change on the natural environment, agriculture, energy production and use, land and water resources, transportation, human health and welfare, human social systems, and biological diversity; Analyze current trends in global change, both human-induced and natural, and project major trends for the subsequent 25–100 years.

1.8.2.2 Highlights from the Fourth National Climate Assessment, 2018

The Fourth NCA was the product of more than 300 scientists in the U.S. federal government. The report is available to the public [62]. The NCA presented highlights from the scientists' findings. Their highlights follow.

"The climate of the United States is strongly connected to the changing global climate. The statements below highlight past, current, and projected climate changes for the United States and the globe.

Global annually averaged surface air temperature has increased by about 1.8°F (1.0°C) over the last 115 years (1901–2016). This period is now the warmest in the history of modern civilization. [...]. These trends are expected to continue over climate timescales.

> Fourth National Climate Report: The climate of the U.S. is strongly connected to the changing global climate.

This assessment concludes, based on extensive evidence, that it is extremely likely that human activities, especially emissions of greenhouse gases, are the dominant cause of the observed warming since the mid-20th century. For the warming over the last century, there is no convincing alternative explanation supported by the extent of the observational evidence.

In addition to warming, many other aspects of global climate are changing, primarily in response to human activities. Thousands of studies conducted by researchers around the world have documented changes in surface, atmospheric, and oceanic temperatures; melting glaciers; diminishing snow cover; shrinking sea ice; rising sea levels; ocean acidification; and increasing atmospheric water vapor.

> Fourth National Climate Report: Global annually averaged surface air temperature has increased by about 1.8°F (1.0°C) over the last 115 years (1901–2016).

For example, the global average sea level has risen by about 7–8 inches since 1900, with almost half (about 3 inches) of that rise occurring since 1993. [...] Global sea level rise has already affected the U.S.; the incidence of daily tidal flooding is accelerating in more than 25 Atlantic and Gulf Coast cities.

Global average sea levels are expected to continue to rise—by at least several inches in the next 15 years and by 1–4 feet by 2100. A rise of as much as 8 feet by 2100 cannot be ruled out. Sea-level rise will be higher than the global average on the East and Gulf Coasts of the U.S.

Changes in the characteristics of extreme events are particularly important for human safety, infrastructure, agriculture, water quality and quantity, and natural ecosystems. Heavy rainfall is increasing in intensity and frequency across the U.S. and globally and is expected to continue to increase. The largest observed changes in the U.S. have occurred in the Northeast.

> Fourth National Climate Report: Global average sea levels are expected to continue to rise—by at least several inches in the next 15 years and by 1–4 feet by 2100.

Heatwaves have become more frequent in the U.S. since the 1960s, while extreme cold temperatures and cold waves are less frequent. [..]. Annual average temperature over the contiguous U.S. has increased by 1.8°F (1.0°C) for the period 1901–2016; over the next few decades (2021–2050), annual average temperatures are expected to rise by about 2.5°F for the U.S., [..].

The incidence of large forest fires in the western U.S. and Alaska has increased since the early 1980s and is projected to further increase in those regions as the climate changes, with profound changes to regional ecosystems.

Annual trends toward earlier spring melt and reduced snowpack are already affecting water resources in the western U.S. and these trends are expected to continue. [...]

The magnitude of climate change beyond the next few decades will depend primarily on the amount of greenhouse gases (especially carbon dioxide) emitted globally. Without major reductions in emissions, the increase in annual average global temperature relative to preindustrial times could reach 9°F (5°C) or more by the end of this century. With significant reductions in emissions, the increase in annual average global temperature could be limited to 3.6°F (2°C) or less.

> Fourth National Climate Report: The magnitude of climate change beyond the next few decades will depend primarily on the amount of greenhouse gases (especially carbon dioxide) emitted globally.

The global atmospheric CO_2 concentration has now passed 400 parts per million (ppm), a level that last occurred about 3 million years ago, when both global average temperature and sea level were significantly higher than today. [..].

The observed increase in carbon emissions over the past 15–20 years has been consistent with higher emissions pathways. In 2014 and 2015, emission growth rates slowed as economic growth became less carbon-intensive. Even if this slowing trend continues, however, it is not yet at a rate that would limit global average temperature change to well less than 3.6°F (2°C) above preindustrial levels" [62].

Perspective: The NCR presents a dire picture of the current and future consequences of climate change. The warnings in the report should resonate in environmental, social, and economic policies enacted by both public and private parties. Regrettably, this resonance has not occurred. When asked in November 2018 by a news reporter about the NCR, U.S. President Donald Trump (R-NY) responded, "I've seen it, I've read some of it, and it's fine, I don't believe it" [63]. Further, no action was taken by the U.S. Congress. The impact of this absence of policymaking will be incalculable.

> Fourth National Climate Report: The global atmospheric CO_2 concentration has now passed 400 ppm, a level that last occurred about 3 million years ago.

1.8.3 OBAMA ADMINISTRATION'S CLIMATE CHANGE POLICIES

How to deal within the U.S. federal government with the political, economic, and health issues arising from climate change became itself a matter of environmental policy as the science of climate change began to amass. EPA was the logical and perhaps only choice, in which to entrust U.S. policy leadership on climate change. This choice was dictated by both matters of practicality as well as legality. EPA, since its establishment by President Richard Nixon (R-CA) administration, had become the U.S.'s central federal resource on environmental policy and implementation of environmental statutes enacted by Congress. Moreover, EPA was the lead federal agency for air pollution policies, principally the CAAct (Chapter 2), and had thereby accrued more than 40 years of practical experience in working on issues of contaminants in the air, the environmental medium of greatest relevance to climate change.

Early in the 21st century, EPA made a policy decision to consider using the existing Clean Air Act (CAAct) as its basis for regulating the emissions of greenhouse gases. This represented a new extension of the act and was quickly met with resistance by some industrial groups and politically conservative organizations that opposed any attempt by the U.S. federal government to develop climate change policies. As EPA proceeded with its plans to develop regulations to control greenhouse gases, the policymaking arena moved from EPA to U.S. federal courts. Without discussing the litany of court cases, three decisions by the U.S. Supreme Court were a key to shaping the U.S. policies on climate change. These decisions are discussed in the following section.

> Early in the 21st century, EPA made a policy decision to consider using the existing CAAct as its basis for regulating the emissions of greenhouse gases.

1.8.3.1   U.S. Supreme Court Decisions

The first of three seminal U.S. Supreme Court decisions that helped shape U.S. policy on climate change dealt with the central issue of whether EPA could regulate greenhouse gases under the provisions of the CAAct (Chapter 2). In 2003, EPA decided that it lacked authority under the CAAct to regulate greenhouse gases, principally carbon dioxide emissions into the atmosphere. Further, the agency had some uncertainty on whether the science base on climate change was sufficiently robust so as to consider using the CAAct in relation to climate change. EPA's policy position of 2003 was challenged in litigation [64].

"Massachusetts and several other states petitioned the EPA (EPA), asking EPA to regulate emissions of carbon dioxide and other gases that contribute to global warming from new motor vehicles. Massachusetts argued that EPA was required to regulate these 'greenhouse gases' by the CAAct—which states that Congress must regulate 'any air pollutant' that can 'reasonably be anticipated to endanger public health or welfare.

The EPA denied the petition, claiming that the CAAct does not authorize it to regulate greenhouse gas emissions. Even if it did, EPA argued, the Agency had discretion to defer a decision until more research could be done on 'the causes, extent and significance of climate change and the potential options for addressing it.' Massachusetts appealed the denial of the petition to the Court of Appeals for the D.C. Circuit, and a divided panel ruled in favor of EPA" [65].

The petitioners appealed the appellate court's decision to the U.S. Supreme Court, with the following two questions forming the central questions put before the court:

1. May EPA decline to issue emission standards for motor vehicles based on policy considerations not enumerated in the CAAct?
2. Does the CAAct give EPA authority to regulate carbon dioxide and other greenhouse gases?

By a 5-4 vote, the Court reversed the D.C. Circuit and ruled in favor of Massachusetts.

"The opinion by Justice John Paul Stevens held that Massachusetts, due to its 'stake in protecting its quasi-sovereign interests' as a state, had standing to sue EPA over potential damage caused to its territory by global warming. The Court rejected EPA's argument that the CAAct was not meant to refer to carbon emissions in the section giving EPA authority to regulate 'air pollution agent[s]'. The Act's definition of air pollutant was written with 'sweeping,' 'capacious' language so that it would not become obsolete. Finally, the court's majority ruled that EPA was unjustified in delaying its decision on the basis of prudential and policy considerations. The Court held that if EPA wishes to continue its inaction on carbon regulation, it is required by the act to base the decision on a consideration of "whether greenhouse gas emissions contribute to climate change." Chief Justice Roberts's dissenting opinion argued that Massachusetts should not have had standing to sue because the potential injuries

> The Supreme Court decided that existing law, the CAAct, could be interpreted as applying to greenhouse gases.

from global warming were not concrete or particularized (individual and personal). Justice Scalia's dissent argued that the CAAct was intended to combat conventional lower-atmosphere pollutants and not global climate change" [64].

The court's decision was decided in April 2007.

Perspective: This was a pivotal decision for U.S. climate change policy, since the court decided that existing law, the CAAct, could be interpreted as applying to greenhouse gases. However, the court only opened the door to EPA's potential use of the CAAct for regulating greenhouse gases (GHG). How EPA chose to pursue that use was up to the agency. EPA's use of its new authority viz. greenhouse gases led to the second of the three major Supreme Court decisions.

Given the path that EPA chose, i.e., the use of the CAAct for extension to cover GHS, one might ask why not amend the CAAct and make explicit its coverage of GHS? Presumably, such a course might obviate the need for subsequent litigation. However, legislative experience has shown that an attempt to amend an act such as the CAAct might lead to unpredictable and undesirable outcomes. The CAAct is a complex law, with four decades of judicial decisions and executive branch policies and attempts to amend the law by Congress would likely open a stampede of parties interested in various changes to current law. In other words, opening the legislative door to amend current laws is a risky proposition, given the vicissitudes of legislative processes.

The second of the three key Supreme Court decisions on climate change followed the 2007 Court decision in Massachusetts v. EPA, as previously described. Following that decision from the Court, EPA developed a series of standards governing greenhouse gas emissions.

"One of these benchmarks set emission standards for vehicles, while another one required stationary sources of greenhouse gases to obtain constructing and operating permits from EPA. The petitioners, who include various state and industry groups, challenged these rules on the grounds that they were based on an improper construction of the CAAct and were arbitrary and capricious because they were based on an inadequate scientific record. The U.S. Court of Appeals for the Federal Circuit dismissed the challenges" [65].

The petitioners appealed the appellate court's decision to the U.S. Supreme Court.

The gist of the case was the question: "Did the EPA permissibly determine that its regulation of greenhouse gas emissions from new motor vehicles under the CAAct also triggered permit requirements for stationary sources of greenhouse gas emissions?" [65].

The Court decided No. Justice Antonin Scalia delivered the opinion for the 9-0 member majority. The Court held that, while the Massachusetts decision found that the CAAct's general definition of "air pollutant" included greenhouse gas emissions, it does **not require** (emphasis added) the EPA to include greenhouse gas emissions every time the act uses the term "air pollutant." Instead, EPA retains its ability to interpret the term in a context-appropriate way depending on where the term was being used. Because the inclusion of greenhouse gases as an "air pollutant" under the permitting scheme would compel EPA to regulate tens of thousands of additional pollution emitters, it would not be reasonable for EPA to interpret this specific instance of "air pollution" to include greenhouse gas emissions. Furthermore, even if EPA were able to interpret this instance of "air pollution" to include greenhouse gases, EPA lacks the authority to modify the threshold limits Congress dictated. Though EPA overstepped its authority in trying to regulate greenhouse gases under this section of the CAAct, the Court held that EPA's decision was within the boundaries of EPA's discretion" [65].

An independent assessment of the Court's decision stated, "The Supreme Court on Monday mostly validated the EPA's plans to regulate major sources of greenhouse gas emissions such as power plants and factories but said the agency had gone too far in interpreting its power. The court's bifurcated opinion on one hand criticized the agency for trying to rewrite provisions of the CAAct. But it nevertheless granted the Obama administration and environmentalists a big victory by agreeing that there are other ways for EPA to reach its goal of regulating the gases that contribute to global warming." "It bears mention that EPA is getting almost everything it wanted in this case," Justice Antonin Scalia said in announcing his opinion from the bench. "It sought to regulate sources that it said were responsible for 86% of all the greenhouse gases emitted from stationary sources nationwide. Under our holdings, EPA will be able to regulate sources responsible for 83% of those emissions" [66].

> The U.S. Supreme Court mostly validated the EPA's plans to regulate major sources of greenhouse gas emissions such as power plants and factories but said the agency had gone too far in interpreting its power.

The Obama administration's commitment to mitigating the causes and effects of climate change has resulted in a decrease in U.S. greenhouse gas (GHG) emissions, which is illustrated in Figure 1.4. Annual emissions have fluctuated in recent years, including a sharp uptick in 2014. But the general trend since 2004 has been down, a trend influenced by power companies' substitution of natural gas for coal as fuel. Overall, U.S. GHG emissions have declined about 13% since 2007, even with the Trump administration's policy of climate change denial and support for the fossil fuel industry. Electricity generation accounts for about 30% of total U.S. emissions. Greenhouse gases from cars, trucks, and the rest of the transportation sector were flat in 2019. Emissions from agriculture and industry increased slightly.

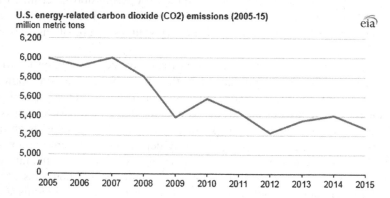

FIGURE 1.4 U.S. greenhouse gas emissions. (U.S. Energy Information Service, 2021, https://www.eia.gov/todayinenergy/detail.php?id=30712.)

The transition of electric power plant fuel from coal to gas is the primary factor in decreased GHG emissions.

Perspective: This second decision by the Supreme Court essentially validated EPA's plan to regulate greenhouse gas emissions from stationary sources, e.g., electrical power plants. This plan was built on EPA's regulatory framework for mobile sources of air pollution emissions, e.g., vehicles. Although the Court's opinion in both the cited cases split along ideological lines, a few decisions arrive from this court with a common voice and opinion. One might interpret this diversity of legal opinion as a desirable product of democratic processes.

The Supreme Court's third decision on climate change was less supportive of an EPA climate change policy, its Clean Power Plan. The Court's decision will be discussed subsequent to a description of the Clean Power Plan.

1.8.3.2 Greenhouse Gas Reporting Program, 2008

In response to the FY2008 Consolidated Appropriations Act, EPA issued the Mandatory Reporting of Greenhouse Gases Rule, which requires reporting of greenhouse gas (GHG) data and other relevant information from large sources and suppliers in the U.S. The purpose of the rule is to collect accurate and timely GHG data to inform future policy decisions. In general, the Rule is referred to as 40 CFR Part 98 (Part 98). Implementation of Part 98 is referred to as the Greenhouse Gas Reporting Program (GHGRP).

Suppliers of certain products that would result in GHG emissions if released, combusted, or oxidized; direct emitting source categories; and facilities that inject CO_2 underground for geologic sequestration or any purpose other than geologic sequestration, are covered in Part 98. Facilities that emit 25,000 metric tons or more per year of GHGs are required to submit annual reports to EPA. Part 98 was published in the *Federal Register* on October 30, 2009.

Categories subject to Part 98 began reporting their yearly emissions with the 2010 reporting year. Additional sources began reporting yearly emissions in September 2012, bringing the total to 41 source categories reporting. In January 2012, EPA made the first year of GHGRP reporting data available to the public through its interactive Data Publication Tool, called Facility Level Information on Greenhouse gases Tool (FLIGHT) [67].

1.8.3.3 Climate Action Plan, 2013

For several reasons, including accrued climate change science, global political pressure, and urging from domestic environmental organizations, the Barack Obama (D-IL) administration became the first U.S. presidential administration to adopt policies on climate change. On June 25, 2013, the Obama administration announced its Climate Action Plan. The Plan states, "While no single step can reverse the effects of climate change, we have a moral obligation to act on behalf of future generations. Climate change represents one of the major challenges of the 21st century, but as a nation of innovators, we can and will meet this challenge in a way that advances our

The Obama administration became the first U.S. presidential administration to adopt policies on climate change.

economy, our environment, and public health all at the same time." That is why the President's comprehensive plan takes action to [68]:

"Cuts Carbon Pollution in America: In 2012, U.S. carbon pollution from the energy sector fell to the lowest level in two decades even as the economy continued to grow. To build on this progress, the Obama administration is putting in place tough new rules to cut carbon pollution—just like we have for other toxins like mercury and arsenic—so we protect the health of our children and move our economy toward American-made clean energy sources that will create good jobs and lower home energy bills. For example, the plan:

- Directs EPA to work closely with states, industry, and other stakeholders to establish carbon pollution standards for both new and existing power plants;
- Makes up to $8 billion in loan guarantee authority available for a wide array of advanced fossil energy and efficiency projects to support investments in innovative technologies;
- Directs DOI to permit enough renewables project—like wind and solar—on public lands by 2020 to power more than 6 million homes; designates the first-ever hydropower project for priority permitting; and sets a new goal to install 100 megawatts of renewables on federally assisted housing by 2020; while maintaining the commitment to deploy renewables on military installations;
- Expands the President's Better Building Challenge, focusing on helping commercial, industrial, and multifamily buildings cut waste and become at least 20% more energy efficient by 2020;
- Sets a goal to reduce carbon pollution by at least 3 billion metric tons cumulatively by 2030— more than half of the annual carbon pollution from the U.S. energy sector—through efficiency standards set over the course of the Obama administration for appliances and federal buildings;
- Commits to partnering with industry and stakeholders to develop fuel economy standards for heavy-duty vehicles to save families money at the pump and further reduce reliance on foreign oil and fuel consumption post-2018; and
- Leverages new opportunities to reduce pollution of highly potent greenhouse gases known as hydrofluorocarbons; directs agencies to develop a comprehensive methane strategy; and commits to protect our forests and critical landscapes.

Prepares the United States for the Impacts of Climate Change. Even as we take new steps to cut carbon pollution, we must also prepare for the impacts of a changing climate that are already being felt across the country. Building on progress over the last 4 years, the plan:

- Directs agencies to support local climate-resilient investment by removing barriers or counterproductive policies and modernizing programs; and establishes a short-term task force of state, local, and tribal officials to advise on key actions the Federal government can take to help strengthen communities on the ground;

- Pilots innovative strategies in the Hurricane Sandy-affected region to strengthen communities against future extreme weather and other climate impacts; and building on a new, consistent flood-risk reduction standard established for the Sandy-affected region, agencies will update flood-risk reduction standards for all federally funded projects;
- Launches an effort to create sustainable and resilient hospitals in the face of climate change through a public-private partnership with the healthcare industry;
- Maintains agricultural productivity by delivering tailored, science-based knowledge to farmers, ranchers, and landowners; and helps communities prepare for drought and wildfire by launching a National Drought Resilience Partnership and by expanding and prioritizing forest—and rangeland—restoration efforts to make areas less vulnerable to catastrophic fire; and
- Provides climate preparedness tools and information needed by state, local, and private-sector leaders through a centralized "toolkit" and a new Climate Data Initiative.

Lead international efforts to address global climate change: Just as no country is immune from the impacts of climate change, no country can meet this challenge alone. That is why it is imperative for the U.S. to couple action at home with leadership internationally. America must help forge a truly global solution to this global challenge by galvanizing international action to significantly reduce emissions, prepare for climate impacts, and drive progress through international negotiations. For example, the plan:

- Commits to expand major new and existing international initiatives, including bilateral initiatives with China, India, and other major emitting countries;
- Leads global sector public financing toward cleaner energy by calling for the end of U.S. government support for public financing of new coal-fired powers plants overseas, except for the most efficient coal technology available in the world's poorest countries, or facilities deploying carbon capture and sequestration technologies; and
- Strengthens global resilience to climate change by expanding government and local community planning and response capacities" [68].

The Climate Action Plan did not extend beyond President Obama's term in office, which ended in 2016. One of the key elements of the Plan is the Clean Power Plan. The U.S. President Donald Trump (R-NY) administration announced in 2017 its disavowal of the Clean Power Plan, as discussed subsequently.

1.8.3.4   Clean Power Plan, 2015
On August 3, 2015, the Obama administration announced its Clean Power Plan (CPP), the administration's linchpin policy on reducing U.S. carbon emissions that contribute to climate change. The Plan targets electricity power generation plants,

with the goal of reducing carbon dioxide emissions by forcing plants to use alternative, less polluting fuels, and cleaner technology for power generation. The 2015 Obama-era rule aimed to reduce greenhouse gas emissions from U.S. power plants, the second-largest source of greenhouse gases in the U.S. The CPP gave 47 U.S. states unique emissions targets while leaving it up to them how to get there. The EPA invoked health provisions of the Clean Air Act to make this rule, arguing that cutting greenhouse gas emissions would also limit other pollutants. That, in turn, would avert 3,600 premature deaths, 90,000 asthma attacks in children, and 1,700 heart attacks each year according to the EPA's analysis.

The Clean Power Plan was established through EPA regulations carbon emission standards for power plants and customized goals for U.S. states to cut the carbon pollution generated within their borders. The Clean Power Plan was the culmination of EPA's efforts to generate an environmental policy that would result in lower carbon emissions. The Plan also served as the Obama administration's principal contribution to global negotiations on climate change. The administration promoted the Plan as an example of one industrialized nation's commitment to mitigating the effects of climate change. A summary provided by the EPA of the Clean Power Plan's key provisions follows [69].

- The CAAct—under section 111(d)—creates a partnership between EPA, states, tribes, and U.S. territories—with EPA setting a goal and states, territories, and tribes choosing how they will meet it.
- The final Clean Power Plan follows that approach. EPA is establishing interim and final CO_2 emission performance rates for two subcategories of fossil fuel-fired electric generating units (EGUs):
- Fossil fuel-fired electric steam generating units (generally, coal- and oil-fired power plants)
- Natural gas-fired combined cycle generating units

The Clean Power Plan established EPA regulations on carbon emission standards for power plants and mandated states to develop customized plans to enforce standards for power plants within their borders.

- To maximize the range of choices available to states in implementing the standards and to utilities in meeting them, EPA is establishing interim and final statewide goals in three forms:
- A rate-based state goal measured in pounds per megawatt hour (lb/MWh);
- A mass-based state goal measured in total short tons of CO_2;
- A mass-based state goal with a new source complement measured in total short tons of CO_2;
- States then develop and implement plans that ensure that the power plants in their state—either individually, together, or in combination with other measures—achieve the interim CO_2 emissions performance rates over the period of 2022–2029 and the final CO_2 emission performance rates, rate-based goals, or mass-based goals by 2030.

- These final guidelines are consistent with the law and align with the approach that Congress and EPA have always taken to regulate emissions from this and all other industrial sectors—setting source-level, source category-wide standards that sources can meet through a variety of technologies and measures.

The Clean Power Plan was the Obama administration's centerpiece policy on mitigating the effects of climate change. Moreover, the Plan was the primary offering of U.S. diplomacy at the Paris Agreement. The Plan's supporters promoted it as a statement of U.S. intent to lead global efforts in climate change policies. But because the Plan transacts into substantive economic and sociopolitical changes in the U.S., the Plan was quickly challenged in federal courts. The litigation once again reached the docket of the U.S. Supreme Court, as described subsequently.

However, the Trump administration issued an executive order on March 28, 2017, entitled "Energy Independence," which is targeted at revoking the Obama administration's Clean Power Plan. The Obama plan would have discouraged U.S. coal production in order to reduce the emission of greenhouse gases emitted when fossil fuels are combusted. Whether electric utilities will increase their use of coal, however, is uncertain, given the alternative of less expensive natural gas supplies and the increasing installation of renewable sources of energy [70].

The third key Supreme Court decision on climate change resulted from litigation by some U.S. states and the energy industry both preceded and followed EPAs' issuance of the Clean Power Plan. Some states did not want the responsibility of preparing plans on how they would implement the Plan. Energy industries argued that the Plan would force costly changes in technology and result in high energy costs for customers. On February 9, 2016, the Supreme Court stayed (i.e., set aside) by 5-4 vote EPA's implementation of the Clean Power Plan pending judicial review by a lower federal court. The Court's decision was not on the merits of the rule. EPA asserts that the Clean Power Plan will be upheld when the merits are considered because the rule rests on strong scientific and legal foundations. EPA further asserts that for the states that chose to continue working on cutting carbon pollution, EPA will continue to provide tools and support [71].

> The Obama administration's Clean Power Plan can be characterized as a policy comprising one part technology and one part politics.

Perspective: The Obama administration's Clean Power Plan can be characterized as a policy comprising one part technology and one part politics. The Plan is one part of technology because it would force major changes in how the U.S. generates much of its energy supply. More specifically, the use of carbonaceous fuels would cease for combustion in electricity generating plants, with replacement by sources of renewable energy, e.g., solar and wind, and fuels for combustion that yield a smaller carbon footprint, e.g., natural gas.

As to the political part of the Clean Power Plan, the U.S. and other countries with strong national economies had been criticized by countries with low- to mid-level economies for attempting to impose unpopular climate control actions on them. As previously discussed, this divide between the "haves" and "still developing" economies had led to stalemates at climate control meetings previous to the Paris conference, yielding little agreement on how to proceed in developing global policies on mitigation of climate change. Because the U.S. had historically been the largest emitter of carbon into the atmosphere, the U.S. found itself in an awkward political stance in global meetings on climate change. This stance changed at the Paris meeting.

> The environmental policy of the Trump administration was a shift from the policy priorities and goals of his predecessor, Barack Obama.

The Clean Power Plan became the U.S. centerpiece as a political action for demonstrating to other nations of U.S. intention to seriously reduce its national carbon footprint on the global environment. As previously commented, the Trump administration has disavowed support for the Clean Power Plan. In April 2019, the EPA sent a replacement plan, which would let states set their own rules, to the White House for budget review. In August 2019, a coalition of 29 states and cities sued to block the Trump administration from easing restrictions on coal-burning power plants, setting up a case that could determine how much leverage the federal government has to fight climate change in the future. It is anticipated that the suit will reach the U.S. Supreme Court for determination [72].

1.8.4 TRUMP ADMINISTRATION'S CLIMATE CHANGE POLICIES

The environmental policy of the U.S. President Donald Trump (R-NY) administration represents a shift from the policy priorities and goals of his predecessor, U.S. President Barack Obama (D-IL) [73]. While Obama's environmental agenda prioritized the reduction of carbon emissions through the use of clean renewable energy, the Trump administration sought to increase fossil fuel use and scrap many environmental regulations, which he has referred to as impediments to business. Trump has pulled the U.S. out of the Paris climate agreement, leaving the U.S. the only nation that has not joined the agreement. Trump campaigned on a pledge to roll back government regulations, and within days of taking office he began to implement his "America First Energy Plan," which does not mention renewable energy, rather the plan emphasizes the use of fossil fuels. President Trump often characterized climate change as a "hoax" that had no basis in science.

1.8.4.1 Affordable Clean Energy, 2019

In June 2019 the EPA announced its Affordable Clean Energy (ACE) rule, the Trump administration's substitute for the Obama administration's Clean Power rule [74]. As stated by the EPA, the ACE rule establishes emissions guidelines for U.S. states to use when developing plans to limit CO_2 at their coal-fired power plants. The move largely gives states the authority to decide how far to scale back emissions, or not to do it all, and significantly reduces the federal government's role in setting standards. Specifically, ACE identifies heat rate improvements as the best system of emission reduction (BSER) for CO_2 from coal-fired power plants, and these improvements can be made at individual facilities. States will have 3 years to submit plans, which is in line with other planning timelines under the Clean Air Act.

> The ACE rule establishes emissions guidelines for U.S. states to use when developing plans to limit CO2 at their coal-fired power plants.

Also contained within the rule are new implementing regulations for ACE and future existing-source rules under Clean Air Act Section 111(d). These guidelines will inform states as they set unit-specific standards of performance. For example, states can take a particular source's remaining useful life and other factors into account when establishing a standard of performance for that source.

The EPA asserted that the ACE will reduce emissions of CO_2, mercury, as well as precursors for pollutants like fine particulate matter and ground-level ozone. The EPA projected the ACE rule would reduce by 2030: CO_2 emissions by 11 million short tons, SO_2 emissions by 5,700 tons, NOx emissions by 7,100 tons, $PM_{2.5}$ emissions by 400 tons, and mercury emissions by 59 pounds. The EPA also projected that the ACE rule will result in annual net benefits of \$120–\$730 million, including costs, domestic climate benefits, and health co-benefits.

The Trump administration's EPA also forecast that with ACE, along with additional expected emissions reductions based on long-term industry trends, CO_2 emissions from the electric sector would fall by as much as 35% less than 2005 levels in 2030 [74]. Attorneys general in California, Oregon, Washington State, Iowa, Colorado, and New York announced they intended to sue to block the ACE rule. On January 19, 2021, the U.S. Court of Appeals for the District of Columbia invalidated the Trump ACE rule [74a]. The court called the rule a "fundamental misconstruction" of the nation's environmental laws, devised through a "tortured series of misreadings" of legal statute. The court's invalidation of the Trump administration's Affordable Clean Energy rule brought to an end that administration's major attempt to forego policy actions on climate change mitigation.

> The Massachusetts plan to reduce carbon emissions from power plants is just one example of states seizing the initiative to force reductions in greenhouse gases.

Rather than follow the Trump administration's policies and actions on limiting carbon emissions from power plants, some U.S. states have enacted their own policies [75]. In 2018, the Massachusetts Department of Environmental Protection finalized rules to require power plants within the state's borders to reduce their emissions annually, amounting to a 7% reduction in greenhouse gas emissions from current levels by 2020 and an 80% reduction by 2050. The Massachusetts plan to reduce carbon emissions from power plants is just one example of states seizing the initiative to force reductions in greenhouse gases. California lawmakers voted to require that state's utilities to use 100% carbon-free power by 2050. Hawaii has a similar requirement. Other states have increased their renewable portfolio standards for clean energy.

1.8.4.2 Withdrawal from the Paris Agreement

On June 1, 2017, U.S. President Donald Trump (R-NY) announced from the White House's Rose Garden, "As of today, the United States will cease all implementation of the nonbinding Paris accord—and the draconian financial and economic burdens the agreement imposes on our country" [76]. The President asserted that the Paris Agreement was an agreement that disadvantages the U.S. to the exclusive benefit of other countries. His administration did not describe the alleged disadvantages. However, there is no way for the U.S.—or any other country that signed it—to formally withdraw from the Paris Agreement until 4 years after it went into effect. On November 4, 2020, the U.S. formally withdrew from the Agreement, completing the Trump administration's exit, becoming the only big country besides Iran and Turkey not signed on to the landmark agreement to limit greenhouse gas emissions.

In response to the Trump administration's rejection of the Paris Agreement, more than 1,000 U.S. companies and institutions, including more than a dozen Fortune 500 businesses, signed onto a statement—"We Are Still In"—saying they're committed to meeting the Paris targets. The statement calls Trump's decision "a grave mistake that endangers the American public and hurts America's economic security and diplomatic reputation." The coalition's numbers climbed past 1,400. Twenty U.S. states that together represent the world's third-largest economy, and more than 200 cities have also committed to the Paris accord through various coalitions [77].

> On June 1, 2017, U.S. President Donald Trump announced, "As of today, the U.S. will cease all implementation of the nonbinding Paris accord."

1.8.4.3 Rescinding of Regulations Relevant to Climate Change

A major news source noted that the Trump administration had rescinded 84 environmental rules through August 2019 [78]. In general, the rules were viewed by the administration as a counter to support of the U.S. fossil fuel industries. Many of these efforts have been challenged in the courts. The following two reversals are particularly relevant to climate change issues.

In August 2018, the Trump administration released its Safer Affordable Fuel Efficient (SAFE) Vehicles Proposed Rule for Model Years 2021–2026. The proposed rule amends certain existing Corporate Average Fuel Economy (CAFE) and greenhouse gas emissions standards for passenger cars and light trucks and establishes new standards, covering model years 2021–2026.

> In application, the SAFE rule would lower fuel economy standards, resulting in more fuel being used by vehicles, producing greater amounts of air pollution.

The new rule, announced in March 2020, withdraws the Obama 2012 rule that required automakers' fleets to average about 54 miles per gallon by 2025. Instead, the fleets would have to average about 40 miles per gallon [79]. According to the EPA plan, the new standard would lead to nearly a billion more tons of CO_2 released and the consumption of about 80 billion more gallons of gasoline over the lifetime of the vehicles built during the terms of the rule. The Trump administration argued that the Obama rule was an economic burden to U.S. automakers.

Another rule change was announced in August 2019 when the EPA proposed a rule to eliminate federal requirements that oil and gas companies install technology to detect and fix methane leaks from wells, pipelines, and storage facilities [78]. As previously discussed, methane is a powerful greenhouse gas. EPA officials said the new methane rule, which would replace one from the Obama administration, is a response to President Trump's calls to trim regulations that impede economic growth or keep the U.S. reliant on energy imports. EPA's economic analysis of the rule estimated that it would save the oil and natural gas industry \$17–\$19 million a year. For comparison, the annual revenue of the U.S. oil industry as a whole typically ranges between \$100 and \$150 billion.

Concerning another Obama administration greenhouse gas rule, in February 2020, the Trump administration's EPA finalized a rule that eliminates leak prevention and repair requirements for hydrofluorocarbons (HFCs), the heat-trapping pollutants used in commercial and industrial refrigeration [80]. Though HFCs currently represent around 1% of total greenhouse gases, their impact on global warming can be hundreds to thousands of times greater than that of CO_2 per unit of mass.

Of policy note, in 2020 the Government Accountability Office (GAO) reported that the Trump administration had set a rock-bottom price on the damages done by greenhouse gas emissions, enabling the government to justify the costs of repealing or weakening dozens of climate change regulations. The GAO found the Trump administration had estimated the harm that global warming will cause future generations to be seven times lower than previous federal estimates, a figure used to justify rollbacks of federal climate change regulations [81].

1.8.5 Biden Administration's Climate Team

The U.S. President national election of November 2020 was won by Joseph R. Biden Jr. (D-DE), replacing Donald J. Trump (R-NY). President-elect Biden stated his four most important issues upon taking office in January 2021 were eliminating COVID-19, improving the U.S. national economy, addressing social justice issues, and dealing with climate change. Biden's statement on climate change was in distinction to the Trump administration's denial of climate change and revoking of federal policies intended to mitigate climate change.

> President-elect Biden stated his four most important issues upon taking office in January 2021 were: eliminating COVID-19, improving the U.S. national economy, addressing social justice issues, and dealing with climate change.

In support of his priority of dealing with climate change, president-elect Biden announced his choices of a climate team. Top lieutenant will be Gina McCarthy, former President Obama's EPA administrator whom Mr. Biden tapped to head a new White House Office of Climate Policy [82]. The group includes Representative Deb Haaland (D-NM), Mr. Biden's choice to lead the Department of the Interior, and Jennifer Granholm, the former governor of Michigan whom Mr. Biden selected to be Energy secretary. Michael Regan, North Carolina's top environmental regulator, was named to lead the EPA and Brenda Mallory, an environmental attorney, will chair the Council on Environmental Quality. Ms. Haaland is the first Native American to serve in a president's cabinet, and Mr. Regan is the first African-American to serve as EPA administrator. Former Secretary of State John Kerry will serve as a climate ambassador on issues with other countries.

1.8.6   BIDEN ADMINISTRATION'S CLIMATE POLICIES

The Biden administration's approach to climate change has taken shape in 2021. On January 20, 2021, President Biden signed an executive order to rejoin the Paris agreement, effective February 19, 2021, thereby reversing the action of the Trump administration. Further, the 10 executive actions Biden took on his first day as president to combat the crisis and reduce emissions are as follows [83]:

- Require limits on methane pollution for oil and gas operations.
- Use the federal government procurement system to work toward 100% clean energy and zero-emissions vehicles.
- Ensure US government buildings and facilities are more efficient and climate-ready.
- Implement the already-existing Clean Air Act and reduce greenhouse gas emissions from transportation by developing new fuel economy standards to ensure all new sales for light- and medium-duty vehicles will be electrified, and annual improvements for heavy-duty vehicles.
- Double down on liquid fuels like advanced biofuels and make agriculture a key part of the solution to the climate crisis.
- Reduce emissions and cut consumer costs through new standards for appliance and building efficiency.
- Require federal permit decisions to consider effects of greenhouse gas emissions and climate change and ensure every federal infrastructure investment reduces climate pollution.
- Require public companies to disclose climate risks and greenhouse gas emissions in their operations and supply chains.
- Protect biodiversity, slow extinction rates, and conserve 30% of America's lands and waters by 2030.
- Permanently protect the Arctic National Wildlife Refuge, establish national parks and monuments, ban new oil and gas permits on public lands and waters, modify royalties to account for climate costs and creating programs to enhance reforestation and develop renewable energy on federal lands and waters to double offshore wind by 2030.

Additionally, Biden appointed John Kerry as his special presidential envoy for climate. Kerry, who was President Barack Obama's Secretary of State, will be a Cabinet-level official in Biden's administration and will sit on the National Security Council. As Secretary of State, he played a key role in negotiating the Paris climate agreement, which was adopted by nearly 200 nations in 2015.

On April 22–23, 2021, President Biden convened a virtual international conference on climate change. Several national leaders participated. Mr. Biden pledged that the U.S. would cut its greenhouse gas emissions by 50% by the year 2030. He further vowed that the U.S. would substantially increase the amount of money the U.S. would offer to developing countries to address climate change [83a]. Canada pledged to cut emissions 40%–45% from 2005 levels by 2030, and Japan committed to cut emissions 46% below 2013 levels by 2030. Russia stated a reduction in emissions by

2050, and China and India representatives repeated previous pledges of China's draw down of carbon emissions to net zero by 2060 and India's pledge to install 150 giga-watts of renewable energy by 2030. Whether these pledges by national leaders can be codified into national or international treaties remains to be determined.

1.8.7 POLICY FRAMEWORKS FOR MITIGATING CLIMATE CHANGE

There are policies other than the U.S. federal government's Clean Power Plan and Affordable Clean Energy Plan that have been proposed or actually implemented as means to reduce the emission of greenhouse gases into the atmosphere. Two of these policies are a tax on carbon and a system of jmj of carbon emissions. Both policies have been adopted by certain policymakers, but not at the U.S. federal level. The passage of time may advance both kinds of policies to a wider congress of policymakers.

1.8.7.1 Carbon Taxes and Emissions Trading Systems

A carbon tax is a form of explicit carbon pricing; it refers to a tax directly linked to the level of CO_2 emissions, often expressed as a value per ton CO_2 equivalent (per tCO_2e).

"Carbon taxes provide certainty in regard to the marginal cost faced by emitters per tCO_2e, but do not guarantee a maximum level of emission reductions, unlike an emissions trading scheme. However, this economic instrument can be used to achieve a cost-effective reduction in emissions. Since a carbon tax puts a price on each ton of GHG emitted, it sends a price signal that gradually causes a market response across an entire economy, creat-ing incentives for emitters to shift to less greenhouse gas-intensive ways of produc-tion and ultimately resulting in reduced emissions" [84].

> A carbon tax is a form of explicit carbon pricing; it refers to a tax directly linked to the level of CO_2 emissions.

Carbon taxes can be introduced as an independent instrument or they can exist alongside other carbon pricing instrument, such as an energy tax. While the experience with direct carbon tax implementation is relatively new, such instruments are being increasingly introduced at a fast pace.

Currently, 14 countries and one Canadian province have adopted somewhat different forms of carbon taxes, according to what form of carbon or energy source is subject to taxation. Listed in Table 1.1 is an overview of existing national jurisdictions that have introduced a direct carbon tax [84]. As an example, the Danish carbon tax covers all consumption of fossil fuels (natural gas, oil, and coal), with partial exemption and refund provisions for sectors covered by the EU Emissions Trading System (ETS), described in a subsequent section of this chapter, energy-intensive processes, exported goods, fuels in refineries, and many transport-related activities. Fuels used for electricity production are also not taxed by the Danish carbon tax, but instead a tax on electricity production applies. As a second example, the carbon tax in British Columbia applies to the purchase or use of fuels within the province. The carbon tax there is revenue neutral; all funds generated by the tax are returned there to citizens through reductions in other taxes [84].

A carbon tax can be a regressive tax, depending on a country's economic structure. Recall that regressive tax is a tax that takes a larger percentage from low-income persons

TABLE 1.1
Countries with Taxes on Carbon [88]

Country	Year of Adoption	Cost
Canada/British Columbia	2008	CAD30 per tCO_2e (2012)
Chile	2014	USD5 per tCO_2e (2018)
Costa Rica	1997	3.5% tax on hydrocarbon fossil fuels
Denmark	1992	USD31 per tCO_2e (2014)
Finland	1990	EUR35 per tCO_2e (2013)
France	2014	EUR7 per tCO_2e (2014)
Iceland	2010	USD10 per tCO_2e (2014)
Ireland	2010	EUR 20 per tCO_2e (2013)
Japan	2012	USD2 per tCO_2e (2014)
Mexico	2012	Mex\$ 10-50 per tCO_2e (2014) (Depending on fuel type)
Norway	1991	USD 4-69 per tCO_2e (2014) (Depending on fossil fuel type and usage)
South Africa	2016	R120/tCO_2 (Proposed tax rate for 2016) (Tax is proposed to increase by 10% per year until end-2019)
Sweden	1991	USD168 per tCO_2e (2014)
Switzerland	2008	USD 68 per tCO_2e (2014)

than from those with high income. Because, in general, persons at or below poverty levels tend to be more energy-dependent in the context of income spent on essential resources and services, income spent on energy supply is disproportionally spent by persons of low income. Therefore, a carbon tax can be a regressive tax for the poor or low-income populations. Governments can make a carbon tax less regressive or neutral by utilizing policies of tax rebates or lump-sum payments to persons adversely impacted by carbon taxes.

> Fourteen countries and one Canadian province have adopted forms of carbon taxes, according to what form of carbon or energy source is subject to taxation.

An important study examined the impact of carbon taxes on the economy [85]. Eleven teams participated in the 2018 Stanford Energy Modeling Forum (EMF) project, examining the economic and environmental impacts of a carbon tax. The studies included "revenue recycling," in which the funds generated from a carbon tax are returned to taxpayers either through regular household rebate checks or by offsetting income taxes, an approach used in British Columbia. The key findings were consistent across the 11 modeling teams. First, a carbon tax is effective at reducing carbon pollution, although the structure of the tax (the price and the rate at which it raises) is important. Second, this type of revenue-neutral carbon tax would have a very modest impact on the economy in terms of gross domestic product (GDP). In all likelihood, it would slightly slow economic growth, but by an amount that would be more than offset by the benefits of cutting pollution and slowing global warming.

Perspective: A carbon tax has the following advantages: Predictable carbon prices, easier to understand, and revenue can be returned via tax cuts and/or used for public goods. Disadvantages include most politicians are hesitant to advocate for any tax, the public distrusts government programs, CO_2 tax revenues might end up being wasted in special interest spending, or priorities that are only peripheral to climate or sustainability [86].

1.8.7.2 Cap and Trade of Greenhouse Gases

The second form of non-regulatory policy purposed for the reduction of greenhouse gases is a cap-and-trade system. As will be discussed in Chapter 2, the cap-and-trade system has been used successfully as a means to reduce the release of certain air pollutants, for example, the elimination of "acid rain" in the eastern U.S. was a consequence of a cap-and-trade system for reducing emissions of sulfur compounds. As will become evident, cap and trade is a policy different from a tax on carbon emissions because sources of greenhouse gas emissions play a more active and direct role in the policy's implementation.

1.8.7.2.1 EU's Emissions Trading System

The EU launched the EU Emissions Trading System (EU ETS) in 2005 as the cornerstone of its strategy for reducing emissions of CO_2 and other greenhouse gases at least cost. In contrast to traditional "command and control" regulation, emissions trading harnesses market forces to find the cheapest ways to reduce emissions. The EU ETS is a cap-and-trade system. It caps the total volume of GHG emissions from installations and aircraft operators responsible for around 45% of EU GHG emissions. The system allows the trading of emission allowances so that the total emissions of the installations and aircraft operators stay within the cap and the least-cost measures can be taken up to reduce emissions [87].

> The EU launched the EU Emissions Trading System in 2005 as the cornerstone of its strategy for reducing emissions of CO_2 and other greenhouse gases at least cost.

The system works by putting a limit on overall emissions from high-emitting industry sectors, a limit that is reduced each year. Within this limit, companies can buy and sell emission allowances as needed. This cap-and-trade approach gives companies the flexibility they need to cut their emissions in the most cost-effective way. The EU ETS covers more than 11,000 power stations and manufacturing plants in the 28 EU member states as well as Iceland, Liechtenstein, and Norway. Aviation operators flying within and between most of these countries are also covered. In total, around 45% of total EU CO_2 emissions are limited by the EU ETS. In 2015, the EU was responsible for around 10% of world greenhouse gas emissions. The EU is one of the major economies with the lowest per capita emissions. EU emissions were reduced by 22% between 1990 and 2017, while the economy grew by 58% over the same period, figures that attest to the effectiveness of the EU ETS.

> The EU launched the EU Emissions Trading System (EU ETS) in 2005 as the cornerstone of its strategy for reducing emissions of CO_2.

The EU ETS is the world's first major carbon market and is the largest. As the first international emissions trading system to address greenhouse gas emissions from companies, the European system accounts for more than three-quarters of the trading volume of the international carbon market and functions as its engine. By putting a price on carbon and thereby giving a financial value to each ton of emissions saved, the EU ETS has placed climate change policy on the agenda of company boards across Europe. A sufficiently high carbon price also promotes investment in clean, low-carbon technologies. By allowing companies to buy credits from emission-saving projects around the world, the EU ETS acts also as a major driver of investment in clean technologies and low-carbon solutions, particularly in developing countries.

1.8.7.2.2   Regional Greenhouse Gas Initiative

The Regional Greenhouse Gas Initiative (RGGI) was the first mandatory cap-and-trade program in the U.S. to limit CO_2 from the power sector [88]. It consists of Connecticut, Delaware, Maine, Maryland, Massachusetts, New Hampshire, New York, Rhode Island, and Vermont. Following discussions initiated by the Governor of New York in 2003, the RGGI was established in 2005 and administered its first auction of CO_2 emissions allowances in 2008. By 2020, the RGGI CO_2 cap is projected to contribute to a 45% reduction in the region's annual power sector CO_2 emissions from 2005 levels, or between 80 and 90 million short tons (tons) of CO_2. The RGGI requires fossil fuel power plants more than 25 megawatts in participating states to obtain an allowance for each ton of CO_2 emitted annually. Power plants within the region may comply with the cap by purchasing allowances from quarterly auctions, other generators within the region, or offset projects.

The RGGI states sell nearly all emission allowances through auctions and invest proceeds in energy efficiency, renewable energy, and other consumer benefit programs. These programs are spurring innovation in the clean energy economy and creating green jobs in the RGGI states [88]. RGGI, Inc. has no regulatory or enforcement authority. All such sovereign authority is reserved within the RGGI states. A 3-year review found $1.4 billion in economic benefits across the nine RGGI states, no harm to electric grid reliability, and long-term benefits for residents [89].

1.8.7.2.3   California Cap and Trade of Greenhouse Gases

California has often been the trendsetter in U.S. sociopolitical affairs. The state ranks first in population (38.8 million in 2013) and third in total area behind Alaska and Texas. In 2014, if California were a separate country, it would have had the eighth largest economy in the world, ahead of Italy, India, and Russia [90]. As a state rich in economic, natural, and culture, policies established by California often are emulated in whole or part by other entities. This is certainly true of the state's leadership in environmental health policymaking. For instance, as observed in Chapter 2, California was the first U.S. state to investigate and respond to emerging problems of air pollution, actions that predated those of the U.S. federal government. The state's response to climate change has correspondingly preceded policies of other U.S. states and the U.S. federal government. In particular, California's cap-and-trade policy for the purpose of lowering emissions of greenhouse gases is noteworthy and described in the following.

In 2013 California launched its cap-and-trade program, which uses a market-based mechanism to lower the state's greenhouse gas emissions. The state's cap-and-trade rules went into effect on January 1, 2013, and applied at that time to large electric power plants and large industrial plants. In 2015, the cap-and-trade rules were extended to fuel distributors (including distributors of heating and transportation fuels). At that stage, the program encompassed around 360 businesses throughout California and nearly 85% of the state's total greenhouse gas emissions. The state forecasts that its emissions trading system will reduce greenhouse gas emissions from regulated entities by more than 16% between 2013 and 2020. It is a central component of the state's broader strategy to reduce total greenhouse gas emissions to 1990 levels by 2020 [91].

> In 2013, California launched its cap-and-trade program, which uses a market-based mechanism to lower the state's greenhouse gas emissions.

Under a cap-and-trade system, companies must hold enough emission allowances to cover their emissions and are free to buy and sell allowances on the open market. California held its first auction of greenhouse gas allowances on November 14, 2012. In November 2017, a record 89.3 million emission allowances were sold. The sale raised approximately $1.3 billion in total, with $860 million going to the California Greenhouse Gas Reduction Fund [92]. In July 2017 California legislators in a bipartisan vote extended the cap-and-trade program through 2030. California's program is second in size only to the EU ETS, based on the amount of emissions covered. California's program will provide salient experience in how an economy-wide cap-and-trade system can function in the U.S.

Perspective: Cap and trade can have the following advantages: Predictable carbon emissions, fewer political obstacles than a tax on emissions, revenue can be returned via rebates and/or used, for public good, and revenue rises as emissions decline. Some of the disadvantages include: Total emissions are capped, but the dollar price is unknown and dependent on many market variables; depending on the scope, method of allocation, and other design elements, too many permits may be issued, and other market imperfections may arise [86].

1.8.7.3 Reforestation Policies

Forests are a vital part of the carbon cycle, both storing and releasing this essential element in a dynamic process of growth, decay, disturbance, and renewal. The "carbon cycle" refers to the constant movement of carbon from the land and water through the atmosphere and living organisms. This cycle is fundamental to life on Earth. On a global scale, forests help maintain Earth's carbon balance. Over the past four decades, forests have moderated climate change by absorbing about one-quarter of the carbon emitted by human activities such as the burning of fossil fuels and the changing of land uses. Carbon uptake by forests reduces the rate at which carbon accumulates in the atmosphere and thus reduces the rate at which climate change occurs [93].

> Carbon uptake by forests reduces the rate at which carbon accumulates in the atmosphere and thus reduces the rate at which climate change occurs.

As observed by the Intergovernmental Panel on Climate Change, supporting natural systems that can soak up carbon is widely accepted as a major component of any climate change mitigation strategy—in addition to deploying clean energy, switching to electric vehicles, and curbing consumption overall. A worldwide assessment of current and potential forestation using satellite imagery estimates that letting saplings regrow on land where forests have been cleared would increase global forested area by one-third and remove 205 billion metric tons of carbon from the atmosphere. That figure is two-thirds of the approximately 300 billion metric tons of carbon humans have emitted to Earth's atmosphere since the dawn of the Industrial Revolution [94].

However, the current rate of deforestation is contributing to climate change, according to research by an international team of scientists [95]. The group conducted a mathematical reproduction of the planet's current atmospheric conditions through computer modeling that used a numerical model of the atmosphere developed by the UK's national meteorological service. The researchers concluded that the current pace of deforestation—some 7,000 km² per year in the case of Amazonia—will in three to four decades result in a massive accumulated loss of forests. In turn, this loss would intensify global warming regardless of all efforts to reduce greenhouse gas emissions.

On a more positive note, a study by Swiss university ecologists reported there is enough room in the world's existing parks, forests, and abandoned land to plant 1.2 trillion additional trees, which would have the CO_2 storage capacity to cancel out a decade of carbon dioxide emissions [96]. Combining forest inventory data from 1.2 million locations around the world and satellite images, the scientists estimated there are 3 trillion trees on Earth—seven times more than previous estimates. And they also found that there is abundant space to restore millions of acres of additional forests, not counting urban and agricultural land.

> Swiss university ecologists reported there is enough room in the world's existing parks, forests, and abandoned land to plant 1.2 trillion additional trees.

1.8.7.4 One Trillion Trees Initiative

The 2020 World Economic Forum, held in Davos, announced the creation of the One Trillion Tree Initiative platform for governments, businesses, and civil society to provide support to the UN Decade on Ecosystem Restoration (2020–2030), led by UNEP and FAO. As of February 4, 2020, 13.6 billion trees have been planted. One country, Ethiopia, made tree planting a national policy and priority [97]. In a historic move for the east African nation, Ethiopia announced in July 2019 a tree-planting initiative to outdo virtually any other country in the world. Based initially at the Gulele Botanical Garden in the capital of Addis Ababa, volunteers began planting 350 million trees spanning the country's breadth. In just 12 hours, the world record was broken, with the purpose of combating the effects of deforestation and climate change. The last country to attempt such a feat was India, who has been reigning champions since 2016 when they planted 49.3 million trees in just 1 day, involving 800,000 volunteers.

> The U.S. chapter of the One Trillion Trees program, launched August 27, 2020, aims to plant in the U.S. at least 855 million trees by 2030.

The U.S. is officially participating in a global program that aims to plant 1 trillion trees worldwide, something that Republicans, including President Trump, have adopted as a way to combat climate change [98]. The U.S. chapter of the One Trillion Trees program, launched August 27, 2020, aims to plant in the U.S. at least 855 million trees by 2030. The goal of the initiative is to try to pull carbon out of the air through reforestation, though scientists have said planting trees isn't a panacea and that the U.S. will also have to significantly reduce its emissions to mitigate climate change impacts.

1.8.7.5 Africa's Great Green Wall Project

The world's most ambitious reforestation project is the Great Green Wall of Africa. The project was conceived in 2007 by the African Union as a 7,000 km (4,350-mile) cross-continental barrier stretching from Senegal to Djibouti that would hold back the advancing deserts of the Sahara and Sahel [99]. Notwithstanding the project's vision, progress has been slow. After an investment of more than $200 million, only 4 million hectares have been planted in the past decade. To achieve the 2030 target, more than twice that area will need to be restored every year at an annual cost of $4.3bn. The results have varied substantially from country to country. Ethiopia, which started reforesting earlier than other nations in the region, is a frontrunner, having reportedly planted 5.5 billion seedlings on 151,000 hectares of new forest and 792,000 of new terraces. Other countries have lagged due to different geographies, levels of governance, and economic development. Renewed international commitment will be required if the Great Green Wall is to become a reality of benefit to global climate change mitigation.

> The world's most ambitious reforestation project is the Great Green Wall of Africa.

1.8.8 CLIMATE CHANGE ECONOMICS

In addition to the impacts of climate change, e.g., ambient air temperature rises, ocean acidification, species migration, stronger natural disasters, there are economic consequences. Some studies on this subject by economists have been commended. Some of their findings are presented here.

A University of California San Diego study found that global warming is costing the U.S. economy about $250 billion per year [100]. Researchers measured the economic harm from CO_2 emissions to the globe's nearly 200 countries. The top three countries with the most to lose from climate change are the U.S., India, and Saudi Arabia, with the world's largest CO_2 emitter, China, also placing in the top five countries with the highest losses. In a different study of economic impacts, the U.S. National Oceanic and Atmospheric Administration (NOAA) reported that weather and climate-related disasters cost the U.S. a record $306 billion in 2017, the third-warmest year on record [101]. The agency said western wildfires and hurricanes Harvey, Maria, and Irma contributed to making 2017 the costliest year on record. The previous record was $215 billion in 2005, when hurricanes Katrina, Wilma, and Rita slammed the U.S. Gulf Coast. In 2020 hurricanes, wildfires, and other disasters across the U.S. caused $95 billion in damage, almost double the amount in 2019 and the third-highest losses since 2010.

> A study found that global warming is costing the U.S. economy about $250 billion per year.

On a global scale, a team of the University of Southern California studied the long-term impact of climate change on economic activity across countries, using a stochastic growth model where labor productivity is affected by country-specific climate variables—defined as deviations of temperature and precipitation from their historical norms [102]. Researchers found a persistent increase in average global temperature by 0.04°C per year, in the absence of mitigation policies, reduces world real GDP per capita by 7.22% by 2100. On the other hand, abiding by the Paris Agreement, thereby limiting the temperature increase to 0.01°C per annum, reduces the loss substantially to 1.07%.

A different kind of economic impact of climate change is presented in the *2018 Report of the Global Commission on the Economy and Climate*. The commission comments that over the next 10–15 years, the world will invest $90 trillion in new infrastructure—more than the total current stock that exists in 2019. If those investments are in line with what's needed to address climate change, it could be part of a shift that delivers $26 trillion in economic benefits between now and 2030. Concerted efforts to stop climate change by 2030 would also create 65 million new jobs and prevent 700,000 premature deaths [103].

1.8.9 CLIMATE CHANGE LITIGATION

As a matter of environmental health and policy, sharply divided opinions and political platforms have resulted in litigation issues concerning climate change. Recall that the U.S. Constitution constructs three equal branches of federal government: executive, legislative, and judicial. Of these three branches, the legislative branch has been the most silent on issues of climate change. Neither of the two major political parties in the U.S., Democrats and Republicans, whose members populate the U.S. Congress have made climate change a legislative priority. And the executive branch during the Obama and Trump administrations took polar opposite attitudes on climate change, as discussed elsewhere in this chapter. An outcome of these two branches' inattention or disagreement on climate change issues has resulted in concerned individuals and public and private organizations turning to the third branch, the judiciary. This has resulted in initiating litigation as a potential resource for addressing, and perhaps redressing, climate change concerns.

As of 2019, at least eight U.S. cities, five counties, and one state are suing some of the world's largest fossil fuel companies for selling products that contribute to global warming while misleading the public about their harms [104]. The progress of this litigation through federal courts is typically rather slow, given the large amount of litigation that confronts federal courts. Of policy note, in early 2018, a few months after the cities of Oakland and San Francisco sued several major oil companies over climate change, attorneys with the U.S. Department of Justice began a series of email exchanges and meetings with lawyers for the oil companies targeted in the litigation [105]. The messages were among 178 pages of emails exchanged by government and industry from February through May 2018 as they worked together to oppose

> As of 2019, at least eight U.S. cities, five counties, and one state are suing some of the world's largest fossil fuel companies.

the cities' lawsuits. The propriety of the U.S. government's interceding in a matter of supporting the fossil fuel industry in litigation against U.S. states became a subject of debate between environmental groups and fossil fuel advocates.

Additionally, the litigation momentum is building in other countries, too. In Canada, the Netherlands, and Ireland, citizens have taken their governments to court to demand more ambitious policies to fight climate change. In one outcome of citizen litigation, the highest court in the Netherlands in December 2019 upheld a landmark ruling that defines protection from the devastation of climate change as a human right and requires the Dutch government to set more ambitious targets to cut greenhouse gas emissions [106]. Hailed as an "immense victory for climate justice," the supreme court rejected the Dutch government's appeal against earlier rulings, handing environmental activists a final victory in a grueling 6-year legal battle. The Dutch government was directed to cut emissions by at least 25% by the end of 2020 from benchmark 1990 levels.

Elsewhere in Europe, litigants from eight countries claim EU institutions are not protecting fundamental rights. Lawyers acting for a group including a French lavender farmer and members of the indigenous Sami community in Sweden launched legal action in 2018 against the EU's institutions for failing to adequately protect them against climate change. A case is being pursued in the Luxembourg-based general court, Europe's second highest, against the European parliament and the council of the European Union for allowing overly high greenhouse gas emissions to continue until 2030 [107]. The court's decision could take years before determination.

Of particular policy note, young environmental activists from Portugal filed in September 2020 the first climate change case at the European Court of Human Rights in Strasbourg, demanding 33 countries make more ambitious emissions cuts to safeguard their future physical and mental well-being [108]. The 33 countries comprised the EU countries, Norway, Russia, Switzerland, Turkey, UK, and Ukraine. The plaintiffs were four children and two young adults. The crowdfunded legal action broke new ground by suing multiple countries both for the greenhouse emissions within their borders and also for the climate impact that their consumers and companies have elsewhere in the world through trade, fossil fuel extraction, and outsourcing. The court's decision is awaited by the plaintiffs.

1.8.10 THE PUBLIC'S ROLE IN CLIMATE CHANGE POLICYMAKING

The public has an important role, indeed responsibility, in addressing the issues attending policymaking actions on climate change. Put simply, influencing the responsible policymakers is the course of action to be taken. The currency to be used is education, communication, and persistence. Whether as an individual or member of a like-minded group, one must be educated on the issues at hand (e.g., portent of climate change on human health) and be prepared to communicate one's knowledge and concerns to policymakers, no matter whether they are in the public or private sector. Persistence in advocacy for a particular policy is usually a prerequisite for a successful outreach to policymakers. "Grassroots" advocacy can be a highly effective impactful resource for making or changing environmental health policies. As but one of many examples of grassroots advocacy are the local campaigns by anti-smoking groups. Their currency of education, communication, and persistence was used throughout the U.S. and other countries to achieve policies of no smoking in public buildings and recreation areas.

1.8.11 YOUTH ACTIVISM AGAINST CLIMATE CHANGE

The year 2019 saw the commencement of youth movements globally in protest of what they deemed too little action to combat global climate change. On September 20, 2019, masses of young people poured into the streets on every continent for a day of global climate protests [109]. Organizers estimated the turnout to be around four million in thousands of cities and towns worldwide. The number of young protesters in Berlin numbered 100,000 participants, with similar numbers in Melbourne and London. In New York City, the mayor's office estimated that 60,000 people had assembled in protest.

A different form of youth protest against climate change inaction occurred in the U.S. when 21 young people litigated the U.S. federal government. The litigants are trying to suspend fossil fuel development as part of their high-profile climate rights case, Juliana v. United States [110]. The lawsuit was filed in 2015, accusing the U.S. government of violating the young plaintiffs' constitutional rights by failing to address climate change and continuing to subsidize fossil fuels.

In November 2018, the suit reached the U.S Supreme Court, which suggested that the government's arguments ought to be considered by the Ninth Circuit Court of Appeals before being reviewed by the nation's highest court. As of 2019, the Juliana lawsuit remains undecided by federal courts [110]. The current wave of litigation is bringing up new legal questions in the context of climate change. For local governments suing fossil fuel companies, the fight is over who pays for the damages stemming from rising average temperatures. In youth lawsuits, the key issue is whether a stable climate is a civil right.

> Fourteen teens from across Canada filed litigation against Canada's federal government over its failure to address the climate crisis. The suit claims the federal government violated the teens' Charter rights.

In Canada, 14 teens from across Canada filed litigation against Canada's federal government over its failure to address the climate crisis. Teens from eight Canadian provinces are the plaintiffs. The lawsuit was filed in October 2019 in federal court in Vancouver. The suit claims the federal government violated the teens' Charter rights by contributing to high levels of greenhouse gas (GHG) emissions and dangerous climate change [111]. Canada's Charter of Rights and Freedoms guarantees Canadians the right to life, liberty, the security of the person, and equal protection and benefit under the law.

In Europe, the European Court of Human Rights has told the governments of 33 industrialized countries to promptly respond to a climate lawsuit lodged by six youth campaigners in September, giving it priority status because of the "importance and urgency of the issues raised." The young people, supported by the Global Legal Action Network (GLAN), allege the countries—which include Germany, the UK, Russia, and Portugal—have failed to enact the emission cuts needed to protect their futures [112].

While there are numerous recent and ongoing climate cases, many also involving young plaintiffs, it is the first of its kind to be brought to Strasbourg. The international court, set up in 1959, deals with alleged violations of civil and political rights set out in the European Convention on Human Rights. According to GLAN, if successful, the 33 countries would be legally bound to tackle overseas contributions to climate change, including that of multinational companies, as well as ramping up emissions cuts.

1.9 SOCIAL JUSTICE ISSUES

As a reminder, "Social justice is a concept of fair and just relations between the individual and society as measured by the distribution of wealth, opportunities for personal activity, and social privileges." Fair and just relations can certainly include the idea that the adverse effects of climate change are equitably shared across populations. The issue of climate change and social justice was addressed by Levy and Patz in 2015 [113]. Their primary findings are summarized in the following.

"The environmental and health consequences of climate change, which **disproportionately affect** low-income countries and poor people in high-income countries, profoundly affect human rights and. social justice (emphasis added). Environmental consequences include increased temperature, excess precipitation in some areas and droughts in others, extreme weather events, and increased sea level. These consequences adversely affect agricultural production, access to safe water, and worker productivity, and, by inundating land or making land uninhabitable and uncultivatable, will force many people to become environmental refugees. Adverse health effects caused by climate change include heat-related disorders, vector-borne diseases, foodborne and waterborne diseases, respiratory and allergic disorders, malnutrition, collective violence, and mental health problems."

"These environmental and health consequences threaten civil and political rights and economic, social, and cultural rights, including rights to life, access to safe food and water, health, security, shelter, and culture. On a national or local level, those people who are most vulnerable to the adverse environmental and health consequences of climate change include poor people, members of minority groups, women, children, older people, people with chronic diseases and disabilities, those residing in areas with a high prevalence of climate-related diseases, and workers exposed to extreme heat or increased weather variability. On a global level, there is much inequity, with low-income countries, which produce the least greenhouse gases (GHGs), being more adversely affected by climate change than high-income countries, which produce substantially higher amounts of GHGs yet are less immediately affected. In addition, low-income countries have far less capability to adapt to climate change than high-income countries" [112].

Perspective: Social justice advocates' agenda will increasingly include climate change as the consequences of climate change become more evident and personally experienced. In particular, minority populations and low-income communities

> Climate change will threaten civil and political rights and economic, social, and cultural rights, including rights to life, access to safe food and water, health, security, shelter, and culture.

will become increasingly dissatisfied with social structures such as community government that fail to address their concerns.

1.10   IMPACT OF COVID-19 ON CLIMATE CHANGE

The COVID-19 pandemic that began in early 2020 essentially locked down North America, Europe, China, India, and other countries (Volume 2, Chapter 1). Individuals were advised, sometimes ordered by government agencies, to stay home in order to reduce person-to-person transmission of the virus and to wear a face mask when near other people. The policy of social isolation resulted in far fewer motor vehicles traversing streets and highways, fewer commercial activities such as sporting events, bars and restaurants, and large group activities, all actions that influence climate change dynamics.

The first half of 2020 saw an unprecedented decline in CO_2 emissions. An international team of researchers found that in the first 6 months of 2020, 8.8% less CO_2 was emitted than in the same period in 2019—a total decrease of 1551 million tons [114]. Overall, a reduction in U.S. greenhouse gas emissions of 10.3% occurred in 2020. The greatest reduction of emissions was observed in the ground transportation sector. Largely because of working from home restrictions, transport CO_2 emissions decreased by 40% worldwide. The researchers opined that their results showed the need for structural and transformational changes in energy production and consumption systems in order to achieve meaningful, lasting reductions of greenhouse gases.

A further impact of COVID-19 on climate change dynamics was the decline of ridership on buses and trains [115]. A year into the coronavirus pandemic, public transit has experienced a dramatic, dire loss of ridership. Riders remain at home or they remain fearful of boarding buses and trains. For instance, the Delhi Metro is ferrying fewer than half of the pre-COVID ridership, and in Rio, unpaid bus drivers have gone on strike. And in New York City, subway traffic is just a third of what it was before the pandemic. As to climate change, public transit offers a relatively simple way for cities to lower their greenhouse gas emissions because fewer automobiles are utilized, resulting in lesser emissions of greenhouse gases.

1.11   HAZARD INTERVENTIONS

Assuming that the projections of the dire portent of climate change are true, interventions to mitigate climate change go beyond necessary to the level of vital. Some persons might argue that interventions would be extraordinarily costly, burdensome to global societies, and a misuse of resources needed for use by an increasing global human population. But upon reflection of what can already be observed and attributable to climate change (e.g., rising sea levels, melting polar ice, shrinking glaciers), the question must be put, "Can humankind afford to be wrong about mitigating climate change?" Given this question, interventions must be implemented. Some of the interventions to lessen the hazard posed by climate change are the following:

- On a global scale, support through sociopolitical means those policies that are based on consensus science, presented in transparent reports, and implemented through diplomatic dialog and resolution.

- On a national scale, support through sociopolitical means those policies and policymakers that advocate for the development and promulgation of policies for climate change mitigation.
- On a personal scale, use objective, transparent sources of climate change information as the basis for personal decisions and policies. This could include choosing consumer products manufactured by sources that have a neutral carbon impact on the environment. Other individual policies might include selecting energy sources that are not carbon-based, e.g., solar power, wind, and geothermal.
- Industrial entities should understand their role in mitigating climate change and redesign or replace carbon-dependent manufacturing processes and products.
- Research on control of CO_2 emissions should be encouraged. For example, researchers discovered a new iron-based catalyst that converts CO_2 into jet fuel [116]. If CO_2, rather than oil, were used to make jet fuel, it could reduce the air travel industry's carbon footprint.

1.12   SUMMARY

This chapter presents Earth's climate change as the single greatest current environmental threat to life on the planet. This assertion is based on both the science of climate change together with observable changes already occurring in global temperature, acidification of seas, impacted ecosystems, and human health morbidity and mortality. While the science of greenhouse gas accumulation and its contribution to global warming is not incontrovertible, nevertheless, it is compelling and leads to a conclusion that global interventions to mitigate further climate change cannot wait. Fortunately, as described in some detail in this chapter, international efforts through the auspices of the UN have produced global agreements on actions to reduce the greenhouse effect of heat-trapping gases emitted into the atmosphere, as well as reducing the carbon footprint of national economies. The passage of time will yield an answer to the degree of success of these and subsequent climate control policies.

REFERENCES

1. Wikipedia. 2020. James Hansen. https://en.wikipedia.org/wiki/James_Hansen#Career.
2. Milman, O. 2020. Disease-bearing ticks thrive as climate change heats up US. *The Guardian*, August 11.
3. Johnston, I. 2016. Global warming set to cost the world economy £1.5 trillion by 2030 as it becomes too hot to work. *The Independent*, July 18.
4. Heeb, G. 2019. Climate events have cost the US economy more than $500 billion over the last 5 years, Fed official says. Markets Insider, November 10.
5. *History.Com*. 2016. Industrial Revolution. http://www.history.com/topics/industrial-revolution.
6. NASA (National Aeronautics and Space Administration). 2016. Climate change: How do we know? http://climate.nasa.gov/evidence/.
7. EPA (Environmental Protection Agency). 2019. Overview of greenhouse gases. https://www.epa.gov/ghgemissions/overview-greenhouse-gases.

8. Pearce, F. 2019. Scientists zero in on trees as a surprisingly large source of methane. *Yale Environment 360*, June 24.

9. University of East Anglia. 2020. Nitrous oxide emissions pose an increasing climate threat, study finds. *ScienceDaily*, October 7.

10. Lindsey, R. 2018. Climate change: Atmospheric carbon dioxide. *NOAA Climate.gov*, August 1.

11. Carrington, D. 2019. Last time CO2 levels were this high, there were trees at the South Pole. *The Guardian*, April 3.

12. Staff. 2020. Global CO_2 emissions in 2019. IEA. https://www.iea.org/articles/global-co2-emissions-in-2019.

13. Carrington, D. 2018. Climate-heating greenhouse gases at record levels, says UN. *The Guardian*, November 22.

14. Tabuchi, H. 2020. Global methane emissions reach a record high. *The New York Times*, July 14.

15. Frazin, R. 2020. Court strikes down Trump administration's methane rollback. *The Hill*, July 16.

16. Redford, T. 2020. Food system causes one third of greenhouse gases. *Climate News Network*, November 13.

17. Rutgers University. 2018. Global sea level could rise 50 feet by 2300, study says. *Science Daily*, October 8.

18. Schwartz, J. 2020. How much will the planet warm if carbon dioxide levels double? *The New York Times*, July 22.

19. Hamers, L. 2018. Antarctica has lost about 3 trillion metric tons of ice since 1992. *Science News*, June 13.

20. McKie, R. 2020. Earth has lost 28 trillion tonnes of ice in less than 30 years. *The Guardian*, August 23.

21. American Geophysical Union. 2018. Melting of Arctic mountain glaciers unprecedented in the past 400 years. *Science Daily*, April 10.

22. Pappas, S. 2019. The world's fastest-thinning glacier identified. *Live Science*, October 31.

23. Beitsch, R. 2020. Climate change a factor in most of the 7,000 natural disasters over last 20 years: UN report. *The Hill*, October 12.

24. Lin, L., and P. Chakraborty. 2020. Slower decay of landfalling hurricanes in a warming world. *Nature* 587(7833):230.

25. Newburger, E. 2020. Climate change is driving widespread forest death and creating shorter, younger trees. *CNBC*, May 28.

26. Sullivan, J. 2018. *The Intergovernmental Panel on Climate Change: 30 Years Informing Global Climate Action*. New York: United Nations Foundation, March 13.

27. IPCC (international panel on climate change). 2018. Global warming of 1.5°C, p 11. https://report.ipcc.ch/sr15/pdf/sr15_spm_final.pdf.

28. Li, D., J. Yuan, R. Kopp. 2020. Escalating global exposure to compound heat-humidity extremes with warming. *Environ. Res. Lett.*, doi:10.1088/1748-9326/ab7d04.

29. Davis, N. 2018. Climate change will make hundreds of millions more people nutrient deficient. *The Guardian*, August 27.

30. WHO (World Health Organization). 2018. Climate change and health. https://www.who.int/news-room/fact-sheets/detail/climate-change-and-health.

31. American Academy of Pediatricians. 2015. American Academy of Pediatrics links global warming to the health of children. Policy statement, October 26. Elk Grove Village: Press Releases Office.

32. Elliott, L. 2016. Climate change puts 1.3bn people and $158tn at risk, says World Bank. *The Guardian*, May 16.

33. Rowling, M. 2015. World Bank warns climate change could add 100 million poor by 2030. http://www.reuters.com/article/us-climatechange-poverty-disaster-idUSKCN0SX0WE20151108.

34. Beitsch, R. 2020. US couild avoid 4.5M early deaths fighting climate change, study. *The Hill*, August 5.
35. Obradovich, N., R. Migliorini, M. P. Paulus, I. Rahwan. 2018. Empirical evidence of mental health risks posed by climate change. *PNAS* 115(43):10953–8.
36. Huizen, J. 2019. What to know about eco-anxiety. *Medical News Today Newsletter*, December 19.
37. Harvey, F. 2018. Climate change soon to cause movement of 140m people, World Bank warns. *The Guardian*, March 19.
38. EPA (Environmental Protection Agency). 2016. Ecosystems: Climate impacts on ecosystems. Washington, D.C.: Office of Public Affairs.
39. Berwyn, B. 2018. Fish species forecast to migrate hundreds of miles northward as U.S. waters warm. *Inside Climate News*, May 16.
40. Tollefson, J. 2019. World's oceans are losing power to stall climate change. *Nature*, September 25.
41. Berwyn, B. 2017. Tree-killing beetles spread into northern U.S. forests as temperatures rise. *Inside Climate News*, August 28.
42. Bhalla, N. 2020. Climate change linked to African locust invasion. *Reuters*, January 29.
43. Cang, A. 2020. Army of 100,000 Chinese ducks ready to fight locust plague. *Time*, February 27.
44. Doyle, A. 2018. Coral reefs at risk of dissolving as oceans get more acidic: study. *Reuters*, February 22.
45. McKenna, P. 2016. Spikes in U.S. air pollution linked to warming climate. *Inside Climate News*, April 20.
46. NOAA (National Oceanic and Atmospheric Administration). 2016. Is sea level rising? Silver Spring: National Ocean Service.
47. NOAA (National Oceanic and Atmospheric Administration). 2015. Temperature and precipitation maps. Silver Spring: National Centers for Environmental Information.
48. Cheng, L., J. Abraham, J. Zhu, et al. 2020. Record-setting ocean warmth continued in 2019. *Advances in Atmospheric Sciences* 37 (2):135–42.
49. Center for Biological Diversity. 2015. Endangered oceans. http://www.biologicaldiversity.org/campaigns/endangered_oceans/index.html.
50. FAO (UN Food and Agriculture Organization). 2016. Climate change. http://www.fao.org/climate-change/en/.
51. EC (European Commission). 2016. EU agriculture and climate change. http://ec.europa.eu/agriculture/climate-change/factsheet_en.pdf.
52. Walthall, C.L., et al. 2012. Climate change and agriculture in the United States: Effects and adaptation. USDA Technical Bulletin 1935. Washington, D.C.: U.S. Department of Agriculture, Office of Public Affairs.
53. Hannam, P. 2016. Soil productivity cut by climate change, making societies more marginal: studies. *The Sydney Morning Herald*, January 28.
54. Bond-Lamberty, B., et al. 2016. Soil respiration and bacterial structure and function after 17 years of a reciprocal soil transplant experiment. *PLoS One* 11(3):e0150599. doi:10.1371/journal.pone.0150599.
55. Njagi, K. 2016. Extreme weather increasing level of toxins in food, scientists warn Thomson Reuters Foundation. *Reuters*, May 21.
56. Rothamsted Research. 2018. Soil cannot halt climate change. *Science Daily*, February 28.
57. Rosch, A. 2014. Over 1,400 endangered species are threatened by climate change, says new 'red list'. *Climate Progress*, November 18.
58. Kaplan, K. 2015. One in six species could be wiped out by climate change, study says. *The New York Times*, May 1.
59. Crugnale, J. 2016. First mammal declared extinct due to climate change, scientists say. https://weather.com/science/nature/news/first-mammal-declared-extinct-climate-change.

60. Jackson, P. 2019. From Stockholm to Kyoto: A brief history of climate change. UN Chronicle. New York: UN Department of Public Information.

61. U.S. Global Change Research Program. 2018. Fourth National Climate Assessment. https://www.globalchange.gov/nca4.

62. Wuebbles, D. J., et al. 2017. Executive summary. In: *Climate Science Special Report: Fourth National Climate Assessment*, volume I. pp 12-34. Wuebbles, D. J., (eds.). Washington, D.C.: U.S. Global Change Research Program. doi: 10.7930/J0DJ5CTG.

63. Holden, E. 2018. Trump on own administration's climate report: 'I don't believe it. *The Guardian*, November 26.

64. Oyez (IIT Chicago-Kent School of Law). 2015. Massachusetts v Environmental Protection Agency. https://www.oyez.org/cases/2006/05-1120.

65. Oyez (IIT Chicago-Kent School of Law). 2015. Utility Air Regulatory Group v. Environmental Protection Agency. Utility Air Regulatory Group v. Environmental Protection Agency.

66. Barnes, R. 2014. Supreme Court: EPA can regulate greenhouse gas emissions with some limits. *The Washington Post*, June 23.

67. EPA (Environmental Protection Agency). 2016. Greenhouse gas reporting program (GHGRP). https://www.epa.gov/ghgreporting/learn-about-greenhouse-gas-reporting-program-ghgrp.

68. White House. 2013. Fact sheet: President Obama's climate action plan. https://www.white-house.gov/the-press-office/2013/06/25/fact-sheet-president-obama-s-climate-action-plan.

69. EPA (Environmental Protection Agency). 2016. Clean power plan toolbox for states. https://www.epa.gov/cleanpowerplantoolbox.

70. Davenport, C. 2017. Trump signs executive order unwinding Obama climate policies. *The New York Times*, March 28.

71. Adler, J. H. 2016. Supreme Court puts the brakes on EPA's clean power plan. *The Washington Post*, February 9.

72. Friedman, L. 2019. States sue Trump administration over rollback of Obama-era climate rule. *The New York Times*, August 13.

73. Wikipedia. 2019. Environmental policy of the Donald Trump administration. https://en.wikipedia.org/wiki/Environmental_policy_of_the_Donald_Trump_administration.

74. EPA (Environmental Protection Agency). 2019. News releases from Headquarters. Air and Radiation (OAR). Washington, D.C.: Office of Air and Radiation, June 19.

74a. Friedman, L. 2021. Court voids 'tortured' Trump climate rollback. *The New York Times*, January 19.

75. Lavelle, M. 2018. Massachusetts can legally limit CO2 emissions from power plants, court rules. *Inside Climate News*, September 6.

76. Roberts, T. 2018. One year since Trump's withdrawal from the Paris climate agreement. *Brookings Planet Policy*, June 1.

77. Gustin, G. 2017. Over 1,400 U.S. cities, states and businesses vow to meet Paris climate commitments. *Inside Climate News*, June 6.

78. Schwartz, J. 2019. Major climate change rules the Trump administration is reversing. *The New York Times*, August 29.

80. Lewis, M. 2020. The EPA rolls back greenhouse gas-lead prevention rule. *Electrek*, February 29.

79. Davenport, C. 2020. Trump administration, in biggest environmental rollback, to announce auto pollution rules. *The New York Times*, March 30.

81. Friedman, L. 2020. G.A.O.: Trump boosts deregulation by undervaluing cost of climate change. *The New York Times*, July 14.

82. Knickmeyer, E. and M. Daly. 2020. Biden introduces his climate team. *The New York Times*, December 19.

83. Sullivan, K. 2020. Here are 10 climate executive actions Biden says he will take on day one. *CNN News*, November 11.

83a. Newburger, A. 2021. Here's what countries pledged on climate change at Biden's global summit. *CNBC,* April 22.

84. World Bank. 2015. Putting a price on carbon with a tax. http://www.worldbank.org/content/dam/Worldbank/document/SDN/background-note_carbon-tax.pdf.

85. Nuccitelli, D. 2018. Comprehensive study: Carbon taxes won't hamper the economy. *The Guardian,* July 16.

86. Carbon Share. 2016. Cap and trade vs. carbon tax. http://www.carbonshare.org/docs/capvscarbontax.pdf.

87. EC (European Commission). 2014. EU ETS system. http://ec.europa.eu/clima/publications/docs/factsheet_ets_en.pdf.

88. Union of Concerned Scientists. 2018. Each country's share of CO_2 emissions. https://www.ucsusa.org/global-warming/science-and-impacts/science/each-countrys-share-of-co2.html.

89. Gearing, D. 2018. Carbon markets pay off for these states as new businesses, jobs spring up. *Inside Climate News,* April 17.

90. Perry, M. J. 2015. If New York is Spain and California Brazil, What is Texas? Consumer Financial Protection Bureau. http://www.newsweek.com/if-new-york-spain-and-california-brazil-what-texas-344702.

91. Center for Climate and Energy Solutions. 2015. California's cap and trade. http://www.c2es.org/us-states-regions/key-legislation/california-cap-trade.

92. Cooper, J. 2017. California nets $860 billion from carbon auction. *AP News,* November 21.

93. Riebeek, H. 2011. *The Carbon Cycle.* Washington, D.C.: NASA.

94. Bastin, J.-F., et al. 2019. The potential for global forest cover. *Science,* 365(6448):76–9. doi:10.1126/science.aax0848.

95. Staff. 2018. Current deforestation pace will intensify global warming, study alerts. *Science Daily,* March 6.

96. Staff. 2019. Planting 1.2 trillion trees could cancel out a decade of CO2 emissions, scientists find. E360 Digest, February 20.

97. Campbell, M. 2019. Ethiopia "breaks" tree-planting record to tackle climate change. Euronews, July 29.

98. Frazin, R. 2020. US officially joins global trillion tree planting initiative. *The Hill,* August 27.

99. Watts, J. 2020. Africa's Great Green Wall just 4% complete halfway through schedule. *The Guardian,* September 7.

100. Staff. 2018. Research forecasts US among top nations to suffer economic damage from climate change. *Science Daily,* September 24.

101. Kusnetz, N. 2018. Climate and weather disasters cost U.S. a record $306 billion in 2017. *Inside Climate News,* January 9.

102. Kahn, M. E., et al. 2019. Long-Term macroeconomic effects of climate change: A cross-country analysis. NBER Working Paper No. 26167, August 2019. National Bureau of Economic Research. https://www.nber.org/papers/w26167.

103. Staff. 2018. The 2018 Report of the Global Commission on the Economy and Climate. https://newclimateeconomy.report/.

104. Irfan, U. 2019. Pay attention to the growing wave of climate change lawsuits. *Vox,* June 4.

105. Hasemyer, D. 2020. Emails reveal U.S. Justice Dept. working closely with oil industry to oppose climate lawsuits. *Inside Climate News,* January 13.

106. Gregory, A. 2019. Landmark ruling that Holland must cut emissions to protect citizens from climate change upheld by supreme court. *Independent,* December 12.

107. Boffey, D. 2018. 'We can't see a future': Group takes EU to court over climate change. *The Guardian,* May 24.

108. Watts, J. 2020. Portuguese children sue 33 countries over climate change at European court. *The Guardian,* September 3.

109. Sengupta, S. 2019. Protesting climate change, young people take to streets in a global strike. *The New York Times*, September 20.
110. Carlisle, M. 2020. A federal court threw out a high profile climate change lawsuit. *Time*, January 19.
111. Mast, M. 2019. Teens to sue federal government over climate crisis. *The Tyree*, October 23.
112. Staff. 2020. European court orders countries to respond to lawsuit from young climate activists. *Deutsche Welle*, November 11.
113. Levy, B. S. and J. A. Patz. 2015. Climate change, human rights, and social justice. *Annl. Global Health* 81(3):310–22.
114. Kelley, A. 2021. US greenhouse gas emissions have plunged during COVID-19 pandemic. *The Hill, Changing America*, January 13.
115. Sengupta, S., G. Abdul, M. Andreoni, and V. Penney. 2021. Riders are abandoning buses and trains that's a problem for climate change. *The New York Times*, March 25.
116. Temming, M. 2020. A new iron-based catalyst converts carbon dioxide into jet fuel. *Science News*, December 22.

2 Air Quality

2.1 INTRODUCTION

This chapter's subject is air pollution, which is justifiably considered the world's single greatest environmental hazard, responsible for one in eight premature deaths globally according to the WHO [1], which in 2020 noted that 97% of cities in low- and middle-income countries with more than 100,000 inhabitants do not meet WHO air quality guidelines. Additionally, a study by academic researchers at Harvard and the University of Birmingham found that more than 8 million people died in 2018 from fossil fuel pollution [2]. Researchers estimated that exposure to particulate matter from fossil fuel emissions accounted for 18% of total global deaths in 2018—a little less than 1 out of 5. In a public health context, 20% of global deaths attributable to any single environmental hazard is quite noteworthy.

Air pollution is global; air contaminants released in Asia and elsewhere will appear in North America and Europe and vice versa. Contained in this chapter are descriptions of the major sources of air pollution, one of which, a U.S. power generating plant, is illustrated in Figure 2.1. To be discussed is the toxicology of the primary air contaminants, associations between air pollution and human and ecological health, domestic and global policies developed for control of air pollution sources, and some interventions for reduction of the hazard presented by polluted ambient (outdoor) and indoor air. Along the course of the chapter's development, some pertinent history will be provided that attends the subject of air pollution.

> Air pollution is responsible for one in eight premature deaths globally according to the WHO.

FIGURE 2.1 Plume of air emissions from a U.S. coal power plant. (EPA, 2016.)

DOI: 10.1201/9781003253358-2

Before proceeding, a definition is in order. The WHO states, *"Air pollution* is the contamination of the indoor or outdoor environment by any chemical, physical, or biological agent that modifies the natural characteristics of the atmosphere" [3]. Noteworthy is the mention of both indoor and outdoor locales where air pollution occurs. Indoor sources are especially consequential as an environmental global hazard, principally a problem of developing countries and economically and culturally poor populations. The WHO's assessment points to a huge surge in disease burden and deaths due to air pollution exposure. Deaths due to air pollution, which include outdoor as well as indoor pollution, have increased four-fold across the globe over the past decade, the latest data show. The total number of deaths due to air pollution is estimated at 8 million every year, which is comparable to the population of New York City. Approximately half of the deaths result from exposure to indoor air pollution [1].

Correspondingly, a definition of *air pollutant* is important:

"Any substance in air that could, in high enough concentration, harm animals, humans, vegetation, and/or materials. Such pollutants may be present as solid particles, liquid droplets, or gases. Air pollutants fall into two main groups: (1) those emitted from identifiable sources, and (2) those formed in the air by interaction between other pollutants more than 100 air pollutants have been identified, which include halogen compounds, nitrogen compounds, oxygen compounds, radioactive compounds, sulphur (sulfur) compounds, and volatile organic chemicals (VOC)" [4].

With these definitions in hand, a history of air pollution is instructive as concerns its sources, consequences, and public policies. Morrison provides a helpful summary of this history [5]. "Contaminated air has been around in one form or another for thousands of years. First, it was wood fires in ancient homes, the effects of which have been found in the blackened lungs of mummified tissue from Egypt, Peru, and Great Britain. And the Romans earn the dubious credit of being perhaps the first to spew metallic pollutants into the air, long before the Industrial Revolution. The residents of ancient Rome referred to their city's smoke cloud as *gravioris caeli* ('heavy heaven') and *infamis aer* ('infamous air'). Several complaints about its effects can be found in classical writings. Roman courts considered civil claims over smoke pollution 2,000 years ago. The empire even tried a very early version of the Clean Air Act (CAAct). In 535, then Emperor Justinian proclaimed the importance of clean air as a birthright [...].

Later, smelting to create lead and copper appeared, fouling medieval air. Analyses of ice cores from the Arctic reveal that extraction and smelting on the Iberian Peninsula, England, and Greece and elsewhere increased lead in the environment by a factor of ten. By 1200, [...] London had been deforested and a switch began to 'seacoal', coal that washed up on beaches. As early as the 1280s, there were complaints about smoke from burning coal. Attempts to ban burning then and 250 years later during the reign of Queen Elizabeth I failed. [...]

Europeans imported air pollution to the New World. Spanish conquistadors mining silver in what is now Bolivia in 1572 used amalgamation, a technique that grinds ore into powder and that shot lead plumes into the air. [...] By the 1600s, smoke from burning coal was damaging the architecture in London and other major cities. The invention and eventually widespread use of the steam engine accelerated pollution. Until then, businesses were artisan shops dispersed throughout a city. But centralized factories on a large scale meant even more air pollution.

The shift to fossil fuels eliminated constraints on urban expansion as factories, powered by steam created by burning coal, attracted new workers. Residents of emerging industrial giants—Birmingham, Leeds, Manchester, Chicago, Pittsburgh, and St. Louis, among others—found acrid smoke stung their eyes and hindered their breathing. Thick fogs, especially in colder weather, blanketed the cities. Societies to campaign against the smoke scourge gradually emerged. Laws were passed in Britain, the U.S., and Germany, but with inconsequential impact. [...] [5].

The 'smoke problem' intensified as new coal-burning industrial cities proliferated from the later 18th century onward. Soon a new source of air pollution, the automobile, appeared. By 1940, Los Angeles had more than a million cars. At the time, no one realized the effect of all that exhaust, so when the city was blanketed with what later became known as smog on July 26, 1943, residents feared it was some kind of chemical attack. Figure 2.2 illustrates a day of smog in downtown Los Angeles in the 1970s. Noteworthy is the density of the smog curtain, obscuring residents' view of the surrounding mountains. Four years later, Los Angeles County established the first air pollution control district in the country. California went on to become a leader in regulating air pollution [...]. But it took two other smog incidents to galvanize action in the U.S. and Great Britain. On October 27, 1948, thick smog began to cover the river town of Donora, Pennsylvania. A storm rolled in 2 days later that cleared the air, but in the aftermath 20 died and 6,000 were sickened. Similarly, on December 5, 1952, a fog enveloped London, killing about 4,000

> The "smoke problem" intensified as new coal-burning industrial cities proliferated from the later 18th century onwards.

people before it dissipated 4 days later. Parliament acted with dispatch, passing the U.K. Clean Air Act in 1956, effectively reducing the burning of coal [...]" [5].

FIGURE 2.2   Air pollution in downtown Los Angeles, CA, 1970s. (Howe, Kravchenko, Van Dyke. Blog. Los Angeles Air Pollution, 2015.)

In the U.S., Congress enacted the first Clean Air Act (CAAct) in 1963. Two years later, national emissions standards for cars were set. But it wasn't until the 1970 CAAct that Congress set the framework for air pollution regulations tied to public health. A subsequent section of this chapter will describe the details and policy implications of the CAAct. Suffice it to say here that this act is the most complex and sweeping in the scope of the U.S. environmental health statutes.

Domestic and global policies on control of air pollution will be described in this chapter, but some words about sources of air pollution are necessary, together with the toxic properties of the main contaminants of air pollution, in order to better appreciate the importance of policies on environmental health air quality.

2.2 SOURCES OF EMISSIONS OF AMBIENT AIR POLLUTANTS

As the foregoing history of air pollution implies, the Industrial Revolution commenced the first major release of ambient air pollution, followed by releases due to the invention and global distribution of internal combustion engines in mobile vehicles. Neither industrial operations nor internal combustion engines were designed for zero release of substances released during operation. The global ambient air pollution problem continues to increase due to factors such as human population growth, international trade increase, industrial operations expansion, energy production increase, and exponential growth in the number of motor vehicles.

> The Industrial Revolution commenced the first major release of ambient air pollution.

In the U.S., as described subsequently in the history of the CAAct, the early focus on sources of ambient air pollution was industrial plants, followed later by knowledge of pollution caused by tailpipe emissions from automobiles and other vehicles powered by internal combustion engines. Although there are copious contemporary sources of air pollution, sources are generally grouped by EPA into two broad categories: *point sources* include items such as factories and electric power plants, and *mobile sources* include cars and trucks, lawn mowers, and airplanes. Mobile sources comprise highway vehicles and non-road equipment [6].

Within these two categories, EPA develops regulations and takes other actions pursuant to requirements of the CAAct.

2.2.1 VEHICLE EMISSIONS

Emissions from internal combustion motor vehicles are a main source of air contaminants. As will be subsequently described, this fact came as a shock to early investigators of sources of air pollution. The main emissions from a motor vehicle's internal combustion gasoline engine are: nitrogen gas (N_2)—Air is 78% nitrogen gas, and most of this pass right through the car engine. Carbon dioxide (CO_2)—This is a product of fuel combustion. The carbon in the fuel bonds with the oxygen in the air. Water vapor (H_2O)—This is another product of combustion. The hydrogen in the fuel bonds with the oxygen in the air. Because the combustion process is never perfect, some smaller amounts of more harmful emissions are also produced in car engines, specifically: carbon monoxide (CO), a poisonous gas that is colorless and odorless; hydrocarbons or volatile organic compounds (VOCs) are a major component of smog produced mostly from evaporated, unburned fuel; nitrogen oxides (NO and NO_2, together called NO_x) are a contributor to smog and acid rain, which is precipitation containing harmful amounts of nitric and sulfuric acids, which also causes irritation to human mucus membranes. Internal combustion engine exhaust is routed through catalytic converters that are designed to reduce all these three harmful pollutants [7].

Diesel engines are another form of internal combustion engine used to power motor vehicles [8]. Diesel engines convert the chemical energy contained in the fuel into mechanical power. Diesel fuel is injected under great pressure into the engine cylinder where it mixes with air and where the combustion occurs by virtue of the pressure. The exhaust gases that are discharged from the engine contain several constituents which are harmful to human health and to the environment. Carbon monoxide (CO), hydrocarbons (HC), and aldehydes are generated in the exhaust as the result of incomplete combustion of fuel. Nitrogen oxides (NO_x) are generated from nitrogen and oxygen under high pressure and temperature conditions in the engine cylinder. Sulfur dioxide (SO_2) is generated from the sulfur present in diesel fuel. The concentration of SO_2 in the exhaust gas depends on the sulfur content of the fuel. Diesel particulate matter (DPM) is a complex aggregate of solid and liquid material. Its origin is carbonaceous particles generated in the engine cylinder during combustion.

> Collectively, cars and trucks account for nearly one-fifth of all U.S. greenhouse gas emissions.

Our personal vehicles are also a major contributor to global warming. Collectively, cars and trucks account for nearly one-fifth of all U.S. greenhouse gas emissions, emitting around 24 pounds of CO_2 and other global warming gases for every gallon of gas [9]. In total, the U.S. transportation sector—which includes cars, trucks, planes, trains, ships, and freight—produces nearly 30% of all U.S. global warming emissions, more than almost any other sector. Notable in terms of CO_2 emissions are sport utility vehicles (SUVs). Growing demand for SUVs was the second-largest contributor to the increase in global CO_2 emissions from 2010 to 2018, an analysis found [10]. In that period, SUVs doubled their global market share from 17% to

> Globally, in the period 2010–2018, SUVs' annual emissions of CO_2 exceeded those of the UK and the Netherlands combined.

39% and their annual emissions rose to more than 700 megatonnes of CO_2, more than the yearly total emissions of the UK and the Netherlands combined.

2.2.2 POWER PLANTS

When coal burns, the chemical bonds holding its carbon atoms in place are broken, releasing energy. However, other chemical reactions *also* occur, many of which carry toxic airborne pollutants and heavy metals into the environment [11]. This form of air pollution includes:

Mercury: Coal plants are responsible for 42% of U.S. mercury emissions, a toxic heavy metal that can damage the nervous, digestive, and immune systems, and is a serious threat to child development. Just 1/70th of a teaspoon of mercury deposited on a 25-acre lake can make the fish unsafe to eat. According to EPA's National Emissions Inventory, U.S. coal power plants emitted 45,676 pounds of mercury in 2014 (the latest year data is available). Global emissions of mercury from anthropogenic sources decreased by 30% from 1990 to 2010, in part from decreasing use of coal, the U.S. Geological Survey (USGS) reported in 2016 [12].

Sulfur dioxide (SO_2): When the sulfur in coal reacts with oxygen, SO_2 combines with other molecules in the atmosphere to form small, acidic particulates that can penetrate human lungs. It is linked with asthma, bronchitis, smog, and acid rain, which damages crops and other ecosystems, and acidifies lakes and streams. U.S. coal power plants emitted more than 3.1 million tons of SO_2 in 2014.

> U.S. coal power plants emitted 197,286 tons of small airborne particles (measured as 10 μm or less in diameter) in 2014.

Nitrogen oxides (NO_x): Nitrogen oxides are visible as smog and irritate lung tissue, exacerbate asthma, and make people more susceptible to chronic respiratory diseases like pneumonia and influenza. In 2014, U.S. coal power plants emitted more than 1.5 million tons.

Particulate matter: Better known as "soot," this is the ashy gray substance in coal smoke and can cause chronic bronchitis, aggravated asthma, cardiovascular effects like heart attacks, and premature death. U.S. coal power plants emitted 197,286 tons of small airborne particles (measured as 10 μm or less in diameter) in 2014.

Other harmful pollutants emitted in 2014 by the U.S. coal power fleet include:

- 41.2 tons of lead, 9,332 pounds of cadmium, and other toxic heavy metals.
- 576,185 tons of carbon monoxide, which causes headaches and places additional stress on people with heart disease.
- 22,124 tons of volatile organic compounds (VOCs), which form ozone.
- 77,108 pounds of arsenic. For scale, arsenic causes cancer in one out of 100 people who drink water containing 50 parts per billion.

Most of these emissions can be reduced through pollution controls—sometimes by a significant amount—though many plants don't have adequate controls installed. Under the Clean Air Act, the Clean Water Act, and other environmental laws, EPA has the responsibility and authority to set and enforce emissions limits for pollutants deemed harmful to human health and the environment, including those emitted by power plants.

Notwithstanding the differences between the Obama administration's policies to reduce coal as a fuel for power plants and the Trump administration's policy to support fossil-fuel commerce, U.S. power plants fueled by coal are closing or switching to natural gas for fuel. While power companies' abandonment of coal for reasons of climate change concerns is one basis for their action, a more persuasive reason is economic. Natural gas in the U.S. has become rather abundant and less expensive than coal and current renewable sources of energy such as wind farms, although advances in solar and wind energy technologies have made them more cost-competitive.

The closure of several U.S. power plants that were major emitters of carbon began occurring in the second decade of 2018. Examples include the Navajo Generating Station in Arizona, closing in 2019. The giant coal plant emitted almost 135 million metric tons of CO_2 between 2010 and 2017, according to U.S. Department of Energy figures. Its average annual emissions over that period were about equivalent to what 3.3 million passenger cars would have emitted into the atmosphere in a single year. Similarly, the Bruce Mansfield power plant, a coal plant in Pennsylvania, emitted nearly 123 million tons between 2010 and 2017. It, too, will be retired by the end of 2019. And in western Kentucky, the Paradise plant emitted some 102 million tons of carbon between 2010 and 2017. The Tennessee Valley Authority closed two of Paradise's three units in 2017 and will close the last unit in 2020 [13].

2.2.3 WILDFIRES

Perhaps as an underappreciated source of air pollution are wildfires. But, given the effects of climate change with warmer ambient air temperatures and season droughts (Chapter 1), wildfires should be recognized as an important source of air pollution for reasons of both health and policymaking. The Intergovernmental Panel on Climate Change (IPCC) notes that in some regions of the globe, changes in temperature and precipitation are projected to increase the frequency and severity of fire events. Pollutants from wildfires can affect air quality for thousands of miles. Forest and bush fires can cause burns, damage from smoke inhalation, and produce other injuries to human and ecosystem health. The Panel also notes that toxic gaseous and particulate air pollutants released from large fires can be accompanied by an increased number of patients seeking emergency services and cause adverse effects on morbidity and mortality [14]. Illustrated in Figure 2.3 is a wildfire that occurred in 2020 in northern California. Noteworthy is the large amount of billowing dark smoke that contains particles and gases of carbon combustion.

> Pollutants from wildfires can affect air quality for thousands of miles.

FIGURE 2.3 Wildland fire in California. (U.S. Forest Service, 2021, https://www.fs.usda.gov/managing-land/fire.)

Wildfires affect the climate by releasing large amounts of CO_2, black carbon, brown carbon, and ozone precursors into the atmosphere. These emissions affect radiation, clouds, and climate on regional and even global scales. Wildfires also affect air quality by emitting substantial amounts of volatile and semi-volatile organic materials and nitrogen oxides that form ozone and organic particulate matter. Direct emissions of toxic pollutants can affect first responders and local residents. In addition, the formation of other pollutants as the air is transported can lead to harmful exposures for populations in regions far away from the wildfires [15].

Three illustrations of the impact of wildfires on air quality will demonstrate the adverse effects on human and ecological health. In 2018 California experienced a major outbreak of wildfires caused in large measure by drought conditions across the state. One major wildfire, the Camp Fire blaze in Butte County, killed at least 77 people, destroyed 10,623 homes, and consumed 150,000 acres [16]. Particulates in the air reached as high as 1,500 µg/m³. The threshold set by the WHO is 25. The smoke from the fire drifted far south, where Sacramento's air was listed as unhealthy and particulates reached a level

> One major wildfire, the Camp Fire blaze in Butte County, California, killed at least 77 people, destroyed 10,623 homes, and consumed 150,000 acres.

of 135.4 µg/m³ according to the U.S. Forest Service. San Francisco had a reading of 55 with San Jose at 76.1 and Stockton at 152.

The short-term health consequences to persons who inhaled air pollution from the Camp Fire wildfire were increased numbers of healthcare emergency acute care patients; long-term health effects are unknown [16]. Similarly, a population-based epidemiologic analysis was conducted for daily cardiovascular and cerebrovascular emergency department (ED) visits and wildfire smoke exposure in 2015 among adults in 8 California air basins [17]. Researchers found smoke exposure to be associated with cardiovascular and cerebrovascular ED visits for all adults, particularly for those older than age 65 years.

The second illustration of wildfires and their air pollution impact are fires in the Arctic [18]. In 2019, the Arctic suffered its worst wildfire season on record, with huge blazes in Greenland, Siberia, and Alaska producing plumes of smoke that can be seen from space. In 2019, the Arctic region recorded its hottest June ever. Since the start of that month, more than 100 wildfires have burned in the Arctic circle. In Russia, 11 of 49 regions experienced wildfires. The amount of CO_2 emitted from Arctic circle fires in June 2019 was larger than all of the CO_2 released from Arctic circle fires in the same month from 2010 to 2018 summed. Unusually warm and dry weather conditions were a prime factor in the outbreak of Arctic wildfires.

Wildfires are costly in terms of loss of life, short-and long-term public health effects, ecosystem destruction, social disruption (homes burned), and economic impact. In 2017 the cost of fighting wildfires exceeded $2 billion of U.S. Forest Service funds [19]. This figure does not include costs accrued by state and local governments, insurance companies, business enterprises, and individuals.

2.2.4 WOODSTOVES

Wood-burning stoves used for home heating and some industrial purposes can be a source of air pollution. The EPA notes,

"Smoke forms when wood or other organic matter burns. The smoke from wood burning is made up of a complex mixture of gases and fine particles (particulate matter, or PM). In addition to particle pollution, wood smoke contains several toxic harmful air pollutants including benzene, formaldehyde, acrolein, and polycyclic aromatic hydrocarbons (PAHs)" [20].

The biggest health threat from smoke is from $PM_{2.5}$. The EPA cautions that wood smoke can affect everyone, but children, teenagers, older adults, people with lung disease, including asthma and chronic obstructive pulmonary disease (COPD), or people with heart diseases are the most vulnerable. Research indicates that obesity or diabetes may also increase risk. New or expectant mothers may also want to take precautions to protect the health of their babies because some studies indicate they may be at increased risk. The agency further notes that using an EPA-certified wood stove and dry, seasoned wood creates less smoke and thereby lessens the health risk from wood-burning air pollution [20].

The public health implications of the use of woodstoves depend on the number and kind of stoves used, together with the quality and quantity of wood burned. One study of the air pollution impact of woodstoves was released in 2020 by the Oregon Department of Environmental Quality [21]. The agency estimated that 12.8 million pounds of particulate pollution matter are released into Oregon's air by woodstoves and chimneys annually. The air pollution estimate doesn't take into account wood-burning effects from commercial or business wood-burning, such as restaurants that have woodstoves or chimneys for ambiances. Annual PM levels emitted from wood smoke in Oregon were close to that released by a wildfire, according to the agency.

2.2.5 AGRICULTURE

Agriculture is a major source of ambient air pollution that is less often discussed in its context as an environmental health hazard. A study reports that emissions from farms outweigh all other human sources of fine-particulate air pollution in much of the U.S., Europe, Russia, and China [22]. The emissions are caused by the use of nitrogen-rich fertilizers and animal waste that combine in the air with industrial emissions to form particulate matter. Agricultural air pollution comes mainly in the form of ammonia, which enters the air as a gas from heavily fertilized fields and livestock waste. It then combines with pollutants from combustion—mainly nitrogen oxides and sulfates emitted from motor vehicles, power plants, and industrial processes—to create fine-particulate matter of micrometer diameter. As will be described in a subsequent section, fine-particulate matter can significantly contribute to acute and chronic illnesses of the respiratory system, particularly

> Emissions from farms outweigh all other human sources of fine-particulate air pollution in much of the U.S., Europe, Russia, and China.

in children, including pneumonia, upper respiratory diseases, asthma, and chronic obstructive pulmonary diseases. In a separate study, investigators estimate that particulate matter may cause at least 3.3 million deaths annually globally [23]. As a matter of hazard intervention, reductions in the air pollutants from stationary and mobile sources will reduce the hazard of agriculturally released particulates.

2.2.6 AVIATION AND MARINE SOURCES OF AIR POLLUTION

The public's perception of sources of air pollution is likely linked to land-based sources such as power plants, internal combustion vehicles, and industrial plants such as steel mills. While these are indeed sources of concern, there are other consequential sources of air pollution that are not land-based. In particular, air pollution released from airplanes and ships is substantial and contributes to the health burden of persons and communities so exposed. This section will provide some details on air pollution linked to the operation of aircraft and ships.

2.2.6.1 Air Pollution from Aircraft

The airline industry is a global economic and sociopolitical power. The fleet of airlines globally annually transport passengers, consumer goods, industrial cargo, and unfortunately in a public health sense, weapons. Few could dispute the importance of aircraft as an essential component of a modern global society. Yet with almost all elements of technology, there are environmental costs attending humankind's reliance on aircraft as an essential mode of transportation. One of the costs is air pollution. Other environmental costs are noise pollution (Volume 2, Chapter 5) and issues of waste (Volume 2, Chapter 2). To be discussed here are the air pollution impacts of aircraft.

Motorized aircraft require fuel to power engines that are used to move air past an aircraft's airfoil, thereby providing lift to the aircraft. Although solar-powered and battery-powered aircraft are in development in the 21st century, their replacement of fuel-powered engines will not occur for many years hence. And with the combustion of fossil fuel as jet engine fuel continuing, so will the release of air pollutants from aircraft.

According to EPA, emissions from U.S. aircraft account for 3% of U.S. national greenhouse gas emissions and 12% of emissions from the U.S. transportation sector [24]. After the industrial sector, transportation produces the most greenhouse gas emissions in the U.S. (26% in 2014), and that percentage is growing faster than any other sector. While the aircraft industry has taken steps to increase fuel efficiency and use of alternative fuels, namely biofuels, the Federal Aviation Administration estimates U.S. aviation emissions will rise by 100% by 2050 under a "business-as-usual" scenario, due primarily to increases in air travel.

> Emissions from U.S. aircraft account for 3% of U.S. national greenhouse gas emissions and 12% of emissions from the U.S. transportation sector.

Future EPA regulations would be part of the global effort to reduce aircraft emissions. In February of 2016, the International Civil Aviation Organization (ICAO), a working group of the United Nations' Committee on Aviation Environmental Protection (CAEP), announced the first ever binding standards for aviation emissions, in the form of a CO_2 cap and fuel efficiency standard. The CAEP standards mandate an overall 4% reduction in cruising fuel consumption across the industry by 2028, compared with 2016 levels. Actual reductions will range from 0% to 11%. If the standards are adopted, all new aircraft designs starting in 2020, and new planes in operation by 2023, must comply. Existing aircraft are exempt from the CAEP regulation [24]. On July 22, 2020, the Trump administration's EPA announced the federal government's first proposal to control planet-warming pollution from airplanes, but the draft regulation would not regulate the airlines' emissions beyond limits they have set for themselves [25].

Perspective: The EPA's decision not to require airlines to further lower greenhouse emissions was no doubt influenced by the difficult economic state of airlines due to reduced passenger loads because of COVID-19. EPA's decision also is in line with the Trump administration's position of denial of climate change.

2.2.6.2　Air Pollution from Ships

Motorized ships release large quantities of pollutants into the air, principally in the form of sulfur, NO_x, and particulate matter (PM). These releases are a consequence of using diesel engines to power most ships, and their fuel oil is largely responsible for the soup of air pollutants released. Pitch black and thick as molasses, "bunker" fuel is made from the dregs of the refining process. It's also loaded with sulfur. As noted by a European NGO, the emissions from ships engaged in international trade in the seas surrounding Europe—the Baltic, the North Sea, the northeastern part of the Atlantic, the Mediterranean, and the Black Sea—were estimated to amount to 1.6 million tonnes of SO_2 and 3 million tonnes of NO_x a year in 2013 [26].

> Pitch black and thick as molasses, "bunker" fuel for ships is made from the dregs of the refining process. It's also loaded with sulfur.

In contrast to the progress in reducing emissions from land-based sources over the last 30 years, shipping emissions of sulfur and NO_x have steadily been increasing as the number of ships has grown. Although SO_2 emissions have decreased due to the increased use of low-sulfur fuels, NO_x-emissions are expected to continue increasing. As a result, within 10 years, the NO_x-emissions from international shipping around Europe are expected to equal or even surpass the total from all land-based sources in the 28 EU member states combined [26].

As a matter of environmental health policy, emissions of air pollutants from international shipping are regulated by the International Maritime Organization (IMO), a UN agency, through the Annex VI to the International Convention on the Prevention of Pollution from Ships (MARPOL), which was originally signed in September 1997 and came into force in May 2005. Annex VI set the first global standards for the sulfur content of marine fuel oils and the emissions of nitrogen oxides (NO_x) from new ship engines [26]. Illustrated in Figure 2.4 is a cargo ship releasing air emissions produced by diesel engines combusting bunker fuel. The ship is entering a port in Texas, resulting in the ship's air pollution wafting over the city.

FIGURE 2.4 Cargo ship emitting air pollution from combustion of diesel fuel. (NOAA, 2021, https://www.noaa.gov/stories/guiding-cargo-ships-safely-through-port.)

Starting January 1, 2020, the IMO will require that all fuels used in ships contain no more than 0.5% sulfur. The cap is a significant reduction from the existing sulfur limit of 3.5% and is well below the industry average of 2.7% sulfur content. Public health experts estimate that once the 2020 sulfur cap takes effect, it would prevent about 150,000 premature deaths and 7.6 million childhood asthma cases globally each year [27]. One city, Vancouver, Canada, experienced in 2015 the air pollution benefits of lower-sulfur marine fuels. Local environmental authorities expected the amount of SO_2 released into the atmosphere to drop 79% from the prior decade due in large part to new restrictions on shipping fuel within 200 nautical miles of the Canadian and U.S. coastlines [28].

> In April 2018, the IMO adopted a policy to curb carbon emissions from ships by at least 50% below 2008 levels by 2050.

The IMO is also taking early steps to reduce another form of harmful emissions: greenhouse gases. In April 2018, the UN agency adopted a historic agreement to curb carbon emissions from ships by at least 50% below 2008 levels by 2050. The nonbinding agreement is expected to spur more investment in clean ship technologies, including fuel cells, biofuels, and advanced sail designs.

In advance of the promulgation of the IMO regulation on limiting sulfur in ship fuel, the Association of Arctic Expedition Cruise Operators (AECO), which represents the majority of expedition cruise operators in the northern polar regions, announced in November 2019 a self-imposed ban by its members on the use and carriage of heavy fuel oil (HFO) in the Arctic [29]. The use of better quality fuel oil will reduce sulfur emissions and other air pollution from marine craft.

2.2.7 METALS PRODUCTION

The various metals in Earth's soil and oceans have been put to myriad uses over time for applications with which all humans are familiar, although not always knowledgeable of their origin. The utility of metals perhaps exceeds that of any other natural resource, ranging from cosmetic jewelry to components of transportation to food containers to weapons of war. Indeed, one metal was so important to global social development that an entire age, the Bronze Age, was named for a metal. Metals are removed from their source as ore, which then requires the removal of a particular metal from the remainder of its ore. The removal process is called smelting. Smelting is the process by which a metal is obtained, either as the element or as a simple compound, from its ore by heating beyond the melting point, ordinarily in the presence of oxidizing agents, such as air, or reducing agents, such as coke. Smelting requires a heating source, usually a fossil fuel, and smelting releases air pollutants that come from the ore being smelted as well as the fuel used to heat the smelting furnaces.

The EPA has developed air quality standards and guidelines for the metals production industry under its authorities in the Clean Air Act [30]. As an example, EPA regulations apply to copper smelters, aluminum production, and chromium electroplating. Large smelting operations can release large amounts of air pollutants. An example is a nickel smelting operation in northern Russia that in 2018 released 1.9 million tons of SO_2 into ambient air, according to a NASA satellite study of 500 major point sources of SO_2 emissions globally [31]. This smelter released more SO_2 air pollution than any other point source anywhere on planet Earth.

2.3 SOURCES OF INDOOR AIR POLLUTION

According to the WHO around 3 billion, people still prepare food using solid fuels (such as wood, crop wastes, charcoal, coal, and dung) and kerosene in open fires and inefficient stoves [32]. Most of these people are poor and reside in low- and middle-income countries. These cooking practices are inefficient and use fuels and technologies that produce high levels of household air pollution with a range of health-damaging pollutants, including small soot particles that penetrate deep into the lungs. In poorly ventilated dwellings, indoor smoke can be 100 times higher than acceptable levels for fine particles (PM_{10}, $PM_{2.5}$). Exposure is particularly high among women and young children, who spend the most time near the domestic hearth. The WHO cites the following health statistics regarding indoor air pollution:

> In poorly ventilated dwellings, indoor smoke can be 100 times higher than acceptable levels for fine particles (PM_{10}, $PM_{2.5}$).

- "Around 3 billion people cook using polluting open fires or simple stoves fueled by kerosene, biomass (wood, animal dung and crop waste) and coal.
- Each year, close to 4 million people die prematurely from illness attributable to household air pollution from inefficient cooking practices using polluting stoves paired with solid fuels and kerosene.

- Household air pollution causes noncommunicable diseases including stroke, ischemic heart disease, chronic obstructive pulmonary disease (COPD), and lung cancer.
- Close to half of deaths due to pneumonia among children less than 5 years of age are caused by particulate matter (soot) inhaled from household air pollution" [32].

A particularly important public health study of indoor air quality was the multinational PURE-AIR study, which was conducted from June 2017 to September 2019 in 120 rural communities in eight countries (Bangladesh, Chile, China, Colombia, India, Pakistan, Tanzania, and Zimbabwe) [33]. Data were collected from 2541 households and from 998 individuals (442 men and 556 women). Gravimetric (or filter-based) 48 h kitchen and personal $PM_{2.5}$ measurements were collected. Surveys of household characteristics and cooking patterns were collected before and after the 48 h monitoring period.

A mean $PM_{2.5}$ kitchen concentration gradient emerged across primary cooking fuels: gas (45 µg/m³), electricity (53 µg/m³), coal (68 µg/m³), charcoal (92 µg/m³), agricultural or crop waste (106 µg/m³), wood (109 µg/m³), animal dung (224 µg/m³), and shrubs or grass (276 µg/m³). Among households cooking primarily with wood, average PM_2 concentrations varied ten-fold (range: 40–380 µg/m³). Similar average $PM_{2.5}$ personal exposures between women (67 µg/m³) and men (62) were observed. Using clean primary fuels substantially lowers kitchen $PM_{2.5}$ concentrations. Importantly, average kitchen and personal measurements for all primary fuel types **exceeded** (emphasis added) WHO's Interim Target-1 (35 µg/m³ annual average), highlighting the need for comprehensive pollution mitigation strategies.

In addition to the problem of household air pollution, primarily in low- to medium-income countries, an indoor air pollution problem can occur for persons working in buildings. Called the "Sick Building Syndrome," commercial office buildings have sometimes been a source of health concern due to buildup of air contaminants [34]. The term "sick building syndrome" (SBS) is used to describe situations in which building occupants experience acute health and comfort effects that appear to be linked to time spent in a building, but no specific illness or cause can be identified. The complaints may be localized in a particular room or zone or may be widespread throughout the building. In contrast, the term "building related illness" (BRI) is used when symptoms of diagnosable illness are identified and can be attributed directly to airborne building contaminants. A 1984 WHO Committee report suggested that up to 30% of new and remodeled buildings worldwide might be the subject of excessive complaints related to indoor air quality [34].

> Close to half of deaths due to pneumonia among children less than 5 years of age are caused by particulate matter inhaled from household air pollution.

While indoor air pollution, as described herein, is a substantive public health concern, this chapter will present air pollution as primarily a matter of ambient air policies and issues. Unless specifically termed "indoor air," air pollution in this chapter will refer to ambient air pollution.

2.4   TOXICOLOGY AND STANDARDS FOR CRITERIA AIR POLLUTANTS

The Clean Air Act (CAAct) is the U.S. federal government's major policy statement on the impacts and control of sources of air pollution. The statute requires EPA to set National Ambient Air Quality Standards (NAAQS) for six common air pollutants. These commonly found air pollutants (also known as "criteria pollutants") are found throughout the U.S. They are carbon monoxide, lead, nitrogen oxides, ozone, particle pollution (often referred to as particulate matter), and sulfur oxides. EPA calls these pollutants "criteria" air pollutants because the agency must set NAAQS for them based on the human health effects and/or environmental impact [35].

The CAAct established two types of national air quality standards. *Primary standards* set limits to protect public health, including the health of at-risk populations such as people with pre-existing heart or lung disease (such as asthmatics), children, and older adults. *Secondary standards* set limits to protect public welfare, including protection against visibility impairment, damage to animals, crops, vegetation, and buildings. The CAAct requires periodic review of the science upon which the standards are based and the standards themselves [36]. Primary air quality standards are legally enforceable under provisions of the CAAct. Secondary standards are advisory and not subject to enforcement by the EPA but can be adopted by U.S. states as guidance or regulations.

> The Clean Air Act (CAAct) is the U.S. federal government's major policy statement on the impacts and control of sources of air pollution.

The human health effects of the Criteria Air Pollutants (CAPs) are summarized in the following sections. Shown in Table 2.1 are the EPA standards in 2016 for the six, taking note that particulate matter has two standards [35]. Additional details regarding each of the standards are available from EPA.

Carbon monoxide (CO) is a colorless, odorless gas emitted from combustion processes. Globally, and, particularly in urban areas, the majority of CO emissions to ambient air come from mobile sources. Breathing CO can cause harmful health effects by reducing oxygen delivery to the body's organs (like the heart and brain) and tissues. At extremely high levels, CO can cause death [37].

Lead emissions vary in source. At the national level, major sources of lead in the air originate from ore and metals processing and piston-engine aircraft operating on leaded aviation fuel. Other sources are waste incinerators, utilities, and lead-acid battery manufacturers. The highest air concentrations of lead are usually found near lead smelters. Inhalation or ingestion of lead distributes throughout the body in the blood and accumulates in the bones. Depending on the level of exposure, lead can adversely affect the nervous system, kidney function, immune system, reproductive and developmental systems, and the cardiovascular system. The effects most commonly encountered in current populations are neurological effects in children and cardiovascular effects (e.g., high blood pressure and heart disease) in adults. Infants and young children are especially sensitive to even low levels of lead, which are associated with behavioral problems, learning deficits, and lowered IQ [38].

TABLE 2.1

EPA's Primary and Secondary 2016 Standards for Alphabetized Pollutants [28]

Pollutant	Primary/ Secondary	Averaging Time	Level	Form
Carbon monoxide	Primary	8 hours 1 hour	9 ppm 35 ppm	Not to be exceeded more than once per year
Lead	Primary and secondary	Rolling 3-month average	0.15 µg/m³	Not to be exceeded
Nitrogen dioxide	Primary	1 hour	100 ppm	98th percentile of 1-hour daily maximum concentrations, averaged over 3 years
	Primary and secondary	1 year	53 ppm	Annual mean
Ozone	Primary and secondary	8 hours	0.070 ppm	Annual fourth-highest daily maximum 8-hour concentration, averaged over 3 years maximum 8-hour daily
Particle pollution (PM₂.₅)	Primary	1 hour	12.0 µg/m³	Annual mean, averaged over 3 years
	Secondary	1 hour	15.0 µg/m³	Annual mean, averaged over 3 years
	Primary and secondary	24 hours	150 µg/m³	98th percentile, averaged over 3 years
Particle pollution (PM₁₀)	Primary and secondary	24 hours	150 µg/m³	Not to be exceeded more than once per year on average over 3 years
Sulfur dioxide	Primary	1 hour	75 ppb	98th percentile of 1-hour daily maximum concentrations, averaged over 3 years
	Secondary	3 hours	0.5 ppm	Not to be exceeded more than once per year

Oxides of nitrogen (NO_x) is the general term for a group of highly reactive gases, all of which contain nitrogen and oxygen in varying amounts. The NO_x are created when fuel is burned at high temperatures, including internal combustion engines. Fossil-fueled electric utilities, motor vehicles, and industrial operations are the primary sources of NO_x. Nitrogen dioxide (NO_2) can irritate the lungs and reduce resistance to respiratory infections such as influenza [39].

Ozone (O_3) is a highly reactive gas that results primarily from the action of sunlight on nitrogen oxides and hydrocarbons emitted in combustion of fuels. Ozone exposure can produce significant decreases in lung function, inflammation of the lungs' lining, respiratory discomfort, and impair the body's immune system, making people more susceptible to respiratory illness, including pneumonia and bronchitis. At sufficient high levels, repeated exposure to ozone for several months can cause permanent structural damage to the lungs. Hospital admissions and emergency room visits increase on days of high ozone pollution in outdoor air [40].

Particulate matter (PM) is a general term that refers to very small, carbonaceous, solid particles; dust; and acid aerosols. The size of particles is directly linked to their potential for causing adverse health effects. Particles less than 10 μm in diameter pose the greatest problems because their inhalation can reach the alveoli of the lungs, with some entering the bloodstream. Particulate matter 2.5 μm or less in diameter ($PM_{2.5}$) is produced by incomplete combustion of fossil

ENFORCEMENT EXAMPLE

In 2018, Pennsylvania's Allegheny County Health Department fined U.S. Steel more than $1 million for ongoing violations of an air-quality agreement at the Clairton Coke Works. A March 2016 agreement was to resolve air quality violations at the Clairton plant but failed [41].

fuels and biomass and constitutes one of the biggest health concerns. One-hundredth the thickness of a human hair, $PM_{2.5}$ can penetrate deep into the lungs and blood stream and is dangerous at any concentration. Numerous health studies have linked particle pollution exposure to a variety of problems, including premature death in people with heart or lung disease, nonfatal heart attacks, irregular heartbeat, aggravated asthma, decreased lung function, and increased respiratory symptoms, such as irritation of the airways, coughing, or difficulty breathing [42]. The International Agency for Research on Cancer (IARC) concluded in 2013 that particulate matter is carcinogenic to humans [43]. People with heart or lung diseases, children, and older adults are the most likely to be affected by particle pollution exposure

Sulfur dioxide emissions occur when sulfur-containing fuels are combusted. Exposure to SO_2 at high levels is associated with breathing difficulties, respiratory illness, reduced pulmonary resistance to infectious agents, and aggravation of the existing cardiovascular disease. The major source of SO_2 emissions is electric utilities.

In addition to the six CAPs, EPA has the authority to regulate the release of other air pollutants, such as air toxics released from electric power plants, and regulate releases of greenhouse gases. These are authorities specified in the CAAct and will be discussed as policy issues later in this chapter.

2.5 PREVALENCE OF AIR POLLUTION

As with any other environmental hazard's impact on public health, prevalence of the hazard determines its severity. In the extreme, the most hazardous pollutant is ineffectual if its prevalence is naught.

2.5.1 PREVALENCE OF AIR POLLUTION IN THE U.S.

Data on trends in air quality and source emissions are collected by EPA [45]. Shown in Table 2.2 are air quality trends

ENFORCEMENT EXAMPLE

Denver, January 22, 2020: Federal and state authorities reached a $3.5 million settlement with Denver-based oil and gas producer KP Kauffman Company Inc., which has agreed to pay $1 million and spend $2.5 million more improving pollution controls at 67 Colorado facilities. The company was accused of violating laws designed to minimize air pollution from storage tanks [44].

TABLE 2.2
Trends in U.S. Air Quality and Source Emissions [39]

Percent Change in Air Quality			Percent Change in Emissions		
	1980 vs. 2014	2000 vs. 2014		1980 vs. 2014	2000 vs. 2014
CO	−85	−60	CO	−69	−46
O_3	−33	−18	Pb	−99	−50
Pb	−98	−87	NO_x	−55	−45
NO_2 (annual)	−60	−43	VOC	−53	−16
NO_2 (1-hr)	−57	−29	PM_{10}	−58	−16
PM_{10} (24-hr)	—	−30	$PM_{2.5}$	—	−33
$PM_{2.5}$ (annual)	—	−35	SO_2	−81	−70
$PM_{2.5}$ (24-hr)	—	−36			
SO_2 (1-hr)	−80	−62			

based on concentrations of the common pollutants. The data show that both U.S. air quality and source emissions have improved nationally since 1980. All six criteria air pollutants (CO, O_3, Pb, NO_2, PM, SO_2) show significant downward trends between 1980 vs. 2014. Particularly noteworthy for children's health is the significant decrease in outdoor air levels of lead, an outcome due to phase-out of lead additives in gasoline, commencing in 1974 and continuing to 1996, attributable to the CAAct Amendments of 1970. Reductions in outdoor air lead levels produced corresponding decreases in blood lead levels in children, thereby lessening lead toxicity in children.

The EPA also tracks trends in greenhouse emissions, per requirements of the United Nations Framework Convention on Climate Change (Chapter 1). An overview of EPA's national greenhouse gas inventory for 1990–2014 revealed that in 2014, U.S. greenhouse gas emissions totaled 6,870 million metric tons of CO_2 equivalents, with U.S. emissions increasing by 1.0% from 2013 to 2014. Recent trends can be attributed to multiple factors: increased fuel use, year-to-year changes in the prevailing weather, and an increase in miles traveled by on-road vehicles. Greenhouse gas emissions in 2014 were 9% below 2005 levels [46].

The overall pattern of air pollution trends in the U.S. indicates improved air quality for the six CAPs and a small decrease in emissions of greenhouse gases. Notwithstanding this EPA report, the American Lung Association's *State of the Air 2019* report contained the following findings pertaining to U.S. air quality [47]:

- "Ozone and short-term particle pollution worsened in many cities in 2015–2017, compared to 2014–2016. Even levels of year-round particle pollution increased in some cities.
- More than 4 in 10 Americans, approximately 43.3% of the population, live in counties that have monitored unhealthy ozone and/or particle pollution.
- The number of people exposed to unhealthy air increased to nearly 141.1 million. That represents an increase from the past two reports: higher than

the 133.9 million in the 2018 report (covering 2014–2016) and the 125 million in the 2017 report (covering 2013–2015).

- Close to 20.2 million people, or 6.2%, live in 12 counties with unhealthful levels for all three measures.
- Still, progress continues, thanks to the tools in the Clean Air Act. While this is a significant spike in areas with unhealthy levels of ozone and particle pollution, it remains still far below the 166 million in the years covered in the 2016 report (2012–2014).
- Los Angeles remains the city with the worst ozone pollution as it has for 19 years of the 20-year history of the report" [47].

> More than 4 in 10 Americans, approximately 43.3% of the population, live in counties that have monitored unhealthy ozone and/or particle pollution.

2.5.2 PREVALENCE OF GLOBAL INDOOR AIR POLLUTION

Although ambient outdoor air pollution can be a hazard to human health, less often is thought given to the consequences of polluted indoor air. While indoor air in households and offices is generally not a problem in industrialized countries, owing to air conditioning codes and outdoor air regulations, household air can be a public health hazard in some instances. According to WHO, almost 3 billion people, mostly in low- and middle-income countries, still rely on solid fuels (wood, animal dung, charcoal, crop wastes, and coal) burned in inefficient and highly polluting household stoves for cooking and domicile heating. Higher-income countries do face health problems associated with household air pollution, mainly from the burning of solid heating fuels (e.g., coal) in rural or mountainous areas, but these countries generally have systems in place and resources to address these problems [48].

The inefficient stoves used for cooking and heating practices produce high levels of household (indoor) air pollution that includes a range of health-damaging pollutants such as fine particles and carbon monoxide. In poorly ventilated dwellings, smoke within and around the home can exceed acceptable levels for fine particles 100-fold. Exposure is particularly high among women and young children, who spend the most time near the domestic hearth. In 2012 alone, no fewer than 4.3 million children and adults died prematurely from illnesses caused by such household air pollution, according to estimates by WHO. In addition to the adverse health effects, the widespread use of kerosene stoves, heaters, and lamps, these practices also result in many serious injuries and deaths from scalds, burns, and poisoning [49].

> In 2012, no fewer than 4.3 million children and adults died prematurely from illnesses caused by household air pollution, according to estimates by the WHO.

In response to the morbidity and mortality associated with household air pollution, WHO developed indoor air quality recommendations and air quality guidelines for household fuel combustion that aim to help public health policymaker, as well as specialists working on energy, environmental, and other issues understand best approaches to reducing household air pollution.

The recommendations include general considerations for policy, a set of four specific recommendations, and a best-practice recommendation addressing linked health and climate impacts. Among the general consideration, or overarching advice, is that policies should promote community-wide action and that the safety of new fuels and technologies must be assessed rather than assumed. The set of four WHO recommendations pertaining to household air pollution are excerpted herein. Recommendation 1: Emission Rate Targets: Emission rates from household fuel combustion should not exceed the following targets (ERTs) for particles with aerodynamic diameters of less than 2.5 μm ($PM_{2.5}$) and carbon monoxide (CO), based on the values for kitchen volume, air exchange, and duration of device use per day set out in the cited reference and which are assumed to be representative of conditions in low- and middle-income countries. Recommendation 2: Policy during transition to technologies and fuels that meet WHO's air quality guidelines: Governments and their implementing partners should develop strategies to accelerate efforts to meet these air quality guidelines emission rate targets (see Recommendation 1). Recommendation 3: Unprocessed coal should not be used as a household fuel (unprocessed coal is that which has not been treated by chemical, physical, or thermal means to reduce contaminants). Recommendation 4: Household use of kerosene (paraffin) is discouraged [49].

> In response to the morbidity and mortality associated with household air pollution, WHO developed indoor air quality recommendations and air quality guidelines for household fuel combustion.

Good practice recommendation: Considering the opportunities for synergy between climate policies and health, including financing, WHO recommends that governments and other agencies developing and implementing policy on climate change mitigation consider action on household energy and carry out relevant assessments to maximize health and climate gains.

The WHO observes that tackling household air pollution will demand significant resources. For governments of low- and middle-income countries, this calls for coordinated efforts by ministries, nongovernmental organizations and the public sector, international development and finance organizations, and others. Consonant with this observation is ongoing research by EPA and NIEHS on the health impacts of indoor air pollution and work supported by the NGO Global Alliance for Clean Cookstoves.

2.6 IMPACTS OF AIR POLLUTION ON HUMAN HEALTH

As observed by the WHO, air pollution is a major environmental risk to health. By reducing air pollution levels, countries can reduce the burden of disease from stroke, heart disease, lung cancer, and both chronic and acute respiratory diseases, including asthma. The lower the levels of air pollution, the better the cardiovascular and respiratory health of the population will be, both in the long and short term. The WHO air quality guidelines provide an assessment of health effects of air pollution and thresholds for health-harmful pollution levels. As a preamble, ambient air pollution in both cities and rural areas was estimated by WHO to cause 3.7 million premature deaths worldwide in 2012. Some 88% of those premature deaths occurred in low- and middle-income countries, and the greatest number in WHO's Western Pacific and South-East Asia regions [50].

The ensuing sections elaborate the effects of air pollution on the morbidity and mortality of exposed populations. Special emphasis will be given to the effects of air pollution on children's health. The following sections present the impacts of air pollution on measures of morbidity. Separately, effects of air pollution on measures of mortality will ensue following the morbidity data.

2.6.1 IMPACTS OF AIR POLLUTION ON MORBIDITY

The effects of polluted air on human health are numerous and significant. The health effects of contaminated air are well known to the U.S. public from news reports and the continuous release of new scientific information. But prior to a summary description of the current associations between air pollution and human health,

> The Harvard six-city air pollution study is the foundation study of the public health impacts of air pollution.

2.6.1.1 Overview of Adverse Health Effects

Although air quality in the U.S. has generally improved since 1980, millions of Americans live in areas where urban smog, particle pollution, and toxic pollutants can pose serious health concerns. Adverse health effects can occur from inhalation exposure to: particulate matter (PM), noxious gases (SO_2, NO_x, CO), ground-level O_3, and other hazardous toxic substances. As summarized by the EPA, air pollution can affect human health in many ways [51]. The agency asserts that numerous health studies have linked chronic exposure to air pollution to a variety of health problems including: (1) aggravation of respiratory and cardiovascular disease; (2) decreased lung function; (3) increased frequency and severity of respiratory symptoms such as difficulty breathing and coughing; (4) increased susceptibility to respiratory infections; (5) effects on the nervous system, including the brain, such as IQ loss and impacts on learning, memory, and behavior; (6) cancer; and (7) premature death.

Some sensitive individuals appear to be at greater risk for air pollution-related health effects, for example, those with pre-existing heart and lung diseases (e.g., heart failure/ischemic heart disease, asthma, emphysema, and chronic bronchitis), diabetics, older adults, and children [51]. People exposed to acute, high levels of certain air pollutants may experience:

- Irritation of the eyes, nose, and throat,
- Wheezing, coughing, chest tightness, and breathing difficulties,
- Worsening of existing lung and heart problems, such as asthma,
- Increased risk of heart attack [52].

Given this overview by EPA of the effects of air pollution on human health, attention will turn to the impacts of air pollution on specific organ systems.

2.6.1.2   Morbidity: Air Pollution and Lung Disease

Concerning adverse health effects of air pollution, what's best known to the public are the deleterious effects of air pollutants on the lungs. These effects are generally well known because of newsmedia reports on lung disease related to air pollution and, more importantly, from weather reports that advise the public when air pollutants have reached hazardous levels. When such conditions occur, persons are advised to remain indoors and reduce their activity levels when outdoors. Also, government agencies promote vehicle use reductions on days when pollution levels are hazardous. The sum of these newsmedia and governmental acts is a general awareness among the U.S. public of the health hazards of air pollution, usually focused on the effects on the lungs.

> It is estimated that about 500,000 lung cancer deaths and 1.6 million COPD deaths can be attributed to air pollution.

A Review by the Forum of International Respiratory Societies' Environmental Committee, concluded, "Although air pollution is well known to be harmful to the lung and airways, it can also damage most other organ systems of the body. It is estimated that about 500,000 lung cancer deaths and 1.6 million COPD deaths can be attributed to air pollution [...]" [53].

Exposure to acute high levels of air pollutants can result in persons' visits to emergency rooms to seek medical care. A research team from George Washington University reviewed emergency room (ER) visits for asthma in 54 countries and the territory of Hong Kong [54]. The team noted that Asthma is the most prevalent chronic respiratory disease worldwide, affecting about 358 million people. The ER data were then combined with an analysis of epidemiological studies globally and global pollution levels as detected by satellites orbiting the Earth (with support from NASA). The researchers found that between 9 and 23 million annual asthma ER visits globally may be triggered by ozone pollution and also estimated that between 5 and 10 million asthma emergency room visits every year (or 4%–9% of total global asthma ER visits) were linked to fine-particulate matter.

2.6.1.3   Morbidity: Air Pollution and Cardiovascular Disease

Regarding the association between air pollution and heart attack, Japanese investigators in 2020 reported a nationwide case-crossover study that examined population-based registry data for out-of-hospital cardiac arrest (OHCA) in Japan from January 1, 2014, to December 31, 2015 [55]. Daily $PM_{2.5}$, CO, NO_2, photochemical oxidants (Ox), and SO_2 exposure on the day of the cardiac arrest or 1–3 days before the arrest were assessed. Daily exposure was calculated by averaging the measurements from all $PM_{2.5}$ monitoring stations in the same prefecture. Over the 2 years, 249,372 OHCAs were identified, with 149,838 presumed of cardiac origin. The median daily $PM_{2.5}$ was 11.98 μg/ m^3. Among other findings, of special significance, each 10 μg/m^3 increase in $PM_{2.5}$ was associated with an increased risk of all-cause OHCA on the same day.

> Between 9 and 23 million annual asthma ER visits globally may be triggered by ozone pollution; between 5 and 10 million asthma ER visits every year were linked to fine-particulate matter.

2.6.1.4 Morbidity: Air Pollution and Cancer

Some recent research findings on associations between air pollution and cancer merit special comment. In 2013, the International Agency for Research on Cancer (IARC), a WHO agency, classified outdoor air pollution as carcinogenic to humans. The agency concluded that there was sufficient evidence that exposure to outdoor air pollution causes lung cancer. IARC also noted a positive association with an increased risk of bladder cancer. Particulate matter, a major component of outdoor air pollution, was evaluated separately and was also classified as carcinogenic to humans [43]. These were persuasive statements from the world's most respected cancer agency.

A later study investigated cancer rates in a cohort of Hong Kong residents. The investigators enrolled 66,280 people who were age 65 or older when initially recruited between 1998 and 2001. Researchers followed the study subjects until 2011, ascertaining causes of death from Hong Kong registrations. Annual concentrations of $PM_{2.5}$ at their homes were estimated using data from satellite readings and fixed-site monitors. After adjusting for confounding factors, results showed that for every 10 $\mu g/m^3$ of increased exposure to $PM_{2.5}$, the risk of dying from any cancer rose by 22%. Increases of 10 $\mu g/m^3$ of $PM_{2.5}$ were associated with a 42% increased risk of mortality from cancer in the upper digestive tract and a 35% increased risk of mortality from accessory digestive organs, which include the liver, bile ducts, gall bladder, and pancreas. For women, every 10 $\mu g/m^3$ increase in exposure to $PM_{2.5}$ was associated with an 80% increased risk of mortality from breast cancer, and men experienced a 36% increased risk of dying of lung cancer for every 10 $\mu g/m^3$ increased exposure to $PM_{2.5}$ [56].

> In 2013 the International Agency for Research on Cancer (IARC) classified outdoor air pollution as carcinogenic to humans.

In another study, air pollution exposure was linked to lung cancer incidence in non-smokers. Though cigarette smoking is the number one cause of lung cancer, about one in 10 people who develop lung cancer have never smoked. In this study, investigators followed more than 180,000 non-smokers for 26 years. Throughout the study period, 1,100 people died from lung cancer. The participants lived in all 50 U.S. states and in Puerto Rico, and based on their zip codes, the researchers estimated exposure units of $\mu g/m^3$ PM. Pollution levels overall averaged 17 units across the study period. After taking into account other cancer risk factors, such as second-hand smoke and radon exposure, the investigators found that for every 10 extra units of PM exposure, a person's risk of lung cancer rose by 15% to 27% [57].

However, cancer incidence can be reduced when air pollution levels decrease. The California Air Resources Board (CARB) reported that Californians' overall cancer risk from toxic air pollution declined 76% over more than two decades, a trend the agency attributes to the state's array of air pollution regulations. State scientists measured the drop from 1990 to 2012 by tracking airborne concentrations of the seven toxic air contaminants that are most responsible for increasing cancer risks. Concentrations of diesel PM, the largest contributor to airborne cancer risk in the state, declined more than 68% in California over the 23-year study period, largely because of California's requirements for cleaner diesel fuels and strict emissions-control rules for diesel trucks adopted in 2008 [58]. In 2020 California adopted a landmark environmental rule, the first in the U.S., requiring more than half of all trucks sold in the state to be zero emissions by 2035 and all of them by 2045. Electric vehicles will likely become California's trucks of the future.

> The California Air Resources Board (CARB) reported that Californians' overall cancer risk from toxic air pollution declined 76% over more than two decades.

2.6.1.5   Morbidity: Air Pollution, Mental Health, and Crime

In addition to research findings on the association between air pollution and cancer, cardiovascular disease, and lung disease, a nascent area of research is mental health. Two investigations are illustrative of this area of research. In one study, King's College London investigators combined high-resolution air pollution exposure estimates and prospectively collected phenotypic data to explore associations between air pollutants of major concern in urban areas and mental health problems in childhood and adolescence. Of the 284 children studied, those who lived in the top 25% most polluted areas at age 12 were found to be three to four times more likely to have depression at 18, compared with those living in the 25% least polluted areas [59].

A second investigation by King's College London researchers used estimates of air and noise pollution levels across London and correlated the estimates with patients' health records for 131,000 patients aged between 50 and 79 at 75 medical practices within London. The subjects' health was tracked for 7 years from 2005, during which period 1.7% of the patients were diagnosed with dementia. Their exposure to air pollution was estimated based on their home postal codes. The results suggested that air pollution elevated the risk of dementia by 7%. The researchers postulated that air pollution is added to the list of risk factors for dementia [60].

Perhaps related to the issue of mental health and air pollution, a 2019 study by a team of Canadian university epidemiologists conducted a cohort study of within-city spatial variations in ambient ultrafine particles (UFPs) across Montreal and Toronto, Canada [61]. The cohort population was 1.9 million adults included in multiple cycles of the Canadian Census Health and Environment Cohorts. UFP exposures (3-year moving averages) were assigned to residential locations using land-use regression models. Researchers followed cohort members for malignant brain tumors between 2001 and 2016. In total, 1,400 incident brain tumors were identified during the follow-up period. Each $10,000/cm^3$ increase in UFPs was positively associated with brain tumor incidence (HR=1.112, 95% CI: 1.042, 1.188) after adjusting for $PM_{2.5}$, NO_2, and sociodemographic factors.

In a different kind of investigation, Colorado State University researchers estimated the effect of short-term air pollution exposure ($PM_{2.5}$ and O_3) on several categories of crime, with a particular emphasis on aggressive behavior [62]. To identify this relationship, investigators combined detailed daily data on crime, air pollution, and weather for an eight-year period across the U.S. Results showed a robust positive effect of increased air pollution on violent crimes, and specifically assaults, but no relationship between increases in air pollution and property crimes. The effects were present in and out of the home, at levels well below Ambient Air Pollution Standards, and $PM_{2.5}$ effects are strongest at lower temperatures. The researchers suggested that a 10% reduction in daily $PM_{2.5}$ and O_3 could save $1.4 billion in crime costs per year, a previously overlooked cost associated with pollution.

2.6.1.6 Morbidity: Air Pollution and Neurodegenerative Diseases

A university study team of public health researchers analyzed the link between $PM_{2.5}$ and neurodegenerative diseases across the U.S. The data of more than 63 million individuals who were 65 years or older were collected between the years 2000 and 2016. One million of the senior adults developed Parkinson's disease while three to four million had Alzheimer's and related dementias. The number of cases was directly related to the increase in $PM_{2.5}$ every year.

Results also showed that air pollution could potentially make existing conditions worse "by accelerating these biological pathways or worsening intermediate processes" [63]. The researchers suggested that air pollution may contribute to neurodegeneration due to oxidative stress, systematic inflammation, and neuroinflammation. A second study added weight to these proffered suggestions. In a 2020 cross-sectional study of 18,178 individuals with cognitive impairment, people living in areas with worse air quality were more likely to have positive amyloid positron emission tomography scan results. That is, higher $PM_{2.5}$ concentrations appeared to be associated with brain amyloid-β plaques, a signature characteristic of Alzheimer's disease [64]. Further research will be needed to validate these findings.

2.6.1.7 Morbidity: Air Pollution and Elderly Patients

In a study that evaluated hospital records for persons exposed to air pollutants, researchers at Harvard University reported in 2019 that short-term exposure to fine-particulate matter ($PM_{2.5}$) led to higher levels of hospitalization for people with illnesses such as septicemia or blood poisoning, kidney failure, urinary tract infections, skin and other tissue infections, and electrolyte disorders often brought on by loss of fluids from vomiting or diarrhea [65]. The population studied was Americans older than age 65 years, through a vast analysis of Medicare records for the 48 continental U.S. states. As the amount of $PM_{2.5}$ increased between the days analyzed, so did the hospital admission rate connected to those disease groups. Researchers defined "short-term" as exposure on the day of hospitalization and compared the pollution levels to the day before.

In a different kind of study of Medicare enrollees, researchers looked at 16 years' worth of data from 68.5 million Medicare enrollees, adjusting for factors such as body mass index, smoking, ethnicity, income, and education. They matched the participants' zip codes with air pollution data gathered from locations across the U.S. [66]. The authors found that an annual decrease of 10 $\mu g/m^3$ in $PM_{2.5}$ pollution would lead to a 6%–7% decrease in mortality risk. The investigators opined that U.S. standards for $PM_{2.5}$ concentrations are not protective enough and should be lowered.

2.6.2 Mortality: Impacts of Air Pollution

While the effects of air pollutants on lungs are, and will remain, significant in terms of the public's health, health evidence emerged that fine-particulate matter (PM) exerts an even greater public health burden as a contributor to cardiovascular and heart disease. Particularly alarming is the association between PM in air and its contribution to sudden heart failure. Research now implicates moderate levels of air pollution can be triggers of fatal heart attacks. It is possible that heart attacks, not lung disease, might be the most serious medical threat posed by ambient air pollution [67].

It is important to cite here the study most often called the foundation investigation of the adverse health effects of air pollution. The 1993 study is generally referred to as the "Harvard six-city air pollution study." In 1993, Harvard School of Public Health researchers published their findings of an association between air pollution and adverse health effects [68]. Prior to their study, some other studies had reported an association between air pollution and adverse health effects, but these studies had not controlled for confounding factors, especially tobacco smoking and potential industrial exposure to air pollutants.

In this prospective cohort study, the effects of air pollution on mortality were estimated, while controlling for individual risk factors. Survival analysis was conducted with data from a 14- to 16-year mortality follow-up of 8,111 adults in six U.S. cities. The six communities were Watertown, Massachusetts; Harriman, Tennessee; selected census tracts of St. Louis; Steubenville, Ohio; Portage, Wisconsin; and Topeka, Kansas. As part of the original study design, ambient (outdoor) concentrations of total suspended particulate matter, sulfur dioxide, ozone, and suspended sulfates were measured in each community at a centrally located air-monitoring station. Researchers found that mortality rates were most strongly associated with cigarette smoking. After adjusting for smoking and other risk factors, air pollution was positively associated with death from lung cancer and cardiopulmonary disease but not with death from other causes considered together. Mortality was most strongly associated with air pollution with fine particulates, including sulfates. The Harvard six-city air pollution study stimulated subsequent epidemiological studies of the adverse human health impacts, laying a foundation of air pollution research globally.

For example, Rossi et al. examined air pollution levels for the years 1980–1989 in Milan, Italy, for association with deaths on days of elevated pollution [69]. Among the findings, a significant association was found for heart failure deaths (7% increase/100 $\mu g/m^3$ increase in total suspended particulate [TSP]). Similarly, Neas et al. analyzed daily mortality rates among Philadelphia, Pennsylvania, residents from 1973 to 1980 [70]. Investigators found that a 100 $\mu g/m^3$ increase in the 48-hr mean level of TSP was associated with deaths due to cardiovascular disease. In another study, investigators examined air pollution levels in Seoul, Korea, and stroke mortality data over a 4-year period [71]. They reported "[t]hat PM(10) and gaseous pollutants are significant risk factors for acute stroke death and that the elderly and women are more susceptible to the effect of particulate pollutants."

Investigators at the University of Southern California investigated a large database in regard to chronic health effects of air pollution [72]. Researchers examined data from 22,906 residents of Los Angeles and adjacent areas. They

Among participants, for each increase of 10 $\mu g/m^3$ of PM in the neighborhood's air, the risk of death from any cause rose by 11%–17%. Ischemic heart disease mortality risks rose by 25%–39% for the 10 $\mu g/m^3$ increase in air pollution.

determined air pollution exposure in 267 different zip codes where participants lived and compiled causes of death for the 5,856 participants who died by year 2000. The effects of exposure to $PM_{2.5}$ were examined across the study areas. Among participants, for each increase of 10 µg/m³ of fine particles in the neighborhood's air, the risk of death from any cause rose by 11%–17%. Ischemic heart disease mortality risks rose by 25%–39% for the 10 µg/m³ increase in air pollution. The investigators believed particulate matter may promote inflammatory processes, including atherosclerosis, in key tissues.

Investigators found an increase in overall mortality associated with each 10 µg/m³ increase in $PM_{2.5}$ modeled as the overall mean or as exposure in the year of death.

In another study, investigators found an increase in overall mortality associated with each 10 µg/m³ increase in $PM_{2.5}$ modeled as the overall mean or as exposure in the year of death [73]. $PM_{2.5}$ was associated with increased lung cancer and cardiovascular deaths. Of note, the investigators' database included $PM_{2.5}$ levels that had decreased because of environmental controls. Findings showed improved overall mortality rates were associated with decreased $PM_{2.5}$. Although further research is needed to clarify the association between air pollution and fatal heart attacks, there is already sufficient data to move forward with public health prevention actions, such as public awareness and physician education campaigns.

In a different kind of mortality study of an criteria air pollutant, investigators analyzed daily data on air pollution, meteorology, and total mortality from 337 cities in 18 countries or regions, covering various periods from 1979 to 2016 [74]. All included cities had at least 2 years of both CO and mortality data. Overall, a 1 mg/m³ increase in the average CO concentration of the previous day was associated with a 0·91% increase in daily total mortality. The pooled exposure–response curve showed a continuously elevated mortality risk with increasing CO concentrations, suggesting no threshold. Findings persisted at daily concentrations as low as 0·6 mg/m³ or less. The authors assert that theirs was the largest international study of the effects on mortality of short-term exposure to ambient air CO levels and recommend further study.

Another study reported the association between overall mortality and air pollution in a U.S. Medicare population [75]. Estimates were made of the risk of death associated with exposure to increases of 10 µg/m³ for $PM_{2.5}$ and 10 ppb for O_3. Increases of 10 µg/m³ in $PM_{2.5}$ and of 10 ppb in O_3 were associated with increases in all-cause mortality of 7.3% and 1.1%, respectively.

Further troubling findings about the adverse health effects of ambient ground-level ozone were published by Bell et al. [76]. Using data from a national air pollution database, investigators estimated a national average relative rate of mortality associated with short-term exposure to ambient ground-level ozone for 95 large U.S. urban communities for the period 1987–2000. Findings showed that a 10-ppb increase in the previous week's ozone was associated with a 0.52% increase in daily mortality and a 0.64% increase in cardiovascular and respiratory mortality.

It is important to cite here the study most often called the foundation investigation of the adverse health effects of air pollution. This study is generally referred to as the "Harvard six-city air pollution study." In 1993 Harvard School of Public Health researchers published their findings of an association between air pollution and adverse health effects [68]. Prior to their study, some other studies had reported an association between air pollution and adverse health effects, but these studies had not controlled for confounding factors, especially tobacco smoking and potential industrial exposure to air pollutants. In this prospective cohort study, the effects of air pollution on mortality were estimated, while controlling for individual risk factors.

Survival analysis was conducted with data from a 14- to 16-year mortality follow-up of 8,111 adults in six U.S. cities. The six communities were Watertown, Massachusetts; Harriman, Tennessee; selected census tracts of St. Louis; Steubenville, Ohio; Portage, Wisconsin; and Topeka, Kansas. As part of the original study design, ambient (outdoor) concentrations of total suspended particulate matter, sulfur dioxide, ozone, and suspended sulfates were measured in each community at a centrally located air-monitoring station. Researchers found that mortality rates were most strongly associated with cigarette smoking. After adjusting for smoking and other risk factors, air pollution was positively associated with death from lung cancer and cardiopulmonary disease but not with death from other causes considered together. Mortality was most strongly associated with air pollution with fine particulates, including sulfates. The Harvard six-city air pollution study stimulated subsequent epidemiological studies of the adverse human health impacts, laying a foundation of air pollution research globally.

2.6.2.1   Mortality: Air Pollution and Global Metrics

Regarding global mortality from air pollution, in 2014 WHO reported that in 2012 around 7 million people died—one in eight of total global deaths—as a result of air pollution exposure. This finding more than doubles previous estimates and confirms that air pollution is now the world's largest single environmental health risk. In particular, the new WHO data revealed a strong link between both indoor and outdoor air pollution exposure and cardiovascular diseases, such as strokes and ischemic heart disease, as well as between air pollution and cancer.

These adverse health effects are in addition to air pollution's role in the development of respiratory diseases, including acute respiratory infections and chronic obstructive pulmonary diseases [77]. For comparison, the 7 million global deaths caused by air pollution approximates the year 2013 population of Hong Kong, China.

> The WHO reported in 2014 that around 7 million people died, one in eight of total global deaths, as a result of air pollution exposure. WHO asserted that air pollution is now the world's largest single environmental health risk.

Concerning global mortality, a study in 2018 found as many as 153 million premature deaths linked to air pollution could be avoided worldwide this century if governments speed up their timetable for reducing fossil fuel emissions [78]. The study is the first to project the number of lives that could be saved, city by city, in 154 of the world's largest urban areas if nations agree to reduce carbon emissions and limit global temperature rise to 1.5°C in the near future rather than postponing the biggest emissions cuts until later, as some governments have proposed. Premature deaths would drop in cities on every inhabited continent, the study shows, with the greatest gains in saved lives occurring in Asia and Africa.

2.6.2.2 Morbidity: Air Pollution and Life Expectancy

Global life expectancy, as impacted by air pollution, was the subject of research by the Energy Policy Institute at the University of Chicago [79]. The Institute's Air Quality Life Index (AQLI) can be used to estimate life longevity as a function of environmental hazards. Globally, the AQLI reveals that particulate pollution reduces average life expectancy by 1.8 years, making it the greatest global threat to human health. By comparison, first-hand cigarette smoke leads to a reduction in global average life expectancy of about 1.6 years. On average, people in India would live 4.3 years longer if their country met the WHO guideline on PM—expanding the average life expectancy at birth there from 69 to 73 years. In the U.S., about a third of the population lives in areas not in compliance with the WHO guideline. Those living in the U.S. country's most polluted counties could expect to live up to 1 year longer if pollution met the WHO guideline.

On a more positive note, a study reported in 2009 by researchers at Brigham Young University and the Harvard School of Public Health found that average life expectancy in 51 U.S. cities had increased nearly 3 years over recent decades, with approximately 5 months of that increase attributable to cleaner air. Investigators evaluated the impact of resulting decreases in particulate pollution on average life spans in cities for which air pollution data were available. In cities that had previously been the most polluted air and cleaned up the most, the cleaner air added approximately 10 months to the average resident's life [80].

> Globally, the AQLI reveals that particulate pollution reduces average life expectancy by 1.8 years, making it the greatest global threat to human health.

2.6.3 IMPACTS OF AIR POLLUTION ON CHILDREN'S HEALTH

The effects of air pollution on children's health is a particularly important subject, as any disease or disability in children reduces their quality of life and that of their parents and can lead to expensive health care costs and issues of social development. The early work on the relationship between air pollution and children's health was focused on lung function and disease. These early studies are retained here because of their historical significance and the investigators' commitment to children's health. In an early study, Gauderman et al. reported on the effect of air pollution on lung development of children 10–18 years of age [81]. Children ($n = 1,759$) recruited from schools in 12 southern California communities served as the study population. Results showed that over the 8-year period of study, deficits in the growth of FEV(1) (forced expiratory volume in 1 second, a measure of lung function) were statistically significant with exposure to NO_2, acid vapor, $PM_{2.5}$, and elemental carbon.

Researchers at the University of Southern California investigated the pollution-asthma link in 208 children who had resided since 1993 in 10 Southern California cities [82]. Air samplers were placed outside the home of each student in order to measure NO$_2$ levels. Further, the distance of each child's home from local freeways, as well as how many vehicles traveled within 150 m of the child's home, was determined. Aerodynamic models were used to estimate traffic-related air pollution levels at each child's home. Results showed a link between asthma prevalence in the children and NO$_2$ levels at their homes. For each increase of 5.7 ppb in average NO$_2$, the risk of asthma increased by 83%. Further, the closer the students lived to a freeway, the higher the students' asthma prevalence. Asthma risk increased by 89% for every 1.2 km (about three-quarters of a mile) closer the students lived to a freeway.

> Of the total number of deaths attributable to the joint effects of household and ambient air pollution worldwide in 2016, 9% were in children.

These cited studies helped set a public health path of research on air pollution and children's health. The problem of air pollution's deleterious impact on children's health was noted by the WHO, which observed that globally in 2016, one in every eight deaths was attributable to the joint effects of ambient and household air pollution—a total of 7 million deaths [83]. Of this number:

- Some 543,000 deaths in children less than 5 years and 52,000 deaths in children aged 5–15 years were attributed to the joint effects of ambient and household air pollution in 2016.
- Together, household air pollution from cooking and ambient air pollution causes more than 50% of acute lower respiratory tract infection (ALRI) in children less than 5 years in lower-to-middle-income countries.
- Of the total number of deaths attributable to the joint effects of household and ambient air pollution worldwide in 2016, 9% were in children.

In 2018 the WHO released a summary report on the agency's review of the data on air pollution and child health [83]. The agency's summary findings include the following: "There is compelling evidence that exposure to air pollution damages the health of children in numerous ways. The evidence summarized in this report is based on a scoping review of relevant studies published within the past 10 years and input from dozens of experts around the world." Quoting from the summary report:

"Adverse birth outcomes: Numerous studies have shown a significant association between exposure to ambient air pollution and adverse birth outcomes, especially exposure to PM, SO$_2$, NO$_x$, O$_3$, and CO. There is strong evidence that exposure to ambient PM is associated with low birth weight. There is also growing evidence that maternal exposure, especially to fine PM, increases the risk of preterm birth. There is emerging evidence for associations between exposure to air pollution and other outcomes, such as stillbirth and infants born small for gestational age.

Infant mortality: There is compelling evidence of an association between air pollution and infant mortality. Most studies to date have focused on acute exposure and ambient air pollution. As pollution levels increase, so too does the risk of infant mortality, particularly from exposure to PM and toxic gases" [83]. Concerning infant mortality, a global health study estimated some 476,000 infants across the world died from the adverse effects of exposure to air pollution in 2019 [84]. The State of Global Air study said nearly two-thirds of those deaths were related to the burning of poor-quality fuels for cooking, that is, indoor air pollution.

Returning to the WHO report on child health: "Neurodevelopment: A growing body of research suggests that both prenatal and postnatal exposure to air pollution can negatively influence neurodevelopment, lead to lower cognitive test outcomes, and influence the development of behavioral disorders such as autism spectrum disorders and attention deficit hyperactivity disorder. There is strong evidence that exposure to ambient air pollution can negatively affect children's mental and motor development.

Childhood obesity: A limited number of studies have identified a potential association between exposure to ambient air pollution and certain adverse metabolic outcomes in children. The findings include positive associations between exposure to air pollution in utero and postnatal weight gain or attained body mass index for age, and an association has been reported between traffic-related air pollution and insulin resistance in children.

> There is robust evidence that exposure to air pollution damages children's lung function and impedes their lung function growth, even at lower levels of exposure.

Lung function: There is robust evidence that exposure to air pollution damages children's lung function and impedes their lung function growth, even at lower levels of exposure. Studies have found compelling evidence that prenatal exposure to air pollution is associated with impairment of lung development and lung function in childhood. Conversely, there is evidence that children experience better lung function growth in areas in which ambient air quality has improved.

ALRI (Acute Lower Respiratory Infection), including pneumonia: Numerous studies offer compelling evidence that exposure to ambient and household air pollution increases the risk of ALRI in children. There is robust evidence that exposure to air pollutants such as $PM_{2.5}$, NO_2, and O_3 is associated with pneumonia and other respiratory infections in young children. Growing evidence suggests that PM has an especially strong effect.

Asthma: There is substantial evidence that exposure to ambient air pollution increases the risk of children developing asthma and that breathing pollutants exacerbates childhood asthma as well. While relevant there are fewer studies on household air pollution, there is suggestive evidence that exposure to household air pollution from the use of polluting household fuels and technologies is associated with the development and exacerbation of asthma in children.

Otitis media: There is clear, consistent evidence of an association in children between ambient air pollution exposure and the occurrence of otitis media, a group of inflammatory diseases of the middle ear. Although relatively a few studies have examined the association between non-tobacco smoke household air pollution and otitis media, there is suggestive evidence that combustion-derived household air pollution might increase the risk of otitis media.

Childhood cancers: There is substantial evidence that exposure to traffic-related air pollution is associated with an increased risk of childhood leukemia. Several studies have found associations between prenatal exposure to ambient air pollution and higher risks for retinoblastomas and leukemia in children. While relatively a few studies have focused on household air pollution and cancer risk in children, household air pollution is strongly associated with several types of cancer in adults and typically contains many substances classified as carcinogens.

Relation between early exposure and later health outcomes: Children exposed to air pollution prenatally and in early life are more likely to experience adverse health outcomes as they mature and through adulthood. Exposure to air pollution early in life can impair lung development, reduce lung function, and increase the risk of chronic lung disease in adulthood. Evidence suggests that prenatal exposure to air pollution can predispose individuals to cardiovascular disease later in life.

Altogether, there is clear, compelling evidence of significant associations between exposure to air pollution and a range of adverse health outcomes. The evidence suggests that the early years, starting in pregnancy, are the best time to invest in a child's health, through action to improve their environment and reduce their exposure to pollutants. This window of time offers a great opportunity: precisely because children are most vulnerable and sensitive to environmental influences in their earliest years, action taken during this critical phase can yield immense health benefits" [83].

Perspective: These WHO data on the health impacts of children's exposure to air pollution present a grim picture of the wide-ranging adverse health effects. It is important here to repeat WHO's admonition: "The evidence suggests that the early years, starting in pregnancy, are the best time to invest in a child's health, through action to improve their environment and reduce their exposure to pollutants." The WHO summary of the adverse health effects of children's exposure to air pollution presents a compelling argument for the development and implementation of policies to control sources of air pollution.

Regarding the benefits to children's respiratory health, University of Southern California researchers measured lung function annually in 2,120 children from three separate cohorts corresponding to three separate calendar periods: 1994–1998, 1997–2001, and 2007–2011 [85]. Mean ages of the children within each cohort were 11 years at the beginning of the period and 15 years at the end. Linear-regression models were used to examine the relationship between declining pollution levels over time and lung-function development from 11 to 15 years of age, measured as the increases in forced expiratory volume in 1 second (FEV1) and forced vital capacity (FVC) during that period. Over the 13 years spanned by the three cohorts, improvements in 4-year growth of both FEV1 and FVC were associated with declining levels of NO_2 and of $PM_{2.5}$ and less than PM_{10} for FEV1 and FVC. In summary, long-term improvements in air quality were associated with statistically and clinically significant positive effects on lung-function growth in children.

2.6.4   EPA's NATA HEALTH DATABASE

Concerning the health risk of individual air pollutants, the EPA has developed a database to assist local, state, tribal, and federal governments involved in air pollution

decision-making. The National-Scale
Air Toxics Assessment (NATA) is a
screening tool that estimates cancer and
other health risks from exposure to toxic
air pollutants, called *air toxics* by EPA
[68]. Air toxics are those air pollutants
known or suspected to be carcinogens
or known to cause other health effects,
such as birth defects or respiratory prob-

The National-Scale Air Toxics
Assessment (NATA) is an EPA screening tool that estimates cancer and
other health risks from exposure to
toxic air pollutants, called air toxics
by EPA.

lems. Risk assessment methods are used by EPA to estimate human health risks
from lifetime exposure to air pollutants at year 1999 levels as the baseline for their
assessment.

The EPA released its first NATA, based on year 1996 air emissions data, in year
2002. In August 2018, EPA released the results of its 2014 National Air Toxics
Assessment (NATA) [68]. NATA is a screening tool intended to help EPA and state,
local, and tribal air agencies determine if areas, pollutants, or types of pollution
sources need to be examined further to better understand risks to public health
NATA helps assess which air toxins and emission source types may pose health
risks. NATA also helps EPA and other agencies determine which places might need
further study to better understand risks.

The version of NATA released in August 2018 is based on emissions for the calendar year 2014. It includes estimates of exposure and risk for 180 air toxics that EPA
regulates under the Clean Air Act. It also estimates exposure and risks for diesel
particulate matter (noncancer effects only). Three findings in the 2018 NATA report
of particular relevance to environmental public health are the following:

- "Nationwide, total emissions of air toxins are declining, and air quality
  monitoring data show that concentrations of many toxins in the air, such as
  benzene, also are trending downward.
- The 2014 NATA estimates that the nationwide average cancer risk from
  air toxics exposure is 30 in 1 million in the U.S. About half of that risk
  comes from the formation of formaldehyde—produced when other pollutants chemically react in the air. This is known as secondary formation and
  comes from emissions from industries, mobile sources, and natural sources.
  The other half of the nationwide cancer risk comes from pollution that is
  directly emitted into the air.
- Despite improvements, some local areas still face challenges. The 2014
  NATA results also indicate that some census tracts may have elevated risks
  of cancer from air toxics exposure (less than 1% of all tracts). Census tracts
  are small subdivisions of a county
  or county equivalent, such as a par-
  ish. Industrial emissions of three
  pollutants—ethylene oxide, chloro-
  prene, and coke oven emissions—
  contribute to most of the risk in
  these tracts" [68].

The 2014 NATA estimates that the
nationwide average cancer risk from
air toxics exposure is 30 in 1 million
in the U.S.

The NATA database and findings from research investigations like those cited in this chapter provide public health and environmental protection authorities with essential data from which to set federal, state, and local air pollution control policies.

2.6.5 AIR POLLUTION AND SOCIAL JUSTICE ISSUES

Researchers in the EPA's National Center for Environmental Assessment found that people of color in the U.S. are much more likely to reside near polluters and breathe polluted air [69]. Specifically, their study found that people in poverty are exposed to more fine-particulate matter than people living above the poverty level. According to the study's authors, "results at national, state, and county scales all indicate that non-Whites tend to be burdened disproportionately to Whites." Previous works have also linked disproportionate exposure to particulate matter and America's racial geography.

A 2016 study found that long-term exposure to the pollutant is associated with racial segregation, with more highly segregated areas suffering higher levels of exposure. A 2012 study found that overall levels of particulate matter exposure for people of color in the U.S. were higher than those for white people. While differences in overall particulate matter by race were significant, differences for some key particles were immense. For example, Hispanics faced rates of chlorine exposure that are more than double those of whites. Chronic chlorine inhalation is known for degrading cardiac function [69]. On a global scale, persons in low- to middle-income countries are more likely to be exposed to hazardous air pollution than those in high-income countries due to fewer resources available to control air pollution sources.

> EPA's National Center for Environmental Assessment found that people of color in the U.S. are much more likely to reside near polluters and breathe polluted air than white residents.

Perspective: The adverse health effects of air pollution are a major domestic and global health problem. Decades of health investigations have elucidated myriad effects on public health of pollutants released into ambient air. While early health studies rightly focused on effects on lung function and lung diseases, subsequent research on fine-particulate matter revealed serious effects associated with cardiovascular disease. Later, research revealed an association between air pollution and lung and other cancers. The disparity between persons of color and whites in the U.S. in terms of exposure to air pollution is a social issue of significant relevance to public health programs, meriting their action, including actions to reduce air pollution levels.

2.6.6 DIESEL ENGINE EXHAUST AND HEALTH IMPACTS

Diesel engines provide power to a wide variety of motor vehicles, heavy equipment, and other machinery used in a large number of industries including mining, transportation, construction, agriculture, maritime, and many types of manufacturing operations. The exhaust from diesel engines contains a mixture of gases and fine particle that can create a health hazard when not properly controlled. Diesel particulate matter is a component of diesel exhaust that includes soot particles made up primarily of carbon, ash, metallic abrasion particles, sulfates, and silicates. Diesel-powered vehicles and equipment account for nearly half of all nitrogen oxides (NO_x) and more than two-thirds of all particulate matter (PM) emissions from U.S. transportation sources [86].

> Diesel-powered vehicles and equipment account for nearly half of NO_x and more than two-thirds of all PM emissions from U.S. transportation sources.

Diesel emissions of NO_x contribute to the formation of ground-level ozone, which irritates the respiratory system, causing coughing, choking, and reduced lung capacity. Ground-level ozone pollution, formed when nitrogen oxides and hydrocarbon emissions combine in the presence of sunlight, presents a hazard for both healthy adults and individuals suffering from respiratory problems. Urban ozone pollution has been linked to increased hospital admissions for respiratory problems such as asthma, even at levels below the federal standards for ozone.

Diesel exhaust has been classified as a potential human carcinogen by the EPA and the International Agency for Research on Cancer. Exposure to high levels of diesel exhaust has been shown to cause lung tumors in rats, and studies of humans routinely exposed to diesel fumes indicate a greater risk of lung cancer. For example, occupational health studies of railroad, dock, trucking, and bus garage workers exposed to high levels of diesel exhaust over many years consistently demonstrate a 20%–50% increase in the risk of lung cancer or mortality [86]. Because of the human health and ecological impacts of diesel exhaust, policy-

ENFORCEMENT EXAMPLE

Sacramento, California, September 15, 2020–Mercedes-Benz and its German parent have been fined $1.5 billion for equipping their diesel cars with illegal software, agreeing to pay $1.5 billion in fines, including $286 million to the state of California, after investigators discovered diesel cars equipped with software enabling the vehicles to evade air-pollution limits [87].

makers have begun instituting controls on the presence of diesel-powered vehicles. In particular, diesel vehicles face a grim future in European cities. Over the next decade, 24 European cities with a total population of 62 million people will ban diesel vehicles, and 13 of those cities will ban all internal combustion cars by 2030 [88]. In 2018, diesel car registrations fell 36% across Europe, and they decreased by more than half since 2015. The exodus away from diesel vehicles has been driven not only by concerns over air pollution and climate change, but also by public anger after officials disclosed in 2015 that Volkswagen had been rigging emissions tests to make its diesel vehicles appear less polluting than they were.

Similar to the Volkswagen illegal tampering with its diesel engine emissions, some U.S. owners and operators of diesel pickup trucks have illegally disabled their vehicles' emissions control technology over the past decade according to a November 2020 report by the EPA [89]. More than half a million diesel pickup trucks have been allowing excess emissions equivalent to 9 million extra trucks on the road, according to EPA. The report said "diesel tuners" will allow the trucks to release more than 570,000 tons of NO_2, a pollutant linked to heart and lung disease and premature death, over the lifetime of the vehicles. The illegal devices are typically purchased via the internet and are produced by small machine shops, making legal actions quite challenging.

2.7   IMPACTS OF AIR POLLUTION ON ECOSYSTEM HEALTH

Research shows that air pollution can affect ecosystems. Air pollutants such as sulfur may lead to excess amounts of acid in lakes and streams and can damage trees and forest soils. Nitrogen in the atmosphere can harm fish and other aquatic life when deposited on surface waters. Ozone damages tree leaves and negatively affects scenic vistas in protected natural areas. Mercury and other heavy metal compounds that are emitted into the air from fuel combustion and deposited on land and in water accumulate in plants and animals, some of which are consumed by people [90]. The Massachusetts Department of Environmental Protection has elaborated on the most significant effects of air pollution on ecosystems, as follows [53]:

> "**Acid rain** is precipitation containing harmful amounts of nitric and sulfuric acids. These acids are formed primarily by NO_x and SO_x released into the atmosphere when fossil fuels are burned. These acids fall on the Earth either as wet precipitation (rain, snow, or fog) or dry precipitation (gas and particulates). Some are carried by the wind, sometimes hundreds of miles. In the environment, acid rain damages trees and causes soils and water bodies to acidify, making the water unsuitable for some fish and other wildlife. It also speeds the decay of buildings, statues, and sculptures.
>
> **Eutrophication** is a condition in a water body where high concentrations of nutrients (such as nitrogen) stimulate blooms of algae, which in turn can cause fish kills and loss of plant and animal diversity. Air emissions of NO_x from power plants, cars, trucks, and other sources contribute to the amount of nitrogen entering aquatic ecosystems.
>
> **Haze** is caused when sunlight encounters tiny pollution particles in the air. Haze obscures the clarity, color, texture, and form of what can be seen. Some haze-causing pollutants (mostly fine particles) are directly emitted to the atmosphere by sources such as power plants, industrial facilities, trucks and automobiles, and construction activities. Others are formed when gases emitted into the air (such as SO_2 and NO_x) form particles as they are carried downwind.

Haze is caused when sunlight encounters tiny pollution particles in the air. Haze obscures the clarity, color, texture, and form of what can be seen.

> **Effects on wildlife:** Toxic pollutants in the air, or deposited on soils or surface waters, can impact wildlife in a number of ways. Like humans, animals can experience health problems if they are exposed to sufficient concentrations of air toxics over time. Air toxics can contribute to birth defects,

reproductive failure, and disease in animals. Persistent toxic air pollutants (those that break down slowly in the environment) are of particular concern in aquatic ecosystems. These pollutants accumulate in sediments and may biomagnify in tissues of animals at the top of the food chain to concentrations many times higher than in the water or air.

Ozone depletion: In the stratosphere, ozone forms a layer that protects life on earth from the sun's harmful ultraviolet (UV) rays. The Earth's ozone layer has been damaged by releases of ozone-depleting chemicals, including chlorofluorocarbons, hydrochlorofluorocarbons, and halons. These substances were formerly used and sometimes still are used in coolants, foaming agents, fire extinguishers, solvents, pesticides, and aerosol propellants. UV can damage sensitive crops, such as soybeans, and reduce crop yields.

Crop and forest damage: Air pollution can damage crops and trees in a variety of ways. Ground-level ozone can lead to reductions in agricultural crop and commercial forest yields, reduced growth and survivability of tree seedlings, and increased plant susceptibility to disease, pests, and other environmental stresses (such as harsh weather). Crop and forest damage can also result from acid rain and from increased UV radiation caused by ozone depletion.

Global climate change: Releases of greenhouse gases (Chapter 1) into Earth's atmosphere are causing climate change and impacting human and ecosystem health as a consequence of global temperature rise, increased number and severity of weather events, and impacts on sea life and food production" [53].

Regarding acid rain, the 1990 amendments to the CAAct contain provisions to control acid rain. A marketplace cap-and-trade policy was implemented by EPA as the principal policy to mitigate acid rain. Subsequent studies have shown a dramatic decrease in the emissions of the two primary pollutants from acid rain. A NASA study found that stringent air regulations and technological improvements have reduced NO_2 emissions by 40% and SO_2 emissions by 80% between 2005 and 2014 [91]. In addition to public health benefits, reductions in acid rain have produced ecological benefits. For example, scientists from the U.S. and Canada reported that the acidity of soils in some parts of the Eastern U.S. and Canada has declined, abating years of acid rain's harm to plants and aquatic life by reversing the depletion of a critical nutrient in soil, calcium. Less acidic soil promotes plants' growth and crop yields [92].

Ground-level ozone (O_3) is another important air pollutant, which damages human health and crops. It is estimated that global losses to soybean, maize, and wheat crops due to ground-level ozone pollution could be US$17–35 billion per year by 2030 [93].

2.8   U.S. POLICIES ON AIR QUALITY

The history of air pollution demonstrates that when societies experience what is perceived as excessive air pollution, policies will eventually flow from this discontent.

Policies can range from decisions made by individuals (e.g., annual vehicle emissions maintenance) to global policies that involve international agreements on controlling air pollution. National policies on air pollution control have occurred globally, with the U.S. CAAct, as amended, serving as an example. While the form and details of air pollution control policies may differ, there is common harmony in implementing actions that will protect public health. Societies afflicted with noxious air pollution do not remain silent. Complaints arise and are made known to policymakers, who in response develop policies that are intended to control the sources of pollution. Without the complaints of persons exposed to objectionable levels of air pollution, little effort would have occurred, because control of sources runs against the grain of economic benefits derived from the sources of pollution. For example, adding pollution controls to automobiles added costs to their manufacture, resulting in increased costs to the consumer. Similarly, requiring pollution controls on electric power plants adds to the cost of electricity paid by the consumer. These costs of pollution control must be considered in light of the benefits derived from decreased effects on human and ecosystem health.

To be described in this section are policies on control of air pollution. Attention is given to the U.S. CAAct, as amended, given this act's key role in controlling air pollution in the U.S.

2.8.1   U.S. PRIMARY POLICY: CLEAN AIR ACT, 1955

The CAAct, as amended, is the central environmental health policy on air pollution control in the U.S. This is a complex, comprehensive law that impacts human and ecosystem health throughout the U.S. and globally, as well. Regarding the latter, the CAAct has served as a model for comparable policies in part or whole in Europe and elsewhere. Further, air pollution in the U.S. eventually traverses international borders and can add to locally generated pollution levels. This section presents a précis summary of the CAAct and its key elements that related to public health and policy issues that attend the act's administration and application.

2.8.1.1   History of the Clean Air Act

Pollution of the air we breathe for life's very existence is a problem likely dating from antiquity, perhaps from the time when humans first came into contact with smoke from fires used for warmth and to ward off predators. One source cites an action in the year 1306 when citizens of London petitioned their government to take action to reduce levels of smoke in ambient air. In response, King Edward I issued a royal proclamation to prohibit artificers (i.e., craftsmen) from burning sea coal (coal that washes up on the seashore), as distinguished from charcoal, in their furnaces [94]. This is an example of government taking action against the adverse effects of air pollution. Given this 14th century example of one government's attempts to improve citizens' air quality, it is not surprising to learn that in the 20th century the U.S. public's concern about air pollution also led to legislative action.

Given this early British experience in air pollution policymaking, it is somewhat ironic that 20th century U.S. federal air pollution control legislation was influenced by a "killer smog"[1] that occurred in London during the winter of 1952, an event in which it was first reported that more than 4,000 people died from breathing polluted air caused by a temperature inversion.[2] However, a reassessment of mortality data for December 1952–February 1953 found that more than 12,000 excess deaths occurred due to acute exposure to heavily contaminated ambient air. The primary constituent in the polluted air was smoke from home heating coal-burning stoves and fireplaces.

U.S. federal air pollution policymaking is indebted to the state of California, which had provided early and sustained leadership on controlling air pollution. Many of the state's concerns were focused on air pollution in Los Angeles. In 1943, the first recognized episodes of smog occurred in Los Angeles, resulting in limited visibility of approximately three blocks and reports of eye irritation, respiratory discomfort, nausea, and vomiting. The source of the pollution was unknown but speculated to be an industrial facility. In 1947, California Governor Earl Warren signed into law the Air Pollution Control Act, which authorized the establishment of an air pollution control district (APCD) in every California county, leading to the creation of the Los Angeles County APCD, the first of its kind in the U.S. [4]. This is an example of a state taking action to control an environmental hazard before similar action was taken by the federal government. In 1952, Dr. Arie Haagen-Smit, a professor of chemistry at the California Institute of Technology, discovered the nature and causes of photochemical smog. He determined that nitrogen dioxides and hydrocarbons in the presence of ultraviolet radiation from the sun form smog, a key component of which is ozone.

As described by Fromson [94], the first serious congressional recognition of the need for air pollution control occurred with the Air Pollution Control Act of 1955. This act provided research and technical assistance for the control of air pollution. The tragic events of London's killer smog in 1952 also raised awareness of the need to address the growing air pollution problem in the U.S. A similar episode of fatal air pollution had occurred during October 23–30, 1948, in Donora, Pennsylvania, where 20 people died and half the city's 12,000 residents became ill from breathing industrial contaminants trapped under a layer of temperature-inverted air [95].

The Air Pollution Control Act of 1955 declared that states had the primary responsibility for air pollution control. The federal government's role was advisory, providing technical services and financial support to state and local governments. The U.S. Department of Health, Education and Welfare (DHEW) was vested with these responsibilities under the Act.

The next major federal air pollution legislation occurred with the enactment of the CAAct Amendments of 1963. Whereas the Air Pollution Control Act of 1955 was limited primarily to research and technical and financial assistance to state and local governments, the CAAct Amendments enhanced federal responsibility for controlling air pollution. At the same time, the act continued Congress's intent that "[t]he

---

[1] The word *smog* was first recorded in 1905 in a newspaper report of a meeting of the Public Health Congress, where Dr. H.A. des Vœux gave a paper entitled "Fog and Smoke" in which he coined the word smog [138].

[2] A temperature inversion, which occurs when a cold layer of air settles under a warmer layer, can slow atmospheric mixing and allow pollutants to accumulate hazardously near ground level.

prevention and control of air pollution at the source is the primary responsibility of state and local governments" [94].

Following passage of the CAAct Amendments of 1963, the attention of Congress and environmental groups turned to air pollution caused by automobile emissions [94]. Given the passage of time, it may be difficult for some persons to comprehend the incredulity that accompanied the discovery in California of automobiles' contribution to air pollution, and, more specifically, smog. The discovery of vehicle emissions as the primary constituents of Los Angeles' smog prompted federal laboratory research that found increased cancer rates in cancer-resistant mice. These findings were the subject of a 1962 report to Congress from Surgeon General Luther L. Terry. The report added weight to the need for further congressional action to control air pollution.

> The Clean Air Act requires EPA to set mobile source limits, ambient air quality standards, standards for new pollution sources, and significant deterioration requirements, and to focus on areas that do not attain standards.

In 1965, Congress enacted the Motor Vehicle Air Pollution Control Act. The act required federal standards to be promulgated for controlling pollutants emitted from automobiles. The emission standards were to be established on the basis of "technological feasibility and economic costs" of controlling automobile emissions [95]. Upon promulgation of the emission standards, manufacturers of new motor vehicles or new motor engines were prohibited from selling or importing a non-conforming product into commerce.

The federal CAAct was enacted by Congress in 1970, heavily amended in 1977, and again substantively amended in 1990 [96]. The Act's titles are listed in Table 2.3 [96]. The effects of unclean air on the public's health remain key motivations for keeping the act enforced. The Act, as amended, is a comprehensive, complex statute that controls air pollution emissions and regulates government, business, and community lifestyles that affect the release of air contaminants into outdoor ambient air.

The CAAct adopted the policy of developing *National Ambient Air Quality Standards* (NAAQS) for individual air contaminants, and then placed most of the responsibility on the states to achieve compliance with the standards. The CAAct established two kinds of national air quality standards. *Primary standards* are based

TABLE 2.3
Clean Air Act's Titles [111]

Title	Name of Title
I	Air pollution prevention and control
II	Emission standards for moving sources
III	General
IV	Acid deposition control
V	Permits
VI	Stratospheric ozone protection

on the protection of human health, including the health of sensitive populations such as children, elderly persons, and persons with infirmities (e.g., asthma). *Secondary air quality standards* set limits to protect public welfare, including protection against decreased visibility, damage to buildings, and deleterious ecological effects.

The 1977 CAAct amendments added special provisions for geographic areas with air cleaner than national standards in order to prevent their deterioration in air quality, and special provisions were added pertaining to *nonattainment areas*, that is, geographic areas that had failed to meet national air quality standards [97]. Under the 1977 amendments to the CAAct, states were required to develop *State Implementation Plans* (SIPs) that would meet the air quality standards by 1982, except for ozone, for which the deadline was 1987 [96].

The 1990 CAAct amendments substantively revised the earlier version of the CAAct. Signed into law on November 15, 1990, by President George H.W. Bush (R-TX), these amendments added comprehensive provisions to regulate emissions of air toxicants, acid rain, and substances thought to be a threat to the ozone layer. In addition, the 1990 amendments added an elaborate permit program and markedly strengthened enforcement provisions and requirements for geographic areas that fail to meet air quality standards (i.e., nonattainment areas), mobile source emissions, and automobile fuels [97]. Of special relevance to children's health, these amendments finally banned the sale in the U.S. of gasoline that contained lead additives, ending one of the 20th century's worst environmental health missteps, the use of tetraethyl lead as a gasoline additive.

The 1990 amendments also changed the way hazardous air pollutants are regulated. The Act, as amended, in effect, recognizes two kinds of outdoor ambient air pollutants: the six *Criteria Air Pollutants* and, basically, everything else. The latter category comprises *Hazardous Air Pollutants* (HAPs). Before 1990, regulation of HAPs was a two-step process. EPA had to first establish that a pollutant was likely to be hazardous at ambient levels. Once this determination was made (and survived an elaborate hearing process), the second step was to choose the emission sources to be regulated.

Congress became increasingly impatient with EPA's science and risk-based approach for regulating HAPs, because under the pre-1990 CAAct, only a handful of HAPs had been regulated. Sharply curtailing EPA's discretion on how to regulate HAPs, Congress specified more than 180 HAPs in the Act. Moreover, with respect to these substances, Congress shifted the burden of proof. "Whereas before, EPA had to go through an elaborate process to prove a compound **guilty** (emphasis added) before it could be regulated, now EPA must go through an elaborate process to prove a compound **innocent** (emphasis added) before it can avoid regulation. Secondly, Congress required that maximum available control technology (MACT) be installed on all sources, regardless of extent of resulting exposure or toxicity. Risk assessment has been related to a residual risk provision that provides for additional action should MACT controls still leave a risk to the maximally exposed individual beyond a relatively stringent level. This shift of the burden of proof requirement of MACT across the board and downgrading of the importance of risk assessment clearly falls within the Precautionary Principle, as does the use of a stringent risk criterion and of the maximally exposed individual rather than the population as the target of concern" [97].

The 1990 CAAct amendments also contained a significant environmental policy now called *cap and trade*, a marketplace incentive that was introduced in Chapter 1. In the CAAct amendments, Congress, concerned that acid rain generation and deposition was causing consequential detrimental environmental effects on ecosystems, directed EPA to implement a marketplace approach to reducing sulfur dioxide emissions, the main ingredient of acid rain.

On March 10, 2005, EPA announced the Clean Air Interstate Rule (CAIR). Through the use of the cap-and-trade approach, CAIR targeted substantial reductions in levels of sulfur dioxide (SO_2) and nitrogen oxides (NO_x) emissions in more than 450 counties in the eastern U.S. and helped them meet EPA's air quality standards for ozone and fine particles. CAIR covered 28 eastern U.S. states and the District of Columbia. States were to achieve the required emission reductions by using one of two compliance options: (1) meet the state's emission budget by requiring power plants to participate in an EPA-administered interstate cap-and-trade system that caps emissions in two stages, or (2) meet an individual state emissions budget through measures of the state's choosing.

Before presenting excerpts of key public health provisions of the CAAct, it is important in a positive public health sense to comment that this complex work of environmental health policy has achieved success in reducing adverse health impacts. Researchers reported in 2018 that the number of premature deaths related to air pollution in the U.S. decreased by 47% between 1990 and 2010, dropping from 135,000 per year to 71,000. The researchers attributed the decline in mortality to stricter regulations and improvements in air quality. This is despite increases in the country's population, energy use, and car travel [98].

2.8.1.2 EPA's Air Quality Index

A good illustration of an intersection between environmental policy and public health practice is EPA's Air Quality Index (AQI) [99]. As developed by EPA, the AQI is a scale of 0 to 500, divided into several color-coded categories. A region's AQI score at any time is based on the highest of five criteria air pollutants: PM, SO_2, CO, NO_2, and ground-level O_3. The intervals and the terms describing the AQI air quality levels are as shown in Table 2.4. Using the state of Georgia as an example, the Ambient Monitoring Program at the Georgia Environmental Protection Division (EPD), Air Protection Branch, is responsible for measuring air pollutant levels throughout the state. When these levels are reported, the EPD utilizes the Air Quality Index (AQI) to gauge their public health importance and make adjustments to applicable air quality programs.

AQI figures inform the public about whether air pollution levels in a particular location are Good, Moderate, Unhealthy for Sensitive Groups, Unhealthy, or Very Unhealthy. In addition, the AQI can inform the public about the general health effects associated with different pollution levels and describe possible precautionary steps to take if air pollution rises into unhealthy ranges. Local newsmedia provide alerts when AQI levels are unhealthful, which helps individuals make health-based decisions in support of daily activities. The AQI construct has become a useful metric globally, although some differences exist in deriving AQIs.

TABLE 2.4
Air Quality Index (AQI) Values and Their Meanings [126]

AQI Values	Levels of Health Concern	Colors
When the AQI is in this range	...air quality conditions are:	.. as symbolized by this color
0–50	Good	Green
51–100	Moderate	Yellow
101–150	Unhealthy for sensitive persons	Orange
151–200	Unhealthy	Red
201–300	Very unhealthy	Purple
301–500	Hazardous	Maroon

2.8.1.3 Clean Air Act Regulations on Greenhouse Gases

As described in Chapter 1, EPA has promulgated regulations under the provisions of the CAAct, as amended, to control air emissions that contribute to climate change. Also discussed in Chapter 1, many of these regulations were subject to litigation, with outcomes generally favorable to EPA. The agency undertook a comprehensive approach to developing standards for greenhouse gas emissions from mobile and stationary sources under the CAAct. Following are the key proposed or completed actions proposed by EPA to implement CAAct requirements for carbon pollution and other greenhouse gases [100]. It must be noted that the following actions were set aside by the Donald Trump (R-NY) administration.

Clean power plan: On August 3, 2015, EPA issued the Clean Power Plan, which put the nation on track to cut harmful pollution from the power sector by 32% below 2005 levels, while also cutting smog- and soot-forming emissions that threaten public health by 20%.

Final greenhouse gas tailoring rule: On May 13, 2010, EPA set greenhouse gas emissions thresholds to define when permits under the New Source Review Prevention of Significant Deterioration (PSD) and Title V Operating Permit programs are required for new and existing industrial facilities. This final rule "tailors" the requirements of these CAAct permitting programs to limit covered facilities to the nation's largest greenhouse gas emitters: power plants, refineries, and cement production facilities.

Timing of applicability of the PSD permitting program to greenhouse gases: On December 23, 2010, EPA issued a series of rules that put the necessary regulatory framework in place to ensure that (1) industrial facilities can get CAAct permits covering their greenhouse gas (GHG) emissions when needed and (2) facilities emitting GHGs at levels below those established in the Tailoring Rule do not need to obtain CAAct permits.

EPA and NHTSA standards to cut greenhouse gas emissions and fuel use for new motor vehicles: EPA and the National Highway Traffic Safety Administration (NHTSA) are taking coordinated steps to enable the production of a new generation of clean vehicles—from the smallest cars to the largest trucks—through reduced greenhouse gas emissions and improved fuel use.

Renewable fuel standard program: EPA is also responsible for developing and implementing regulations to ensure that transportation fuel sold in the U.S. contains a minimum volume of renewable fuel. By 2022, the Renewable Fuel Standard (RFS) program is estimated to reduce greenhouse gas emissions by 138 million metric tons.

Landfill air pollution standards: On August 14, 2015, EPA issued two proposals to further reduce emissions of methane-rich gas from municipal solid waste (MSW) landfills. The proposals would require new, modified, and existing landfills to begin capturing and controlling landfill gas at emission levels nearly a third lower than current requirements.

Oil and natural gas air pollution standards: EPA has proposed a suite of requirements that provide greater certainty about CAAct permitting requirements for the oil and natural gas industry. The proposals are part of EPA's strategy to reduce emissions of the methane and smog-forming volatile organic compounds from this rapidly growing industry.

Geologic sequestration of carbon dioxide: Geologic sequestration is the process of injecting CO_2 from a source, such as a coal-fired electric generating power plant, into a well thousands of feet underground and sequestering the CO_2 underground indefinitely. EPA has finalized requirements for geologic sequestration, including the development of a new class of wells,

Greenhouse gas endangerment findings: On December 7, 2009, Administrator Lisa Jackson signed a final action, under Section 202(a) of the CAAct, finding that six key well-mixed greenhouse gases constitute a threat to public health and welfare, and that the combined emissions from motor vehicles cause and contribute to the climate change problem.

Greenhouse gas reporting program: The Greenhouse Gas Reporting Program collects greenhouse gas data from large emission sources across a range of industry sectors, as well as suppliers of products that would emit greenhouse gases if released or combusted.

Perspective: The preceding actions that were developed by the Obama administration were in concert with that administration's commitment to combating the causes of climate change. Having this commitment set aside by the Trump administration is a demonstration of the impact of national political systems that have rather different environmental values. Ultimately, it is voters in a democratic system of government that must decide which environmental commitment to choose.

2.8.1.4 The Mercury and Air Toxics Standards Rule, 2011

The Mercury and Air Toxics Standards (MATS) was established in 2011. It was the first set of federal rules to limit hazardous pollution from coal-burning and oil-burning plants, targeting toxicants like mercury [101]. Coal power plants are the biggest emitters of mercury. According to the EPA's projections, these rules would save upward of 17,000 lives per year in the U.S. The Center for American Progress found that MATS reduced mercury emissions from power plants by 81% since going into effect. However, the energy industry and some U.S. states filed lawsuits to block MATS. One suit, Michigan v. EPA, fought its way to the Supreme Court. In 2015, the court ruled 5-4 in favor of Michigan, deciding that the EPA has to weigh the costs to industry of an environmental regulation against the benefits to society [101].

In 2018, the Trump administration's EPA announced a reinterpretation of MATS, concluding that the current regulation as it stands is too costly and does not meet the "appropriate and necessary" standard [101], therefore setting the rule aside. As justification, the EPA said complying with MATS could cost power producers upward of $9.6 billion per year, but the benefits could only be quantified up to $6 million. This is a much lower tally than what the Obama administration had calculated when it issued MATS. They had estimated $80 billion in benefits because they included the health side benefits of limiting toxic chemical pollution. Environmental groups accused EPA of changing the cost-benefit analysis of MATS in order to make it easier to challenge its implementation. Interestingly, the electric power industry reversed its position of opposition to MATS and counseled EPA in 2018 to continue enforcing the MATS rule, stating that almost all coal-fired power plants had complied with the Obama-era standards in 2016 [102]. However, the EPA did not change its reinterpretation of MATS.

2.8.1.5   The Clean Power Plan, 2016

The Clean Power Plan, announced by President Obama in August 2015, "set the first-ever limits on carbon pollution from U.S. power plants, the largest source of the pollution in the country that's driving dangerous climate change. The U.S. Environmental Protection Agency issued the final Clean Power Plan under the Clean Air Act, the nation's fundamental air pollution law. The plan sets flexible and achievable standards that give each state the opportunity to design its own most cost-effective path toward cleaner energy sources. This historic step to rein in power plant pollution would speed America's transition away from fossil fuels, protect our health, prevent future generations from experiencing the worst effects of climate change, and position the U.S. as a global climate leader.

Enforceable carbon pollution limits would kick in starting in 2022 and ramp up into full effect by 2030. Of course, power companies would have to act sooner than 2022 to prepare and respond to additional clean energy incentives in the plan. According to EPA projections, by 2030, the Clean Power Plan would cut the electric sector's carbon pollution by 32% nationally, relative to 2005 levels. In 2030 alone, there would be 870 million fewer tons of carbon pollution. This is like canceling out the annual carbon emissions from 70% of the nation's cars or avoiding the pollution from the yearly electricity use of every home in America.

Economists believe that in 2030, the Clean Power Plan could save the country $20 billion in climate-related costs and deliver $14 to $34 billion in health benefits. The shift to energy efficiency and cleaner power will also save the average American family $85 on its electricity bills in 2030" [103].

2.8.1.6   Affordable Clean Energy Rule, 2019

The Trump administration announced on June 19, 2019, their Affordable Clean Energy (ACE) rule—"replacing the prior administration's overreaching Clean Power

Plan (CPP) with a rule that restores the rule of law and empowers states to continue to reduce emissions while providing affordable and reliable energy for all Americans. The ACE rule establishes emissions guidelines for states to use when developing plans to limit CO_2 at their coal-fired power plants. Specifically, ACE identifies heat rate improvements as the best system of emission reduction (BSER) for CO_2 from coal-fired power plants, and these improvements can be made at individual facilities. States will have 3 years to submit plans, which is in line with other planning timelines under the Clean Air Act.

Also contained within the rule are new implementing regulations for ACE and future existing-source rules under Clean Air Act Section 111(d). These guidelines will inform states as they set unit-specific standards of performance. For example, states can take a particular source's remaining useful life and other factors into account when establishing a standard of performance for that source. ACE will reduce emissions of CO_2, mercury, as well as precursors for pollutants like fine-particulate matter and ground-level ozone" [104].

"In 2030, the ACE rule is projected to reduce CO_2 emissions by 11 million short tons. reduce SO2 emissions by 5,700 tons, reduce NO_x emissions by 7,100 tons, reduce $PM_{2.5}$ emissions by 400 tons, and reduce mercury emissions by 59 pounds. EPA projects that ACE will result in annual net benefits of $120 to $730 million, including costs, domestic climate benefits, and health co-benefits. With ACE, along with additional expected emissions reductions based on long-term industry trends, we expect to see CO_2 emissions from the electric sector fall by as much as 35% below 2005 levels in 2030" [104].

2.8.1.7 Corporate Average Fuel Economy Standards, 1975

Automobile emissions of air pollutants are correlated with the amount of fuel combusted in internal combustion engines. Quite simply, the less fuel combusted, the lesser amount of pollution emitted. Fuel economy is therefore relevant for air pollution policies. In the U.S. the Corporate Average Fuel Economy (CAFÉ) standards are regulations, first enacted by Congress in 1975, after the 1973–1974 Arab Oil Embargo, to improve the average fuel economy of cars and light trucks produced for sale in the U.S. [105]. CAFE standards are established by the National Highway Transportation Safety Administration (NHTSA) and regulate how far vehicles in the U.S. must travel on a gallon of fuel. NHTSA sets CAFE standards for passenger cars and for light trucks (collectively, light-duty vehicles) and separately sets fuel consumption standards for medium- and heavy-duty trucks and engines. NHTSA also regulates the fuel-economy window stickers on new vehicles. The EPA enforces the fuel standards through its vehicles' emissions regulations.

The fuel economy standards were set by the Obama administration in 2012 with the voluntary acceptance of automakers [106]. In particular, there was an agreement between the Obama administration and the automobile industry that fuel economy standards would gradually increase the average miles per gallon requirements for cars to 54.5 mpg by 2025. The increase in vehicle fuel efficiency was a component of the Obama administration's climate change program, with a focus on using less fossil fuel (coal, oil) in order to lessen the production of corresponding greenhouse gases. However, that agreement with automobile manufacturers was set aside by the Trump administration in 2018.

2.8.1.8   The Safer Affordable Fuel-Efficient Vehicles Rule, 2018

In 2018, the Trump administration set aside the Obama administration's CAFÉ standards for vehicle fuel efficiency. Their Safer Affordable Fuel-Efficient (SAFE) Vehicles Rule would amend existing Corporate Average Fuel Economy (CAFE) and tailpipe CO_2 emissions standards for passenger cars and light trucks and establish new standards covering model years 2021–2026. The Trump administration has said it plans to roll back the Obama-era standard to about 37 miles per gallon. The proposal would retain the model year 2020 standards for both programs through model year 2026. The Trump administration's NHSTA asserts that if the new rule is adopted, the proposed rule's preferred alternative would save more than $500 billion in societal costs and reduce highway fatalities by 12,700 lives (over the lifetimes of vehicles through Model Year 2029) [105]. The SAFE rule is undergoing judicial review in 2019 in response to litigation by environmental organizations and the matter of CAFÉ standards remains a subject of political and judicial debate.

California and 13 other states have vowed to keep enforcing the stricter rules (i.e., the Obama rules), potentially cleaving the U.S. auto market into two markets. In reaction to the political debate at the federal government level concerning CAFÉ standards, one U.S. state, California in conjunction with the Canadian province Quebec, has acted independently of federal fuel efficiency standards. In 2019, with car companies facing the prospect of having to build two separate lineups of vehicles, they opened secretive talks with California regulators in which the automakers — Ford Motor Company, Volkswagen of America, Honda, and BMW—won rules that are slightly less restrictive than the Obama standards and that they can apply to vehicles sold nationwide Under the agreement, the four automakers, which together make up about 30% of the U.S. auto market, would face slightly looser standards than the original Obama rule: Instead of reaching an average 54.5 miles per gallon by 2025, they would be required to achieve about 51 miles per gallon by 2026 [107].

An interesting policy note derives from the California cap-in-trade policy. The U.S. Department of Justice sued the state in October 2019, arguing that California's program to reduce vehicle emissions, done in collaboration with the Canadian province of Quebec, was akin to an international treaty beyond the scope of the powers of a state. A federal judge in California denied the Trump administration, asserting that the California-Quebec program was not an international treaty [108].

Perspective: The bruhaha between the Trump administration and some U.S. states in regard to vehicle fuel efficiency standards will become a matter of serious litigation, but the ultimate decision about vehicles' fuel standards will be settled in the commercial market of vehicle sales. Specifically, if customers of automotive vehicles demand fuel-efficient products, automobile manufacturers will produce them irrespective of what federal and state governments stipulate. One has only to look at the replacement of fossil-fueled vehicles with electric vehicles to appreciate the policy impact of climate change concern as a stimulus of changes in marketplaces.

2.8.1.9   Air Quality Policies: Obama and Trump

Distinct differences in environmental policies, including air quality, exist between the Obama and Trump U.S. presidential administrations. The differences were a product of each administration's political philosophy and each's realpolitik. The Obama

administration had garnered strong political support from environmental organizations during the 2012 presidential election. This support had a strong influence on
the Obama administration's commitment to programs of climate change mitigation.
Similarly, in terms of the influence of support during the 2016 presidential campaign,
the Trump administration had the strong support of corporations and private individuals connected with the U.S. fossil fuel industry. A consequence of that support
was the Trump administration's platform of repealing policies that regulated actions
of the fossil fuel industry, particularly coal. In this section are short summaries of
environmental health policies that differed between the two administrations.

- Clean Power Plan: This was the centerpiece of the Obama policy of U.S.
  contribution to a global program of climate change mitigation through
  reduction of greenhouse gas emissions from power plants. In 2016 the
  Trump administration shelved the plan and withdrew the U.S. from the
  Paris Climate Accord (Chapter 1).
- The Trump administration rescinded the Obama administration's fuel
  economy and global warming standards for cars, choosing to freeze emissions standards at their 2020 level [109]. New cars and light trucks built
  in 2020–2026 would average 37 mpg under the proposed freeze. Obama's
  rule would have resulted in 54.5 mpg 2026. The Trump administration
  justified the rollback as a matter of safety concerns, stating that the rule
  change would save 1,000 lives per year. In distinction, EPA technical
  staffers expected the revision to actually increase highway deaths by
  17 per year [110].
- In 2019 the Trump administration announced its intention to propose a new
  formula for calculating the human health benefits that come from reducing
  air pollution from coal-fueled power plants. The target was the Mercury and
  Air Toxics Standard (MATS)—adopted during the Obama administration.
  The new proposal would negate regulating hazardous air pollution from
  power plants under the Clean Air Act, including coal-fueled plants [111].
- The Trump administration in 2018 reinterpreted language in the Emergency
  Planning and Community Right-to-Know Act so as to amend notification
  regulations to let industrial agricultural operations forego reporting ambient air emissions, including some toxicants, from animal waste at their
  farms [112].
- The Obama-era EPA supported a tighter 0.7 ppb threshold for ambient air
  levels of O_3. The 0.7 ppb cross-state threshold was rooted in a 2015 rule
  that tightened the standards to provide increased public health protection
  against health effects. In October 2018 the Trump administration permitted
  U.S. states to adopt an O_3 standard of 1 ppb, which means a state can emit
  43% more ozone pollution across state lines than before [113].
- In July 2019 the EPA finalized a rule that would ease the air pollution permitting process for certain power plants and manufacturers, specifically for
  obtaining permits under the Clean Air Act, known as New Source Review
  (NSR). The NSR process kicks in both for new facilities and when power
  plants install new equipment or make changes that would significantly

increase air pollution. The Sierra Club said the proposed rule would allow industries to avoid installing modern pollution controls on their dirty facilities. EPA argues it will help ease the process for facilities looking to install new equipment [114].

- In December 2016, EPA finalized a determination that the existing Cross-State Air Pollution Rule Update for the 2008 Ozone National Ambient Air Quality Standards (NAAQS) (CSAPR Update) fully addresses the Clean Air Act's "good neighbor" requirements for 20 states with respect to the 2008 ground-level ozone standard. EPA asserted that there will be no remaining nonattainment or maintenance areas for the 2008 Ozone NAAQS in the eastern U.S. [115]. In October 2019 a federal appeals court struck down the Trump administration's CSAPR, determining that the rule was not permissible under the CA Act [116].

2.8.1.10   The Montreal Protocol, 1987

The ozone layer is a long-standing natural feature of the stratosphere, the part of the atmosphere that begins about six miles above the earth. The ozone layer filters out dangerous ultraviolet radiation from the Sun that can cause skin cancer, cataracts, reduced agricultural productivity, disruption of marine ecosystems, and damage many life forms. The Montreal Protocol of 1987 is a global agreement to protect the stratospheric ozone layer by phasing out the production and consumption of ozone-depleting substances (ODS) [117]. The U.S. ratified the Montreal Protocol in 1988 and has joined four subsequent amendments. Global implementation of the Protocol has led to the phasing out of the production and consumption of ODS such as chlorofluorocarbons (CFCs) and halons.

As noted by the U.S. Department of State, full implementation of the Montreal Protocol is expected to result in avoidance of more than 280 million cases of skin cancer, approximately 1.6 million skin cancer deaths, and more than 45 million cases of cataracts in the U.S. alone by the end of the century, with even greater benefits worldwide [117]. Measurement data acquired by the UN indicates the ozone layer is showing signs of continuing recovery and is likely to heal fully by 2060. In particular, recovery from the holes and thinning caused by aerosol chemicals has progressed at a rate of about 1%–3% a decade since 2000, meaning the ozone layer over the northern hemisphere and mid-latitudes should heal completely by the 2030s, if current rates are sustained. Over the southern hemisphere and in the more problematic polar regions, recovery will take longer, until the middle of the 21st century in the former and about 2060 in the latter case [118].

2.8.1.11   Kigali Amendment to the Montreal Protocol, 2016

The need for the Amendment emerged from the 1987 Montreal Protocol process, which controls ozone-depleting substances. With hydrofluorocarbons' (HFCs) use as an alternative to ozone-depleting substances in cooling equipment, their role in warming the atmosphere became a greater concern. In 2016, the Parties to the Montreal Protocol adopted the agreement on HFCs at the close of the 28th Meeting of the Parties (MOP 28) in Kigali, Rwanda. Governments agreed that it would enter into force on January 1, 2019, provided that at least 20 Parties to the Montreal Protocol had ratified it. On November 17, 2017, Sweden and Trinidad and Tobago ratified their agreement bringing the number of Parties above the required threshold.

On January 3, 2019, the Kigali Amendment entered into force, following ratification by 65 countries. The UN Environment Programme announced the entry into force and noted that it will help reduce the production and consumption of HFCs, potent greenhouse gases (GHGs), and thus avoid global warming by up to 0.4°C in the 21st century. Under the Amendment, all countries will gradually phase down HFCs by more than 80% over the next 30 years and replace them with more environmentally friendly alternatives. A specified group of developed countries will begin the phase-down in 2019. Several developing countries will freeze HFC consumption levels in 2024, followed by additional countries in 2028. As of August 2019, the Trump administration has not ratified the Kigali Amendment, owing to the administration's denial of climate change.

2.9 EMERGING TRANSPORTATION AND TRAFFIC CONTROL POLICIES

Environmental health policymakers, concerned about the adverse health effects of urban air pollution, have begun in the early 21st century to implement motor vehicle transportation policies intended to reduce levels of ambient air pollution. Put simply, there is no future for internal combustion engines. Rather, the future is with electric vehicles (EVs), which will be powered by batteries, fuel cells, or perhaps solar systems. These assertions are based on policies stated by some national governments and by some vehicle manufacturers. For example, France and Britain committed in July 2017 to ban the sales of all gasoline- and diesel-powered cars by 2040, motivated largely by health concerns about air pollution. Then China, the world's largest auto market, announced in September 2017 that it will set a deadline for automakers to stop selling internal combustion engine vehicles and correspondingly set emissions targets for automakers. Of special note is the creation by 10 large corporations of EV100, a coalition of industrial groups that will advocate for the adoption of EVs throughout their industries. The 10 corporations include utilities and an international delivery company. Concerns for the adverse impact of vehicle emissions on human and ecological health, together with issues of climate change, are the environmental health policy foundation for these technological changes in transportation.

A different form of policy has occurred in regard to reducing air pollution in central urban areas such as downtown centers of major cities. Several cities have implemented traffic control policies. Three examples of traffic restriction policies adopted by four European capitals will illustrate the nature of this kind of environmental health policy implemented in 2019 [119,120]. Madrid restricted access to its city center to gas-powered vehicles made prior to 2000 and diesel vehicles made prior to 2006. Paris banned cars made prior to 1997 from entering it city center on weekdays. Oslo restricted its city center to all vehicles. Amsterdam has announced an ambitious slate of car-mitigation goals, including a ban on all gas- and diesel-powered cars in the city by 2030. Each of these cities' policies allows delivery trucks and other commercial vehicles to conduct commercial activities.

2.10 COSTS AND BENEFITS OF U.S. AIR QUALITY

The economic costs of air pollution in the U.S. are huge. A 2007 study estimated the total damage costs associated with emissions of some air pollutants (PM, NO_x, NH_3, SO_2, VOCs) in the U.S. at between \$71 and \$277 billion per year (0.7%–2.8% of GDP [121]. A separate study used air pollution emissions data for the years 2002, 2005, 2008, and 2011 to estimate monetary damages due to air pollution exposure for $PM_{2.5}$, SO_2, NO_x, NH_3, and VOC from electric power generation, oil and gas extraction, coal mining, and oil refineries. In 2011, damages associated with emissions from these sectors totaled \$131 billion (in year 2000 \$), with SO_2 emissions from power generation being the largest contributor to social damages. The investigators noted that damages have decreased significantly since 2002, even as U.S. energy production increased, suggesting that, among other factors, policies that have driven reductions in emissions have reduced damages [122].

The foregoing numbers illustrate the economic burden of air pollution in the U.S. But there are costs to regulating the sources of air pollution emissions. The 1990 CAAct amendments (§812) require EPA to estimate the costs and benefits of the CAAct. In response, EPA has estimated that the total direct compliance costs of the CAAct from 1970 to 1990 were \$500 billion, while the total monetized benefits exceeded \$22,000 billion [123]. In 1999, EPA released a prospective study on the anticipated costs and benefits of the 1990 CAAct Amendments from 1990 to 2010. The EPA study assumed a significant decrease in air pollutants over this period, estimating a net benefit as being \$510 billion, with an expectation that benefits will again exceed direct compliance costs by approximately four to one [124]. A later estimate by EPA found that 25 years after enactment of the Clean Air Act, the agency estimated that the health benefits exceeded the cost by 32:1, saving 2 trillion dollars, and was characterized as one of the most effective public health policies of all time in the U.S. Emissions of the major pollutants particulate matter, sulfur oxides, nitrogen oxides, carbon monoxide, volatile organic compounds, and lead had been reduced by 73% between 1990 and 2015 [125].

> EPA: 25 years after enactment of the Clean Air Act, the health benefits exceeded the cost by 32:1, saving 2 trillion dollars.

The costs of air pollution control are essentially apportioned across the economic sectors of the U.S. public. Businesses increase the price they charge for their products, government authorities charge motorists for the cost of vehicle emissions inspections, and taxes are increased to pay for government inspectors and allied personnel who are charged with enforcing air pollution regulations. Some will argue that these kinds of "hidden costs" are somehow unfair and without merit. Such arguments find a hearing in the court of cost-benefit analysis, where analysts attempt to associate the costs of regulatory impacts (e.g., more stringent air pollution regulations) against the benefits to society (e.g., improvements in the public's health). This kind of analysis is a most difficult calculus because of the many uncertainties in economic models used in the analysis and limited data on health benefits. As a matter of environmental health policy, current cost-benefit analysis must be improved by enriching the databases on associations between environmental hazards and their consequences to the public's health. Having these kind of data benefits policymaking, in general, and advances the possibility of using the Precautionary Principle more effectively.

2.11   GLOBAL ECONOMIC IMPACT OF AIR QUALITY

The burden to humankind of air pollution is not confined to impacts on human and ecosystem health. Air pollution also brings economic consequences to nations and individuals. Economists who have estimated these costs often characterize them as "staggering." The Organization for Economic Cooperation and Development (OECD) has provided estimates of the costs of air pollution. According to OECD, outdoor air pollution could cause 6–9 million premature deaths and represent a global economic cost of around **$2.6 trillion** (emphasis added) annually by 2060 unless action is taken. The economic cost would rise with a surge in related annual healthcare bills to $176 billion from $21 billion in 2015 and with lost work days rising to 3.7 billion from 1.2 billion. A reduction in crop yields as a result of polluted air would also weigh on most countries' economies, according to the OECD [126].

> The global cost of air pollution specifically from burning oil, gas, and coal was estimated to be $2.9 trillion.

A different analysis of the economic cost of air pollution was performed by the Centre for Research on Energy and Clean Air (CREA) and Greenpeace Southeast Asia. These were the first to assess the global cost of air pollution specifically from burning oil, gas, and coal [127]. The global cost of air pollution caused by fossil fuels was estimated at $8bn a day, or about 3.3% of the world's GDP. The global cost for 2018 was estimated to be $2.9 trillion. Among countries taking the biggest economic hit each year were China ($900bn), the U.S. ($610bn), India ($150bn), Germany ($140bn), Japan ($130bn), Russia ($68bn), and Britain ($66bn).

2.12   COVID-19 AND AIR QUALITY

The COVID-19 pandemic that began in early 2020 essentially locked down North America, Europe, China, India, and other countries (Chapter 11). Individuals were advised, sometimes ordered by government agencies, to stay home in order to reduce person-to-person transmission of the virus. This policy resulted in far fewer motor vehicles traversing streets and highways, thus reducing air pollution emissions. Studies of air quality found NO_2 pollution over northern China, Western Europe, and the U.S. decreased by as much as 60% in early 2020 as compared to the same time in 2019, with PM 2.5 decreasing by 35% in northern China. However, the drop in NO_2 pollution caused an increase in surface O_3 levels in China, owing to the fact that O_3 can be destroyed by NO_x, so O_3 levels can increase when NO_2 pollution decreases [128].

Regarding the public health implications, a study by the Centre for Research

> Improvement in air quality over April 2020 of the COVID-19 lockdown led to 11,000 fewer deaths from pollution in the UK and elsewhere in Europe.

on Energy and Clean Air estimated the improvement in air quality over April 2020 of the COVID-19 lockdown led to 11,000 fewer deaths from pollution in the UK and elsewhere in Europe [129]. The researchers estimated that significant decreases in road traffic and industrial emissions resulted in 1.3 million fewer days of work absence, 6,000 fewer children developing asthma, 1,900 avoided emergency room visits, and 600 fewer preterm births. Compared with the same period in 2019, levels of NO_2 decreased by 40% while $PM_{2.5}$ decreased by 10%, which means that people without COVID-19 can breathe easier. Similar estimates for North America and Asia have not been performed but would be expected to also indicate a lessened mortality attributable to lower ambient air pollution levels.

In a separate work, researchers at the Harvard School of Public Health examined the relationship between COVID-19 mortality rates and long-term air pollution [130]. In a nationwide, cross-sectional study using county-level data, COVID-19 death counts were collected for more than 3,000 U.S. counties in the U.S. (representing 98% of the population) up to April 22, 2020. County-level long-term average of $PM_{2.5}$ was the exposure measure. Findings indicated that an increase of only 1 µg/m^3 in $PM_{2.5}$ was associated with an 8% increase in the COVID-19 death rate. The researchers concluded that a small increase in long-term exposure to $PM_{2.5}$ leads to a large increase in the COVID-19 death rate, a finding that gives weight to the public health importance of maintaining air quality standards.

A policy matter related to the issue of $PM_{2.5}$ exposure and COVID-19 enhancement is EPA's December 7, 2020, decision on regulating $PM_{2.5}$. On that date, in one of the final policy moves of the Trump administration, the EPA completed a regulation that keeps in place the current rules on $PM_{2.5}$, instead of strengthening them, notwithstanding the agency's science staff that had recommended tightening the $PM_{2.5}$ rule [131]. EPA administrator Andrew Wheeler announced the rule on a video call with reporters, joined by the governor and the deputy attorney general of West Virginia, who have urged President Trump to loosen rules on coal pollution.

2.13 HAZARD INTERVENTIONS

There are multiple interventions that, if implemented, can reduce the hazard of air pollution to human and ecosystem health. Controlling the release of air pollutants from stationary and mobile sources via national policies is vital. National, state/province, and local laws and ordnances are required, given the breadth of pollution sources and the severity of adverse health and welfare consequences of uncontrolled air pollution. The CAAct, as amended, is an example of such a national environmental health policy. However, such policies are impotent if not enforced, which has occurred in some developing countries.

OECD: Outdoor air pollution could cause 6 to 9 million premature deaths and represent a global economic cost of around $2.6 trillion a year by 2060 unless premeditative action is taken.

Some specific interventions have been offered by WHO as examples of successful policies in transport, urban planning, power generation, and industry that reduce air pollution [50]:

- **For industry:** clean technologies that reduce industrial smokestack emissions; improved management of urban and agricultural waste, including capture of methane gas emitted from waste sites as an alternative to incineration (for use as biogas);
- **For transport:** shifting to clean modes of power generation; prioritizing rapid urban transit, walking and cycling networks in cities as well as rail interurban freight and passenger travel; shifting to cleaner heavy-duty diesel vehicles and low-emissions vehicles and fuels, including fuels with reduced sulfur content;
- **For urban planning:** improving the energy efficiency of buildings and making cities more compact, and thus energy efficient;
- **For power generation:** increased use of low-emissions fuels and renewable combustion-free power sources (like solar, wind, or hydropower); cogeneration of heat and power; and distributed energy generation (e.g. mini-grids and rooftop solar power generation);
- **For municipal and agricultural waste management:** strategies for waste reduction, waste separation, recycling, and reuse or waste reprocessing; as well as improved methods of biological waste management such as anaerobic waste digestion to produce biogas, are feasible, low-cost alternatives to the open incineration of solid waste. Where incineration is unavoidable, then combustion technologies with strict emission controls are critical.

Regulation of motor vehicle traffic is a common intervention used by cities when motor vehicle traffic contributes to ambient air pollution levels of health concern. The alternate day's policy is frequently the chosen policy. Examples of cities that have used this particular intervention are: Mexico City, Madrid (Spain), Milan and Rome (Italy), Sarajevo (Bosnia and Herzegovina), Beijing (China), New Delhi (India), Paris (France), and others [132–134].

Trees are another form of intervention, given their capacity to uptake CO_2 and some other air pollutants. Indeed, the conversion of CO_2 to O_2 is a life-sustaining force of nature. While scientists have known for years that trees work as natural filters, no city or state has replanted forests to reduce smog. However, the large-scale planting of trees along the edges of a city could help to reduce air pollution that forms ozone, or smog. The city of Houston, Texas, is considering returning American elm, green ash, and other native trees to about 1,000 acres of grasslands near the city for the purpose of decreasing ground-level ozone [135]. On a smaller scale, trees have been planted atop an apartment building in Turin, Italy for air pollution reduction. Specifically, a new five-story apartment building is covered with 150 trees, each surrounded by custom-shaped terraces. The apartment building is a large-scale version of a treehouse. The trees help filter pollution from traffic on busy nearby streets, absorbing around 200,000 liters of CO_2 emissions every hour in one of Europe's most polluted cities. As the seasons change, the leaves help shade the apartments or bring in warmth, creating a microclimate for the building [136].

Regulation of motor vehicle traffic is a common intervention used by cities when motor vehicle traffic contributes to ambient air pollution levels of health concern.

There are hazard interventions that an individual can adopt as policy. These include:

- Use social media to support applicable air pollution policies.
- Advocate for elected policymakers who favor environmental policies. Don't support platitudinous individuals whose deeds and words are incongruous.
- Reside in areas distant from major thoroughfares to avoid vehicle emissions.
- Purchase motor vehicles that emit little to no air pollutants.
- Use public transportation whenever and wherever possible.
- Advocate planting of new trees and maintenance of trees in general.
- Don't practice behaviors that produce air pollution, e.g., smoking.
- Utilize EPA and similar databases to ascertain the sources of air pollution in your area.
- Use common household plants for effective removal of VOCs in indoor air [137].

2.14 SUMMARY

Described in this chapter are the policies and impacts of global air pollution. As characterized by WHO, air pollution is the single greatest environmental hazard to humanity, with an estimated 8 million premature deaths caused annually by exposure to polluted air. This figure comprises both ambient air pollution and indoor air pollution, the latter condition accounting for 4 million premature deaths annually. The effects of air pollution on morbidity and mortality were described, with children, elderly persons, and persons with existing infirmities among the most at health risk from exposure to air pollution. As to the impact of poor air quality on the public's health, a considerable and impressive body of health data has accrued over time. These data associate specific air contaminants with adverse health effects on the heart, lungs, and cardiovascular system. Air pollution also can cause deleterious effects on ecosystems, which are magnified by climate change. As illustrated in this chapter, industrialized nations have developed and implement policies to control emissions from stationary and mobile sources of air pollution.

In the U.S. the CAAct, as amended, is the most complex and comprehensive of the federal environmental statutes. Its provisions affect daily the whole of the U.S. population. For instance, the act controls the emissions of air contaminants from sources that range from internal combustion engines to emissions from electricity generating plants. This means that the provisions of the CAAct affect anyone who uses an automobile or relies upon electricity for personal or business purposes. No other federal environmental statute has such a broad sweep of societal impacts. The CAAct contains a strong commitment to federalism, requiring the states to enforce many of the act's provisions such as issuing permits to facilities that emit air contaminants into ambient air. There is also a strong framework of quality standards that are linked to the emission standards. For geographic areas that do not meet air quality standards, the act authorizes such penalties as an area's potential loss of highway transportation funds.

The European Commission has issued directives and regulations that are referred to the EU's Member States for implementation. This arrangement produces variation in how individual Member States implement national policies to control air pollution. And having policies on air pollution control without serious implementation is meaningless. As illustrated in this chapter, countries that are striving to increase national economic development plans can find themselves in conflict with air pollution goals.

REFERENCES

1. WHO (World Health Organization). 2020. WHO global ambient air quality database (update 2018). Geneva: Office of Director-General, Media Centre.
2. Burrows, L. 2021. Fossil fuel air pollution responsible for more than 8 million people worldwide in 2018. Press Release. Harvard University, February 9.
3. WHO (World Health Organization). 2016. Health topics: Air pollution. Geneva: Office of Director-General, Media Centre.
4. *Business Dictionary*. 2016. Air pollutant. http://www.businessdictionary.com/definition/air-pollutant.html.
5. Morrison, J. 2016. Air pollution has been a problem since the days of ancient Rome. *The Smithsonian.Com*. http://www.smithsonianmag.com/history/air-pollution-has-been-a-problem-since-the-days-of-ancient-rome-3950678/?no-ist.
6. EPA (Environmental Protection Agency). 2016. Stationary sources of air pollution. Washington, D.C.: Office of Air and Radiation.
7. Nice, K. and C. W. Bryant. 2019. Pollutants produced by a car engine. https://auto.howstuffworks.com/catalytic-converter1.htm.
8. Staff. 2019. What are diesel emissions? Nett Technologies Inc. https://www.nettinc.com/information/emissions-faq/what-are-diesel-emissions.
9. Staff. 2019. Car emissions & global warming. Union of Concerned Scientists. https://www.ucsusa.org/clean-vehicles/car-emissions-and-global-warming.
10. Kommenda, N. 2019. If SUV drivers were a nation, they would rank seventh in the world for carbon emissions. *The Guardian*, October 25.
11. Staff. 2019. Coal and air pollution. Union of Concerned Scientists. https://www.ucsusa.org/clean-energy/coal-impacts.
12. Staff. 2016. Global mercury emissions down 30 percent as coal use drops: USGS. *Reuters*, January 13.
13. Storrow, B. 2019. America's mega-emitters are starting to close. *E&E News*, August 16.
14. IPCC (Intergovernmental Panel on Climate Change). 2007. Climate change 2007: Working Group II: Impacts, adaptation and vulnerability. https://www.ipcc.ch/publications_and_data/ar4/wg2/en/ch8s8-2-6-3.html.
15. NOAA (National Oceanic and Atmospheric Administration). 2019. The Impact of wildfires on climate and air quality. Washington, D.C.: NOAA ESRL Chemical Sciences Division.
16. Sullivan, B. K. 2018. 'Insane' California air topped World Health standard by 60 times. *Bloomberg News*, November 19.
17. Wettstein, Z. S., S. Hoshiko, J. Fahimi, et al. 2018. Cardiovascular and cerebrovascular emergency department visits associated with wildfire smoke exposure in California in 2015. *J. Am. Heart Assoc.* 7(8):e007492.
18. Vaughan, A. 2019. Huge Arctic fires have now emitted a record-breaking amount of CO_2. *New Scientist*, July 25.
19. Zuckerman, L. 2017. Cost of fighting U.S. wildfires topped $2 billion in 2017. *Reuters*, September 14.

20. EPA (Environmental Protection Agency). 2019. Wood smoke and your health. https://www.epa.gov/burnwise/wood-smoke-and-your-health#what%20is%20ws?
21. Samayoa, M. 2020. Oregon DEQ data shows pollution from woodstoves is on a wildfire's scale. *Oregon Public Broadcasting*, January 16.
22. Bauer, S. E., K. Tsigaridis, R. Miller. 2016. Significant atmospheric aerosol pollution caused by world food cultivation. *Geophysical Research Letters* 43(10):5394–400.
23. Lelieveld, J., et al. 2015. The contribution of outdoor air pollution sources to premature mortality on a global scale. *Nature* 525:367–71. doi:10.1038/nature15371.
24. EESI. 2016. EPA releases endangerment finding for aircraft, good news for biofuels. Washington, D.C.: Environmental and Energy Study Institute, July 29.
25. Davenport, C. 2020. E.P.A. proposes airplane emission standards that airlines already meet. *The New York Times*, July 2.
26. AirClim. 2019. Air pollution from ships. http://www.airclim.org/air-pollution-ships.
27. Gallucci, M. 2018. At last, the shipping industry begins cleaning up its dirty fuels. *Yale Environment 360*, June 28.
28. Pynn, L. 2015. International law dramatically reduces sulphur dioxide emissions in Metro Vancouver. *Vancouver Sun*, February 12.
29. Sevunts, L. 2019. Environmental groups welcome ban on dirty fuel by Arctic cruise operators. *Radio Canada International*, November 14.
30. EPA (Environmental Protection Agency). 2019. Clean Air Act standards and guidelines for the metals production industry. Washington, D.C.: Office of Air and Radiation.
31. Nilsen, T. 2019. Norilsk tops world's list of worst SO_2 polluters. *The Barents Observer*, August 21.
32. WHO (World Health Organization). 2019. Household air pollution and health. https://www.who.int/news-room/fact-sheets/detail/household-air-pollution-and-health.
33. Shupler, M., P. Hystad, A. Birch, et al. 2020. Household and personal air pollution exposure measurements from 120 communities in eight countries: results from the PURE-AIR study. *The Lancet Planetary Health* 4(10): E451–E62.
34. EPA (Environmental Protection Agency). 1991. Indoor air facts no. 4. Sick building syndrome. Washington, D.C.: Office of Air and Radiation.
35. EPA (Environmental Protection Agency). 2016. Criteria air pollutants. Washington, D.C.: Office of Air and Radiation.
36. EPA (Environmental Protection Agency). 2016. Process of reviewing the national ambient air quality standards. Washington, D.C.: Office of Air and Radiation.
37. EPA (Environmental Protection Agency). 2016. Carbon monoxide. Washington, D.C.: Office of Air and Radiation.
38. EPA (Environmental Protection Agency). 2016. National ambient air quality standards (NAAQS) for lead (Pb). https://www.epa.gov/lead-air-pollution/national-ambient-air-quality-standards-naaqs-lead-pb.
39. EPA (Environmental Protection Agency). 2016. Nitrogen dioxide. https://www3.epa.gov/airquality/nitrogenoxides/index.html.
40. EPA (Environmental Protection Agency). 2016. 2015 National ambient air quality standards (NAAQS) for ozone. https://www.epa.gov/ozone-pollution/2015-national-ambient-air-quality-standards-naaqs-ozone.
41. Goldstein, A. 2018. Health department fines U.S. Steel's Clairton coke works $1 million. *Pittsburg Post-Dispatch*, June 28.
42. EPA (Environmental Protection Agency). 2016. 2012 National ambient air quality standards (NAAQS) for particulate matter (PM2.5). https://www.epa.gov/pm-pollution/2012-national-ambient-air-quality-standards-naaqs-particulate-matter-pm25.
43. IARC (International Agency for Research on Cancer). 2013. Press release 221: IARC: Outdoor air pollution a leading environmental cause of cancer deaths. Geneva: Director-General, Media Centre.

44. AP (The Associated Press). 2020. Colorado energy company to pay $3.5M in air pollution case. *The News & Observer*, January 22.
45. EPA (Environmental Protection Agency). 2016. Air quality - national summary. Washington, D.C.: Office of Air and Radiation.
46. EPA (Environmental Protection Agency). 2016. U.S. greenhouse gas inventory report: 1990–2014. Washington, D.C.: Office of Air and Radiation.
47. ALA (American Lung Association). 2019. State of the air – 2019. https://www.lung.org/assets/documents/healthy-air/state-of-the-air/sota-2019-full.pdf.
48. WHO (World Health Organization). 2016. Household (indoor) air pollution. Geneva: Office of Director-General, Media Centre.
49. WHO (World Health Organization). 2016. Indoor air quality guidelines: Household fuel. Geneva: Office of Director-General, Media Centre.
50. WHO (World Health Organization). 2014. Ambient (outdoor) air quality and health. Bulletin 313. Geneva: Office of Director-General, Media Centre.
51. EPA (Environmental Protection Agency). 2015. Criteria air pollutants. Washington, D.C.: Office of Air and Radiation.
52. Massachusetts Department of Environmental Protection. 2016. Health & environmental effects of air pollution. Boston: Department of Public Health.
53. Schraufnagel. D. E., et al. 2019. Air pollution and noncommunicable diseases: A review by the Forum of International Respiratory Societies' Environmental Committee, Part 2: Air pollution and organ systems. *Chest* 155(2):417–26. doi:10.1016/j.chest.2018.10.041.
54. Marusic, K. 2018. Air pollution causes up to 33 million ER visits for asthma annually. *Environ. Health News*, October 25.
55. Zhao, B., F. H. Johnston, F. Salimi, et al. 2020. Short-term exposure to ambient fine particulate matter and out-of-hospital cardiac arrest: A nationwide case-crossover study in Japan. *The Lancet* 4(1): PE15–E23.
56. AACR (American Association for Cancer Research). 2016. Exposure to particulate air pollutants associated with numerous types of cancer. News brief, April 29. Philadelphia: Office of Media Affairs.
57. Grens, K. 2011. Air pollution tied to lung cancer in non-smokers. *Reuters*, October 28.
58. Barboza, T. 2015. How strict California rules on emissions led to lower cancer risk. *Los Angeles Times*, September 21.
59. Roberts, S., et al. 2019. Exploration of NO_2 and $PM_{2.5}$ air pollution and mental health problems using high-resolution data in London-based children from a UK longitudinal cohort study. *Psychiatry Res.* 272:8–17.
60. Harvey, F. 2018. Air pollution linked to much greater risk of dementia. *The Guardian*, September 18.
61. Weichenthal, S., T. Olaniyan, T. Christtidis, et al. 2019. Within-city spatial variations in ambient ultrafine particle concentrations and incident brain tumors in adults. *Epidemiology* 31:177–183.
62. Burkhardt, J., J. Bayhama, A. Wilson, et al. 2019. The effect of pollution on crime: Evidence from data on particulate matter and ozone. *J. Environ. Econ. Manag.* 98:102267.
63. Shi, L., X. Wu, M. D. Yazdi, et al. 2020. Long-term effects of PM2.5 on neurological disorders in the American Medicare population: A longitudinal cohort study. *The Lancet Planetary Health.* doi:10.1016/S2542–5196(20)30227-8.
64. Iaccarino, L., R. La Joie, O. H. Lesman-Segev, et al. 2020. Association between ambient air pollution and amyloid positron emission tomography positivity in older adults with cognitive impairment. *JAMA Neurol.* 78:197–207. Published online November 30.
65. Banerjee, N. 2019. Study links short-term air pollution exposure to hospitalizations for growing list of health problems. *Inside Climate News*, November 27.

66. NEHA. 2020. Study provides evidence on link between air pollution, early death. *Siasat Daily*, June 29.

67. EPA (Environmental Protection Agency). 2018. 2014 national air toxics assessment: Fact sheet. Washington, D.C.: Office of Air and Radiation.

68. Dockery, D. W., C. A. Pope 3rd, X. Xu, et al. 1993. An association between air pollution and mortality in six U.S. cities. *N. Engl. J. Med.* 329(24):1753–9.

69. Rossi, G., et al. 1999. Air pollution and cause-specific mortality in Milan, Italy, 1980–1989. *Arch. Environ. Health* 54(3):158–64.

70. Neas, L. M., J. Schwartz, D. Dockery. 1999. A case-crossover analysis of air pollution and mortality in Philadelphia. *Environ. Health Perspect.* 107(8):629–31.

71. Hong, Y. C., et al. 2002. Effects of air pollutants on acute stroke mortality. *Environ. Health. Perspect.* 110(2):187–91.

72. Jerrett, M., et al. 2004. Spatial analysis of air pollution and mortality in Los Angeles. *Epidemiology* 16(6):727–36.

73. Laden, F., et al. 2006. Reduction in fine particulate air pollution and mortality. *Am. J. Respir. Crit. Care Med.* 173(6):667–72.

74. Chen, K., S. Breitner, K. Wolf, et al. 2021. Ambient carbon monoxide and daily mortality: a global time-series study in 337 cities. *Lancet Planet. Health* 5(4):E191–9.

75. Di, Q., Y. Wang, A. Zanobetti, et al. 2017. Air pollution and mortality in the Medicare population. *N. Engl. J. Med.* 376:2513–22.

76. Bell, M. L., et al. 2004. Ozone and short-term mortality in 95 US urban communities, 1987–2000. *JAMA* 292(19):2372–8.

77. WHO (World Health Organization). 2014. 7 Million premature deaths annually linked to air pollution, news release, March 25. http://www.who.int/mediacentre/news/releases/2014/air-pollution/en/.

78. Duke University. 2018. Cutting carbon emissions sooner could save 153 million lives. *Science Daily*, March 19.

79. Staff. 20018. Air pollution reduces global life expectancy by nearly two years. *UChicago News*, November 19.

80. Anonymous. 2017. Study: Cleaner air adds 5 months to US life span. *MedicalXpress News*, January 21. https://medicalxpress.com/news/2009-01-cleaner-air-months-life-span.html.

81. Gauderman, W. J., et al. 2004. The effect of air pollution on lung development from 10 to 18 years of age. *N. Engl. J. Med.* 351:1057–67. doi:10.1056/NEJMoa040610/ehp.124–A23.

82. Gauderman, W. J., et al. 2005. Childhood asthma and exposure to traffic and nitrogen dioxide. *Epidemiology* 16(6):737–43.

83. WHO (World Health Organization). 2018. Air pollution and child health: Prescribing clean air. https://www.who.int/ceh/publications/air-pollution-child-health/en/.

84. Staff. 2020. State of Global Air/2020. Impacts on Newborns. https://www.stateofglobalair.org/health/newborns.

85. Gauderman, W. J., R. Urman, E. Avol, et al. 2015. Association of improved air quality with lung development in children. *N. Engl. J. Med.* 372:905–13.

86. Union of Concerned Scientists. 2008. Diesel engines and public health. https://www.ucsusa.org/clean-vehicles/vehicles-air-pollution-and-human-health/diesel-engines.

87. Kasler, D. 2020. Your diesel Mercedes-Benz might be illegally polluting. How carmaker is paying for cheating. *The Sacramento Bee*, September 15.

88. Staff. 2019. Diesel vehicles face a grim future in Europe's cities. *Yale Environment 360*, August 5.

89. Davenport, C. 2020. Illegal tampering by diesel pickup owners is worsening pollution, E.P.A. says. *The New York Times*, November 25.

90. EPA (Environmental Protection Agency). 2015. Ecosystems and air quality. Washington, D.C.: Office of Air and Radiation.

91. Coin, G. 2015. Pollution that causes acid rain drops dramatically in Eastern US: Study. *Syracuse.com*, September 22.

92. Shekhtman, L. 2015. Soil health improving in US and Canada due to acid rain decline. *The Christian Science Monitor*, November 4.

93. UNEP (United Nations Environment Program). 2014. UNEP year book 2014 emerging issues update. Air pollution: world's worst environmental health risk. New York: Office of Director-General.

94. Fromson, J. 1970. A history of federal air pollution control. In *Environmental law review–1970*, 214. Albany: Sage Hill Publishers.

95. CARB (California Air Resources Board). 2000. California's air quality history - Key events. Sacramento: Air Resources Board.

96. McCarthy, J. E., et al. 1999. Clean air act. Summaries of environmental laws administered by the EPA. Washington, D.C.: Congressional Research Service.

97. Randle, R. V. and M. E. Bosco. 1991. Air pollution control. In *Environmental Law Handbook*, p. 524. Arbuckle, J. G., et al. (ed.). Rockville: Government Institutes, Inc.

98. Zhang, Y., et al. 2018. Long-term trends in the ambient $PM_{2.5}$- and O_3-related mortality burdens. *Atmos. Chem. Phys.* 18:15003–16.

99. EPA (Environmental Protection Agency). 2014. Air quality index. EPA-456/F-14-002. Research Triangle Park: Office of Air and Radiation.

100. EPA (Environmental Protection Agency). 2016. Climate change: Regulatory initiatives. Washington, D.C.: Office of Air and Radiation.

101. Irfan, U. 2018. The EPA wants to make it harder to ratchet down toxic chemicals from power plants. *Vox*, December 28.

102. Reilly, S. 2018. In about-face, utilities urge EPA to keep mercury rule. *E & E News*, July 11.

103. NRDC. 2017. What is the Clean Power Plan? Washington, D.C.: Natural Resources Defense Council, September 29.

104. EPA (Environmental Protection Agency). 2019. EPA finalizes Affordable Clean Energy rule, ensuring reliable, diversified energy resources while protecting our environment. Washington, D.C.: Office of Air and Radiation, June 19.

105. NHSTA (National Highway Safety Transportation Administration). 2018. Corporate Average Fuel Economy. https://www.nhtsa.gov/laws-regulations/corporate-average-fuel-economy.

106. Guess, M. 2016. EPA reaffirms 54.5mpg target fuel economy by 2025; automakers turn to Trump. *Ars Technica*, December 1.

107. Davenport, C. and H. Tabuchi. 2019. Automakers, rejecting Trump pollution rule, strike a deal with California. *The New York Times*, July 25.

108. Beitsch, R. 2020. Federal judge sides with California in cap-and-trade suit. *The Hill*, March 13.

109. Cama, T. and M. Green. 2018. Trump moves to roll back Obama emission standards. *The Hill*, August 2.

110. Irfan, U. 2019. Trump says his mileage rules make cars safer. His EPA was worried they will kill more people. *Vox*, August 21.

111. Rest. K. 2019. Mercury is toxic. Andrew Wheeler's proposed rollback is even worse. *Blog. Union of Concerned Scientists*, January 8.

112. Hand, M. 2018. A new EPA rule would allow factory farms to avoid reporting air pollution. *Civil Eats*, November 2.

113. Marsh, R., G. Wallace, E. Kaufman. 2018. EPA quietly telling states they can pollute more. *CNN/Environment News*, November 2.

114. Beitsch, R. 2019. EPA proposes easing air pollution permitting process. *The Hill*, August 1.

115. EPA (Environmental Protection Agency). 2018. Final cross-state air pollution rule update. Washington, D.C.: Office of Air and Radiation.
116. Lavelle, M. 2019. Court: Trump's EPA can't erase interstate smog rules. *Inside Climate News*, October 2.
117. U.S. Department of State. 2018. The Montreal Protocol on Substances That Deplete the Ozone Layer. Washington D.C.: Office of Environmental Quality and Transboundary Issues.
118. Harvey, F. 2018. Ozone layer finally healing after damage caused by aerosols, UN says. *The Guardian*, November 5.
119. Bendix, A. 2019. 15 major cities around the world that are starting to ban cars. *Business Insider*, January 12.
120. O'Sullivan, F. 2019. Armed with a street-design tool called the knip, the Dutch capital is slashing car access in the city center, and expanding public transit hours. *City Lab*, October 15.
121. Muller, N. Z. and R. Mendelsohn. 2007. Measuring the damages of air pollution in the United States. *J. Environ. Econ. Manage.* 54:1–14.
122. Jaramilloa, P. and N. Z. Mullerb. 2016. Air pollution emissions and damages from energy production in the U.S.: 2002–2011. *Energy Policy* 90:202–11.
123. EPA (Environmental Protection Agency). 1999. Benefits and costs of the Clean Air Act. http://www.epa.gov/oar/sect812/copy.html.
124. Ryan, H. S. and K. M. Thompson. 2003. When domestic environmental policy meets international trade policy: Venezuela's challenge to the U.S. gasoline rule. *Hum. Ecol. Risk Assess.* 9:811–27.
125. American Thoracic Society. 2019. Dramatic health benefits following air pollution reduction. *Science Daily*, December 6.
126. OECD (Organization for Economic Cooperation and Development). 2014. The cost of air pollution. http://www.oecd.org/env/the-cost-of-air-pollution-9789264210448-en.htm.
127. Staff. 2020. Global cost of air pollution $2.9 trillion a year: NGO report. *News 24*, March 2.
128. American Geophysical Union. 2020. COVUD-19 lockdowns significantly impacting global air quality. *Science Daily*, May 11.
129. Watts, J. 2020. Clean air in Europe during lockdown 'leads to 11,000 fewer deaths.' *The Guardian*, April 30.
130. Wu, X., R. C. Nethery, B. M. Sabath, et al. 2020. Exposure to air pollution and COVID-19 mortality in the United States: A nationwide cross-sectional study. *MedRxiv*. doi:10.1101/2020.04.05.20054502.
131. Davenport, C. 2020. Trump administration declines to tighten soot rules, despite link to covid deaths. *The New York Times*, December 7.
132. Hinckley, S. 2016. How Mexico City plans to fight air pollution. *The Christian Science Monitor*, April 3.
133. *CBS News*. 2015. Smog solutions: How 6 cities are attempting to deal with dangerous air pollution, December 29.
134. Willsher, K. 2015. Paris to stop traffic when air pollution spikes. *The Guardian*, November 3.
135. Tresaugue, M. 2014. Planting trees could be a weapon in the battle against smog. *Houston Chronicle*, September 12.
136. Peters, A. 2015. This tree-covered apartment building cleans a polluted city block. http://www.fastcoexist.com/3043820/this-tree-covered-apartment-building-cleans-a-polluted-city-block.
137. Kaplan, S. 2016. A surprising simple solution to bad indoor air quality: Potted plants. *The Washington Post*, August 24.
138. *American Heritage Dictionary of the English Language*, 3rd edit. 1996. Boston: Houghton Mifflin Co.

3 Water Quality and Security

3.1 INTRODUCTION

Water to drink, air to breathe, and food to consume are the three vital necessities for life. Absent any of these three and life is impossible. A human can go for more than 3 weeks without food (Mahatma Gandhi survived 21 days of complete starvation), but water is a different story. Unlike food, the maximum time a person can go without water is said to be a week to 10 days before dehydration causes loss of life. That time estimate would certainly be shorter in difficult conditions like extreme heat. At least 60% of the adult body is made of water, and every living cell in the body needs it to keep functioning. Water acts as a lubricant for our joints, regulates our body temperature through sweating and respiration, and helps to flush waste [1]. But as will be described in this chapter, water fouled with pollutants can seriously harm human and ecological health.

> The maximum time a person can go without water is said to be a week to 10 days before dehydration causes loss of life.

Regarding access to water, our primordial ancestors likely used surface water sources for their drinking water, relying on springs, streams, and rainwater. As human civilizations grew, wells were dug to access groundwater, along with using surface water resources. Over time, water for drinking was joined by water used for transportation and use in industrial and agricultural operations. As human civilizations grew in number and complexity some sources of water became contaminated with human wastes, chemicals, biological agents, and other unhealthful pollutants. Shown in Figure 3.1 is an example of a water pollution point source that could cause adverse human and ecological health effects, depending on the content of the discharged water.

Described in this chapter are the domestic and global problems of water quality and security and attendant policies meant to prevent the problems. In the context of this chapter, water pollution refers to the quality and security of surface and groundwater sources. Water security will be discussed in terms of water contamination as well as water availability. To be described in some detail are problems and forms of water contamination. Regarding environmental health policies, to be discussed are the two primary U.S. federal policies pertaining to water quality: the Clean Water Act and the Safe Drinking Water Act. While both acts deal with water issues, the former act pertains to the control of contaminants in waterbodies such as rivers, lakes, and wetlands. The second act pertains to the control of contaminants in drinking water supplies. Prior to discussing water quality and security issues, descriptions

DOI: 10.1201/9781003253358-3

FIGURE 3.1 Water pollution discharged from a point source. (NOAA Ocean Service Education [2].)

of the adverse effects of contaminated water on human and ecological health will be presented. The chapter commences with a discussion of the essential value of water as an environmental resource.

3.2 WATER AS A VITAL ENVIRONMENTAL RESOURCE

The vital nature of water is evident by its many uses by humankind. The various uses by the U.S. population will display water's vital role in social well-being. Uses of water in the U.S. are compiled by the U.S. Geologic Survey (USGS). Every 5 years, data at the county level are compiled into a national water-use data system and state-level data are published in a national circular, *Estimated Use of Water in the United States* [3].

According to the USGS, freshwater and saline water (seawater and brackish coastal water) withdrawals in the U.S. in 2015 totaled approximately 322 billion gallons per day, which was 9% less than in 2010 [3]. This decline occurred even as the U.S. population increased by 4%. Water conservation was a factor in this decline. Thermoelectric power generation and irrigation were the dominant uses of this water. The majority of water withdrawals, 61.5% or 198 billion gallons per day, came from surface water sources such as lakes, rivers, and streams. Approximately one-fourth of freshwater withdrawals, totaling 82 billion gallons per day, was from groundwater. The remaining 13% came from saline sources.

Public supply refers to water that is used for domestic purposes—drinking, washing clothes and dishes, bathing, watering lawns, etc.—and public services such as pools, wastewater treatment, and firefighting. According to the USGS, 283 million Americans, 87% got their

> Freshwater and saline water (seawater and brackish coastal water) withdrawals in the U.S. in 2015 totaled approximately 322 billion gallons per day.

drinking water from public supplies in 2015. The remaining 13% relied primarily on private wells. Nevada had the highest self-supplied domestic water per capita use.

Bringing all this closer to home and down to the individual level, according to the EPA, the average U.S. household uses more than 300 gallons of water per day. In a typical residential dwelling, 70% of this usage occurs indoors—primarily in the bathroom [4]. Water-flushed toilets are a major source of household water usage.

3.3 WATER SECURITY: QUALITY

Water security is defined by the UN as "the capacity of a population to safeguard sustainable access to adequate quantities of acceptable quality water for sustaining livelihoods, human well-being, and socio-economic development, for ensuring protection against water-borne pollution and water-related disasters, and for preserving ecosystems in a climate of peace and political stability" [5]. Pollution typically refers to chemicals or other substances or materials in concentrations greater than would occur under natural conditions. Major water pollutants include pathogens, nutrients, heavy metals, organic chemicals, oil, and sediments; heat, which raises the temperature of the receiving water, can also be a pollutant. Pollutants are typically the cause of major water quality degradation around the world [5]. The global state of water security was characterized in 2019 by the UN as follows [6]:

- 2.1 billion people lack access to safely managed drinking water services.
- 4.5 billion people lack safely managed sanitation services.
- 340,000 children under five die every year from diarrheal diseases.
- Water scarcity already affects four out of every 10 people.
- 90% of all natural disasters are water-related.
- 80% of wastewater flows back into the ecosystem without being treated or reused.
- Around two-thirds of the world's transboundary rivers do not have a cooperative management framework.
- Agriculture accounts for 70% of global water withdrawal.
- About 75% of all industrial water withdrawals are used for energy production [6].
- According to one estimate, in 2016 two-thirds of the global population (4.0 billion people) live under conditions of severe water scarcity for at least 1 month of the year. Nearly half live in India and China. Half a billion people in the world face severe water scarcity all year round [7].
- Given the essential value of water as an environmental resource, problems of water pollution become important for human and ecological well-being. Following will be a discussion of the major forms of water pollutants and two other factors, algal contamination and drought, that impact water security.

3.3.1 BIOLOGICAL POLLUTANTS IN WATERBODIES

Biological pollutants consist of various pathogens that pollute surface and groundwater water sources. Pathogens in water, especially surface waters, have existed from the time that humans and other animals first excreted their body wastes into rivers, lakes, streams, and oceans. Bacteria and viruses in urine and feces contaminated water sources and impacted water security. Specific pathogens cause specific diseases, for example, cholera and diarrhea. As human societies increased in number and complexity, wastewater management became necessary and continues as a global necessity. Chlorination or ozonation of water supplies has been generally effective in purifying water of bacterial contaminants. But as noted by the UN, the most significant sources of water pollution are inadequate treatment of human wastes and inadequately managed and treated industrial and agricultural wastes.

> Unsafe water causes 4 billion cases of diarrhea annually and results in 2.2 million deaths, mostly of children less than 5 years old.

3.3.2 CHEMICAL POLLUTANTS IN WATERBODIES

Chemical pollutants have been constituents of water from the time that humans began working with metals and other natural resources, producing materials such as metal weapons that led to waste being discarded into bodies of water. The industrial revolution of the 18th and 19th centuries and the chemical age that followed World War II were social developments that led to large amounts and an increased variety of chemicals being discarded into lakes, rivers, oceans, and other bodies of water. In many parts of the world, this form of environmental pollution continues. Releases of chemical contaminants into waterbodies are controlled under national and regional policies, as described in subsequent sections of this chapter.

Chemical contaminants in U.S. waterbodies reflect the industrial, agricultural, and personal activities of the country's population. Two surveys by the U.S. Geologic Survey (USGS) have identified water pollutants of contemporary origin. In one survey, conducted from 2011 to 2014, USGS discovered insecticides known as neonicotinoids in more than half of both urban and agricultural streams sampled across the U.S and Puerto Rico [8]. Neonicotinoids are a class of pesticides that have properties of nicotine and have been suggested as a factor in the decline in numbers of bees and other pollinators.

In a second USGS survey, the agency investigated pharmaceutical contaminants in a sample of streams in the southeastern U.S. [9]. The USGS noted that pharmaceuticals are a growing aquatic-health concern and largely attributed to wastewater treatment facility discharges. Five biweekly water samples from 59 small streams in the southeastern U.S. were analyzed for 108 pharmaceuticals and degradants using high-performance liquid chromatography and tandem mass spectrometry. The antidiabetic metformin was detected in 89% of samples and at 97% of sites.

> USGS: Pharmaceuticals are a growing aquatic-health concern and largely attributed to wastewater treatment facility discharges.

At least one pharmaceutical was detected at every site (median of 6, maximum of 45), and several were detected at ≥10% of sites at concentrations reported to affect multiple aquatic end points. The results highlight a fundamental biochemical link between global human-health crises like diabetes and aquatic ecosystem health [9]. In recognition of the growing problem of pharmaceuticals in the nation's waterbodies, the EPA announced in 2016 that the agency would be considering a rule for the management of hazardous waste pharmaceuticals by healthcare facilities (including pharmacies) and distributors. The rule would prohibit the flushing of more than 6,400 tons of hazardous waste pharmaceuticals annually by banning healthcare facilities from flushing hazardous waste pharmaceuticals down the sink and toilet [10].

This final pharmaceutical rule became effective on August 21, 2019. The rule is summarized by EPA as the following, "Some pharmaceuticals are regulated as hazardous waste under the Resource Conservation and Recovery Act (RCRA) when discarded. This final rule adds regulations for the management of hazardous waste pharmaceuticals by healthcare facilities and reverse distributors. Healthcare facilities (for both humans and animals) and reverse distributors will manage their hazardous waste pharmaceuticals under this new set of sector-specific standards in lieu of the existing hazardous waste generator regulations. Among other things, these new regulations prohibit the disposal of hazardous waste pharmaceuticals down the drain and eliminate the dual regulation of RCRA hazardous waste pharmaceuticals that are also Drug Enforcement Administration (DEA) controlled substances. The new rules also maintain the household hazardous waste exemption for pharmaceuticals collected during pharmaceutical take-back programs and events, while ensuring their proper disposal" [11].

A different drinking water pollutant with potential adverse health effects are the "forever chemicals." The chemicals, resistant to breaking down in the environment, are known as perfluoroalkyl substances, or PFAS. Some have been linked to cancers, liver damage, low birth weight, and other health problems. An environmental organization, the Environmental Working Group (EWG), estimated in 2018, using EPA data, that 110 million Americans may be contaminated with PFAS [12]. However, a follow-up study released in 2020 by the EWG alleged that the PFAS contamination is far worse than previously estimated with some of the highest levels found in Miami, Philadelphia, and New Orleans.

Of tap water samples taken by EWG from 44 sites in 31 states and Washington, D.C., only one location, Meridian, Mississippi, which relies on 700 feet deep wells, had no detectable PFAS. Only Seattle and Tuscaloosa, Alabama, had levels less than 1 part per trillion (PPT), the limit EWG recommends. EPA has recommended that water contain no more than 70 parts per trillion (ppt) of PFAS, although it is not a water quality standard, leading several U.S. states to enact laws requiring lower levels of PFAS for drinking water. In February 2020 EPA announced it would regulate "forever chemicals" [13]. EPA's decision to regulate PFAS starts a two-year period for the agency to determine what the new mandatory maximum contamination level (MCL) should be. Once that is formally proposed, the agency has another 18 months to finalize its drinking water requirement. In the interim, U.S. state standards will remain in effect.

Of policy note, in December 2020 as the EPA worked to limit the importation of any product with PFAS inside or out, the Trump administration's White House Office of Management and Budget (OMB) significantly weakened EPA guidance, barring importation of only those products with a PFAS coating on the outside, in effect weakening the EPA proposed regulation and increasing the adverse health risk as a product's use erodes a coating, exposing PFAS as a food or water contaminant.

In addition to biological and chemical contaminants of water sources, two current factors that are deleteriously impacting water security are algal contamination and drought, as discussed in the following sections.

3.3.3 HARMFUL ALGAL BLOOMS IN WATERBODIES

The formation of algae blooms in waterbodies has become a major global concern for water security. Algae are plants or plantlike organisms that contain chlorophyll and other pigments that trap light from the Sun. This light energy is then converted into food molecules via photosynthesis. Algae can be either single-celled or large, multicellular organisms. They can occur in freshwater or salt water (most seaweeds are algae) or on the surfaces of moist soil or rocks. Algal blooms occur when colonies of algae reproduce in large numbers. Algal blooms can be toxic as the algae produce toxins. Harmful algal blooms can be green, blue, red, or brown.

The EPA observes that harmful algal blooms (HABs) are a major environmental problem in all 50 U.S. states [14]. Known as red tides, blue-green algae, or cyanobacteria, harmful algal blooms can have severe impacts on human health, aquatic ecosystems, and the economy. Some produce dangerous toxins in fresh or marine water but even nontoxic blooms hurt the environment and local economies. Harmful algal blooms can produce extremely dangerous toxins that can sicken or kill people and animals, create dead zones in the water, raise treatment costs for drinking water, and hurt industries that depend on clean water. Climate change might lead to stronger and more frequent algal blooms as a consequence of warmer water.

> Known as red tides, blue-green algae, or cyanobacteria, harmful algal blooms can have severe impacts on human health, aquatic ecosystems, and the economy.

Concerning HABs, one source observes, "Several decades ago relatively few countries appeared to be affected by HABs, but now most coastal countries are threatened, in many cases over large geographic areas and by more than one harmful or toxic species. Many countries are faced with a bewildering array of toxic or harmful species and impacts, as well as disturbing trends of increasing bloom incidence, larger areas affected, more fisheries resources impacted, and higher economic losses. The causes behind this expansion are debated, with possible explanations ranging from natural mechanisms of species dispersal to a host of human-related phenomena such as pollution, climatic shifts, increased numbers of observers, and transport of algal species via ship ballast water" [15]. Illustrated in Figure 3.2 is a toxic algal bloom in Lake Erie in the summer of 2017. The algal blooms occurred in the southwest corner of Lake Erie.

In order to form a water pollutant, HABs need sunlight, slow-moving water, and nutrients (nitrogen and phosphorus). Nutrient pollution from human activities makes the problem worse, leading to more severe blooms that form more often. The primary sources of nutrient pollution are [15]:

FIGURE 3.2 Satellite image of Lake Erie on September 23, 2017. The bright green areas show algae blooms. (NASA. 2018. Smaller summer harmful algal bloom forecast for western Lake Erie. July 12.)

- **Agriculture:** Animal manure, excess fertilizer applied to crops and fields, and soil erosion make agriculture one of the largest sources of nitrogen and phosphorus pollution in the country.
- **Stormwater:** When precipitation falls on our cities and towns, it runs across hard surfaces—like rooftops, sidewalks, and roads—and carries pollutants, including nitrogen and phosphorus, into local waterways.
- **Wastewater:** Our sewer and septic systems are responsible for treating large quantities of waste, and these systems do not always operate properly or remove enough nitrogen and phosphorus before discharging into waterways.
- **Fossil Fuels:** Electric power generation, industry, transportation, and agriculture have increased the amount of nitrogen in the air through the use of fossil fuels.
- **Households:** Fertilizers, yard and pet waste, and certain soaps and detergents contain nitrogen and phosphorus, and can contribute to nutrient pollution if not properly used or disposed of. The amount of hard surfaces and type of landscaping can also increase the runoff of nitrogen and phosphorus during wet weather.

More than 100,000 miles of rivers and streams, about 2.5 million acres of lakes, reservoirs, and ponds, and more than 800 square miles of bays and estuaries in the U.S. have poor water quality because of nitrogen and phosphorus pollution [14]. As noted by the National Oceanic and Atmospheric Administration (NOAA) in 2016, HABs and hypoxic events (severe oxygen depletion) are some of the most scientifically complex and economically damaging coastal and lake issues. Almost every state in the U.S. now experiences some kind of HAB event and the number of hypoxic (i.e., oxygen-deficient) waterbodies in the U.S. has increased 30-fold since the 1960s with more than 300 coastal systems now impacted. A 2006 study found that the economic impacts from a subset of HAB events in U.S. marine waters averaged $82 million/ year (2005 dollars) [2]. As but one of several examples, in August 2020, Iowa water authorities closed the Des Moines River as a drinking water source for 500,000 central Iowans because of toxins from algae. The Des Moines River's levels of a toxin that comes from blue-green algae, microcystin, have recently been more than ten times the federal recommendation for drinking water. Other sources of drinking water will replace that supplied by the Des Moines River.

During a bloom, algae can produce toxins that can render water unsafe and cause fish mortality, or can impact human health through the consumption of contaminated seafood, skin contact, and swallowed water during recreational activities. Toxins are usually released when an algal bloom dies off. Ingestion of or even just exposure to these toxins has been associated with many human health issues, ranging from diarrhea to cancer, as well as with pet and wildlife deaths [2]. As oceans warm, algae blooms have become more widespread, creating toxins that get ingested by sardines and anchovies, which in turn get ingested by sea lions, causing damage to the brain that results in epilepsy. Sea otters also face risk when they consume toxin-laden shellfish.

> A 2006 study found that the economic impacts from a subset of HAB events in U.S. marine waters averaged $82 million/year.

In 1998, Congress recognized the severity of these threats and authorized the Harmful Algal Bloom and Hypoxia Research and Control Act. The Harmful Algal Bloom and Hypoxia Research and Control Amendments Act of 2004, and 2014 reaffirmed and expanded the mandate for NOAA to advance the scientific understanding and ability to detect, monitor, assess, and predict HABs and hypoxia events. However, Congress has not appropriated funds in support of the law, requiring NOAA and other federal agencies to provide monetary support from existing agency funds, an indication of the long-standing political dogma that fiscal appropriations are a policymaker's statement of priority.

In June 2019, NOAA projected a Massachusetts-sized dead zone would form in the Gulf of Mexico, driven by a vast algae bloom, forming thick undulating mats of green, fed by fertilizer runoff from the upper Midwest. As the bloom decays, it sucks oxygen out of the water. As a result, as NOAA stated, "habitats that would normally be teeming with life become, essentially, biological deserts" [16]. HABs of this magnitude would constitute a major hazard to marine life and accompanying ecosystems.

Research has shown a cause-and-effect relationship between HABs and climate change [17]. Warmer air and water temperatures promote an accelerated growth of the blooms, fueled primarily by farm runoff (fertilizer and manure) in which phosphorus becomes available to the algae, a growth nutrient for them. An effect of HABs is the generation of methane and CO_2 as the algae die. In a study released in March 2018, researchers affiliated with the University of Minnesota found that, globally, lakes and manmade "impoundments" like reservoirs emit about one-fifth the amount of greenhouse gases emitted by the burning of fossil fuels. The majority of that atmospheric effect came from methane [17].

Algae control is the action taken to manage algae problems. The long-term management of algae should at least involve reducing nutrient inflow into the waterbody.
However, long-term nutrient reduction requires extensive changes in policies and
human activities and therefore takes many years before significant improvement in
water quality can be seen. There are some short-term treatment options available
for managing algae problems such as aeration, chemical or biological additives,
or ultrasound technology [18]. In 2020, an NOAA chemist demonstrated the use
of a combination of nanobubble ozone technology (NBOT) machines that take in
oxygen from the air, generate ozone, and release ozone-filled nanobubbles into the
water. Since nanobubbles do not float, the bubbles remain underwater and release
ozone, a strong oxidant that damages the cell wall of algae and breaks down the
chemical bonds of their toxins, destroying the algae [19]. Tests of the NBOT equipment successfully removed all algae in small lakes, with tests on large lakes planned
for 2020.

As a matter of public health concern and subsequent policy, two U.S. states,
Oregon and Ohio, that experienced toxins in drinking water supplies released from
algae blooms have implemented policies that require public water suppliers to test for
algae toxins [20]. Other U.S. states are likely to develop similar drinking water rules
as the number of HABs increases.

Having now discussed various forms of pollution of waterbodies, it is important to
discuss the ramifications of water pollution on human and ecological health.

3.4   IMPACTS OF CONTAMINATED WATERBODIES
ON HUMAN HEALTH

As noted in 2010 by the Secretary, Department of Health and Human Services,
"contamination of water can come from both point (e.g., industrial sites) and non-
point (e.g., agricultural runoff) sources. Biological and chemical contamination
significantly reduces the value of surface waters (streams, lakes, and estuaries)
for fishing, swimming, and other recreational activities, and can cause disease in
humans. For example, during the summer of 1997, blooms of *Pfiesteria piscicida*
were implicated as the likely cause of fish kills in North Carolina and Maryland.
The development of intensive animal feeding operations (e.g., large scale swine
farms) has worsened the discharge of improperly or inadequately treated wastes,
which presents an increased health threat in waters used either for recreation or for
producing fish and shellfish" [21].

Two surveillance systems have provided relevant human health data on U.S. water
quality. One system is a disease surveillance system operated by the CDC. The other
system, which is maintained by EPA, provides data on water quality measurements.
Turning first to the CDC system, since 1971, CDC, EPA, and the Council of State
and Territorial Epidemiologists have maintained a collaborative surveillance system for collecting and periodically voluntarily reporting data on occurrences and
causes of waterborne-disease outbreaks (WBDOs). Tabulation of recreational water-
associated outbreaks was added to the surveillance system in 1978. This surveillance
system is the primary source of data concerning the scope and effects of waterborne
disease outbreaks in the U.S. [22].

During 2000–2014, 493 outbreaks associated with treated recreational water caused at least 27,219 cases and eight deaths. Among the 363 outbreaks with a confirmed infectious etiology, investigations of the 363 outbreaks identified 24,453 cases; 21,766 (89%) were caused by Cryptosporidium, 920 (4%) by Pseudomonas, and 624 (3%) by Legionella. At least six of the eight reported deaths occurred in persons affected by outbreaks caused by Legionella. Hotels were the leading setting, associated with 157 (32%) of the 493 outbreaks [23].

In addition to CDC's waterborne disease surveillance system, EPA is required under section 305(b) of the CWAct, as amended, to report to Congress on the Nation's water quality conditions. Under section305(b), states, territories, and interstate commissions must assess their water quality biennially and report those findings to EPA. These entities must compare their monitoring results to the water quality standards they have set for themselves. In the year 2017 report to Congress, EPA communicated that 46% of U.S. river and stream miles are in poor biological condition; 21% of the nation's lakes are hypereutrophic (i.e., with the highest levels of nutrients, algae, and plants); 18% of the nation's coastal and Great Lakes waters are in poor biological condition; 14% are rated poor based on a water quality index; and 32% of the nation's wetland area is in poor biological condition [24]. The leading causes of inadequate water quality included bacteria, nutrients, metals (primarily mercury), and siltation. EPA cites the following conditions as the primary sources of water degradation: runoff from agricultural lands, sewage treatment plants, and hydrological modifications such as dredging of channels. As a matter of environmental health policy, these statistics indicate that the states, territories, and other governmental entities have a substantial challenge if water quality is to be improved.

> Approximately 50 trillion gallons of raw sewage in the U.S. must be treated every day.

The problem of inadequate sewage treatment is particularly important, given the huge volume of sewage that must be treated in order to prevent waterborne diseases. As noted previously in this chapter, approximately 50 trillion gallons of raw sewage in the U.S. must be treated every day. Unfortunately, many of the sewage systems in the U.S. are old and inadequately designed. To be more specific, some sewage-carrying pipes are almost 200 years old, with 100-year old pipes not uncommon. Moreover, many older municipalities, primarily in the northeastern U.S. and the Great Lakes region, have sewage collection systems designed to carry both sewage and stormwater runoff. Such combined systems can overflow during heavy rainfall, resulting in raw sewage becoming mixed with stormwater, which can bypass sewage treatment plants [25].

The EPA estimated in 2004 that 1.3 trillion gallons of raw sewage are dumped annually due to combined sewer overflows. The agency also estimates that 1.8–3.5 million persons in the U.S. become ill annually from swimming in waters contaminated by sanitary sewage overflows [cited in 22]. To prevent this kind of public health problem will require repairing and upgrading the sewage collection and treatment systems in the U.S. There is government and private sector consensus that there is a funding gap of $1 trillion for water infrastructure [22]. Regrettably, gathering political support for repair and upgrading of municipal infrastructures can be difficult, sewage systems in particular. There is a tendency to pass infrastructure repairs to succeeding governments. Only when emergencies occur, such as the aftermath of hurricanes or the release of large amounts of pollutants in an area or under a court order, do political bodies become energized.

The state of water quality and security in the U.S. is assessed by EPA under provisions of the CWAct of 1972, as amended, as described in a subsequent section of this chapter. Data on water quality are provided to the U.S. Congress under provisions of the same act. The data for 2004 are shown in Tables 3.1 and 3.2. Although somewhat dated, these data remain relevant and representative of the water situation in the U.S. The data in these tables are compiled by EPA from water quality reports from states, territories, and tribes. Specifically, to assess water quality, states, tribes, and other jurisdictions compare their monitoring results to the water quality standards they have set for their waters. Water quality standards consist of three elements: the designated uses (such as drinking, swimming, or fishing) assigned to waters; criteria (such as chemical-specific thresholds that should not be exceeded) to protect those uses; and an anti-degradation policy intended to keep waters that do meet standards from deteriorating from their current condition [26].

Reflection on the data in Table 3.1 [26] provides a troubling characterization, given that 55% of U.S. rivers, 70% of lakes, 78% of estuaries, and 53% of wetlands were assessed in 2016 as "impaired." The leading causes of impairment are listed in Table 3.2 [26]. One notes that causes of impairment differ according to the kind of waterbody. For example, pathogens, sediment, and nutrients are the three leading causes of water impairment for rivers and streams. For rivers and streams, the three leading sources of water impairment are agriculture, unknown, and hydromodification. Looking across the four columns of Table 3.2, mercury and nutrients, are factors common to three of

> EPA reports that 44% of U.S. rivers, 64% of lakes, and 30% of estuaries are assessed as "impaired."

TABLE 3.1
EPA's Summary of Water Quality in U.S. Water Sources, 2016 [23]

Waterbody Type and Statistics			Condition of Assessed Waters (% of Assessed)		
Waterbody Type	Total Size	Amount Assessed (% of Total)	Good	Good But Threatened	Impaired
Rivers & Streams (miles)	3.5 m	1.107 m (31.6%)	487,299 (44.0%)	5,550 (0.5%)	614,153 (55.5%)
Lakes (acres)	41.7 m	18.51 m (44.4%)	5,470,004 (29.6%)	34,621 (0.2%)	13,009,273 m (70.3%)
Bays and Estuaries (mi²)	87,791	35,094 (40.0%)	7,611 (21.7%)	Not rated	27,486 (78.3%)
Wetlands (acres)	107.7 m	1.232 m (1.1%)	574,907 (46.6%)	Not rated	657,653 (53.4%)

TABLE 3.2

Leading Causes and Sources of Impairment in Assessed Waterbodies, 2016 [23]

Rivers and Streams	Lakes	Bays and Estuaries	Wetlands
Causes:	**Causes:**	**Causes:**	**Causes:**
Pathogens	Mercury	Mercury	Nutrients
Sediment	Nutrients	PCBs	Mercury
Nutrients	PCBs	Pathogens	Metals other than Hg
Sources:	**Sources:**	**Sources:**	**Sources:**
Agriculture	Atmospheric deposition	Atmospheric deposition	Agriculture
Unknown/Unspecified	Agriculture	Municipal discharges	Atmospheric deposition
Hydromodification	Natural sources	Other sources	Industry

the four classifications of waterbodies. The data in these two tables provide valuable guides for targeted actions of prevention and policymaking.

3.5   IMPACTS OF CONTAMINATED WATERBODIES ON ECOSYSTEM HEALTH

The United Nations Environment Programme (UNEP) has aided in the coordination of programs of water quality protection. In doing so, the agency has accumulated data on the associations between water quality and impacts on ecosystems. The UNEP observes that "over the past decades, the water quality of surface waters and groundwaters has improved over many parts of the world, particularly in industrialized countries, but also in some parts of middle- and lower-income countries. This has been one of the good news stories of environmental management, achieved by widely introducing wastewater treatment and other water quality management measures. Yet there is important unfinished business. Investing in wastewater treatment, assuring access to safe water, preventing water pollution, and restoring aquatic ecosystems are examples of important unfinished business that require the attention of policymakers and water experts" [27].

> The EPA estimates that 1.8–3.5 million persons in the U.S. become ill annually from swimming in waters contaminated by sanitary sewage overflows.

The UNEP also observes that globally many rivers and other parts of the freshwater system are faced with new threats to their water quality. In emerging and developing countries, water quality is threatened by the increasing discharge of untreated or inadequately treated municipal wastewater as well as by diffuse sources of pollutants from agricultural, urban, and other areas that degrade surface and groundwater.

Regarding ecosystems, water quality degradation poses health risks and undermines ecosystem services provided by surface and subsurface waters. Wastewater loadings deplete dissolved oxygen, increase turbidity, and have other negative effects on freshwater ecosystems thus jeopardizing the services they provide. Impacts might include diminishing stocks of freshwater fish for food, declining aquatic biodiversity, deteriorating water quality for industrial and agricultural use, and higher treatment costs for municipal water supply.

> Water quality degradation poses health risks and undermines ecosystem services provided by surface and subsurface waters.

In countries undergoing rapid economic development, a new threat is caused by the increasing discharge of toxic organic chemicals, heavy metals, and other substances to surface waters. Some of these substances might accumulate in freshwater ecosystems or infiltrate groundwater and thereby pose a long-term risk to human health and aquatic ecosystems.

In industrialized countries, as well as in some developing ones, an increasing threat is the discharge to surface waters of unidentified and unmonitored residues from medicines, e.g., pharmaceuticals) and new chemical products (e.g., cosmetics). Since conventional wastewater treatment might not be able to remove these substances, they may find their way into freshwater systems. Some of these substances might act as endocrine disruptors and be otherwise harmful to people and the environment. Other factors that contribute to water quality degradation and corollary effects on ecosystems include:

- Water quality degradation in developing and rapidly industrializing countries is often associated with the growth of mining, manufacturing, and industrial activities. Wastewaters are often discharged without adequate treatment directly or indirectly to different types of waterbodies (rivers, groundwater aquifers, wetlands, etc.). Better waste management practices are required.
- An important factor contributing to water quality degradation worldwide is unsustainable land use and agriculture. In agricultural regions, the main source of water contamination is the seasonal runoff of pesticides and fertilizers from cropland and pastureland. Other possible sources of land-based water pollution are deforestation and intensive animal husbandry. Urban areas are also a major source of diffuse water pollutants. Better land management and planning could help minimize these problems.
- Climate change is expected to have an increasing impact on worldwide water quantity and quality. Global climate change is expected to have an increasing influence on not only water quantity but also water quality. Where long-term precipitation diminishes, it is likely that stream flow may decrease and along with it the self-purifying capacity of rivers and lakes [27].

Recent reports link pharmaceuticals discarded into water sources as causing ecosystem effects. For example, a review study concluded that recent studies have revealed that pharmaceuticals, both human and veterinary, disperse widely in aquatic and terrestrial environments with uptake into a range of organisms. Pharmaceuticals are designed to have biological actions at low concentrations rendering them potentially potent environmental contaminants. In some cases, the effects can be dramatic, such as the near extinction of three species of vulture in India after eating the carcasses of livestock that had been treated with the anti-inflammatory diclofenac. However, effects can be more subtle but still have potentially significant impacts. Changes

Pharmaceuticals, both human and veterinary, disperse widely in aquatic and terrestrial environments with uptake into a range of organisms.

to behavior of fish and birds after exposure to low concentrations of psychiatric drugs can alter foraging patterns, activity levels, and risk-taking [28].

A long-term, whole-lake experiment was conducted at the Experimental Lakes Area in northwestern Ontario, Canada, using a before-after-control-impact design to determine both direct and indirect effects of the synthetic estrogen used in the birth control pill, 17α-ethynyl estradiol (EE2). Recruitment, i.e., the number of fish surviving to enter a fishery, of fathead minnow failed, leading to a near-extirpation of this species both 2 years during and 2 years following EE2 additions. Body condition of male lake trout and male and female white sucker declined before changes in prey abundance, suggesting direct effects of EE2 on this endpoint [29].

3.6   U.S. WATER QUALITY POLICIES

Protection of water quality and quantity in the U.S. has evolved over the country's history, commencing with wells and cisterns established for use in individual households, water obtained from surface sources such as rivers and lakes used by villages and small towns, and water drawn from surface sources and aquifers for servicing large cities. Commensurate with demands for water came policies for the protection of water quality. These policies were largely the province of towns, municipalities, and states. Needless to say, considerable differences existed between states, in particular, in policies for the protection of water quality. It was not until the mid-20th century that the U.S. federal government assumed a role in water quality policymaking. Although protection of public health was the anchor for federal water policies, there was a corollary issue of reducing the differences across states in how water policies were developed and implemented. As will be described herein, the U.S. has two primary water quality protection policies, the Clean Water Act (CWAct) of 1972 and the Safe Drinking Water Act (SDWAct) of 1974.

The U.S. has two primary water quality protection policies, the CWAct of 1972 and the SDWAct of 1974.

3.6.1   THE CLEAN WATER ACT, 1972

The federal CWAct of 1972 is the principal U.S. law that addresses water pollution in U.S. waterbodies. The CWAct is a policy that does not possess the complexities of the federal Clean Air Act (Chapter 2), and as such, has been subject to less controversy and litigation. However, both of these major U.S. policies on water and air share common features of regulating sources of pollution, requiring permits to pollute water or air, and sharing enforcement responsibilities with U.S. states.

3.6.1.1 History

Water suitable for human consumption and other uses has historically been of public health importance. Indeed, Hippocrates, the father of medicine, emphasized circa 400 BCE the importance of boiling and straining water for health purposes [30]. Modern programs of water quality protection can be dated from the late 19th century, when chlorine was found to be an effective water disinfectant when added in low concentrations to drinking water. In 1902, Belgium became the first country to make continuous use of chlorine as an additive to drinking water supplies. Chlorination of public drinking water supplies in the U.S. dates to 1908, when the Boonton reservoir supply, Jersey City, New Jersey, was chlorinated, triggering a series of lawsuits that were ultimately decided by courts in favor of water chlorination as a means for water purification against pathogens [30].

Chlorination of public drinking water supplies must be considered as a "modern" public health triumph. The notion of adding a human poison, chlorine—even at very low concentrations—to a vital resource, drinking water, must have seemed foolhardy to many persons in the early 20th century. However, as time passed, the marked reduction in waterborne diseases such as cholera, typhus, and dysentery demonstrated the public health benefits of water chlorination and overcame residual public opposition.

> Chlorination of public drinking water supplies in the U.S. dates to 1908, when the Boonton reservoir supply, Jersey City, New Jersey, was chlorinated.

Prior to the enactment of federal water quality statutes, states bore the responsibility for dealing with water quality problems, including issues of sanitation and drinking water quality. According to one source, many of the states' water pollution control policies from the late 19th century through the first half of the 20th century comprised two steps:

1. "First, common-law cases involving adverse effects of pollution upon public health or fish and wildlife resources were brought to court.
2. Second, statutory regulatory authority was then given to state health or fish and wildlife agencies. Sometimes these two authorities were combined and extended by a water pollution control board" [31].

This approach by states led to considering each case of water pollution as an individual matter, subject to informal negotiations between polluters and state officials and attendant negotiations, all of which took considerable time in general, resulting in litigation if the parties could not agree on a pollution control strategy [31]. Little of this kind of informal approach to pollution control was apparent to the general public unless a particular court action attracted newsmedia attention. Other problems with a state-by-state approach to water quality control include different water quality standards between states and the migration of some polluting industries to states with fewer stringent water standards and controls. Problems with state-based pollution controls contributed to pressure to develop federal water quality standards and regulations in the mid-20th century.

The U.S. federal government's involvement with water pollution control dates to the turn of the 20th century, when water pollution control regulations were included in the Rivers and Harbors Act of 1899. This act authorized the regulation of industrial discharges of pollution into waters that might cause navigation problems [32]. Later, the Public Health Service Act expressed the first federal policy on the disposal of human wastes, which authorized the Public Health Service to provide technical advice and assistance to communities and for federal research on sanitary waste disposal methods. Over time, more comprehensive, focused federal water quality legislation was enacted by Congress, as described in this chapter.

> U.S. federal government's involvement in water pollution control dates back to 1899 when water pollution control regulations were included in the Rivers and Harbors Act.

The principal federal law now governing pollution of the Nation's waterways is the Federal Water Pollution Control Act, more commonly called the CWAct [32]. The original purpose of the act was to establish a federal program to award grants to states for the construction of sewage treatment plants. Although originally enacted in 1948, the act was completely revised by amendments in 1972, giving the CWAct most of its current shape. The 1972 legislation declared as its objective the restoration and maintenance of the chemical, physical, and biological integrity of U.S. waters. Two goals were established: zero discharge of pollutants by 1985 and, as an interim goal and where possible, water quality that is both "fishable" and "swimmable" by mid-1983. While those dates have passed, the goals remain, and efforts to attain them are continuing [32].

The CWAct contains a number of complex elements of overall water quality management. Foremost is the requirement in section 303 that states must establish ambient water quality standards for waterbodies, consisting of the designated use or uses of a waterbody (e.g., recreational, public water supply, or industrial water supply) and the water quality criteria that are necessary to protect the use or uses. Through permitting, states or EPA impose wastewater discharge limits on individual industrial and municipal facilities in order to ensure that water quality standards are attained. However, Congress recognized in the CWAct that in many cases pollution controls implemented by industry and municipalities would be insufficient, due to pollutant contributions from other unregulated sources [32].

> The Clean Water Act established a program of grants to construct sewage treatment plants and a regulatory and standards program to control discharges of chemical and microbial contaminants into U.S. waters.

At the heart of the CWAct is a system of permits, called the National Pollutant Discharge Elimination System (NPDES), that determines how much pollution can be released into surface (e.g., rivers) and underground water supplies. Each source of pollution must comply with permits specific to the source. Permits are therefore tailored to the size of the pollution source, the toxicity or hazard of individual pollutants, the technology available to reduce pollution levels, and the quality and size of the waterway receiving the pollution discharges [33]. Unfortunately, according to one environmental organization that in 2000 studied the status of 6,700 permits for major facilities included in the NPDES, about 25% of the permits were not current [34]. That is, more than 1,690 polluting facilities were operating without current discharge permits. Given the purpose of pollution discharge permits, it is important to the public's health that they be kept current.

In 2005, EPA's Inspector General reported that there remains a large backlog of NPDES permits requiring renewal [35]. According to the report, 1,120 major permit facilities, 9,386 individual minor, and 6,512 general minor permit facilities need permit renewals.

In a third analysis of CWAct permits, in 2006 the U.S. Public Interest Research Group reported findings similar to those from the EPA Inspector General [36]. The group's research showed more than 62% of industrial and municipal facilities in the U.S. discharged more pollution into U.S. waterways than the CWAct permits allowed. The investigation covered the period between July 2003 and December 2004. The average facility discharged pollution in excess of its permit limit by more than 275%, or almost four times the legal limit.

Reflection on these three reports of problems with CWAct permits indicates a significant weakness in the permitting policy. Permits to discharge pollution into environmental media are ineffective without a commitment to enforce them.

The CWAct has forced the development and use of technologies to reduce the quantities of pollutants released into waterways. The CWAct gave industries until 1977 to install *best practicable control technology* (BPT) to clean up waste discharges. Later amendments to the CWAct (Table 3.3) required a greater level of pollutant cleanups, generally requiring that by 1989 industry utilize the *best available technology* (BAT) that is economically feasible. Failure to meet statutory deadlines can lead to enforcement action, although compliance extensions of as long as 2 years are available for industrial sources utilizing innovative or alternative technology.

Control of pollution discharges has been the key focus of water quality programs. In addition to the BPT and BAT national standards, states are required to implement

TABLE 3.3
Clean Water Act and Major Amendments [29]

Year	Act
1948	Federal Water Pollution Control Act
1956	Water Pollution Control Act
1961	Federal Water Pollution Control Act Amendments
1965	Water Quality Act
1966	Clean Water Restoration Act
1970	Water Quality Improvement Act
1972	Federal Water Pollution Control Amendments
1977	Clean Water Act
1981	Municipal Wastewater Treatment Construction Grants Amendments
1987	Water Quality Act
2000	BEACH Act

control strategies for waters expected to remain polluted by toxic chemicals even after industrial dischargers have installed the best available cleanup technologies required under the CWAct, as amended. Development of management programs for these post-BAT pollutant problems was a prominent element in the 1987 CWAct amendments and is a key continuing aspect of the CWAct's implementation [37,38].

3.6.1.2 Clean Water Act Amendments

In addition to the 1972 amendments, several other important amendments to the CWAct have occurred, as listed in Table 3.3. Amendments enacted in 1977, 1987, and 2000 are particularly relevant for public health purposes.

The 1977 amendments to the act focused on toxic pollutants. The CWAct of 1977 established the basic structure for regulating discharges of pollutants into waters of the U.S. Further, §404 established a program to regulate the discharge of dredged and fill material into U.S. waters, including wetlands. The basic premise of section 404 is that no discharge of dredged or fill material can be permitted if a practicable alternative exists that is less damaging to the aquatic environment or if the Nation's waters would be significantly degraded. Regulated activities are controlled by a permit review process. An *individual permit*, which is the responsibility of EPA, is usually required for potentially significant impacts. However, for discharges thought to have minimal impact, the U.S. Army Corps of Engineers can grant *general permits*, which are issued for particular categories of activities (e.g., minor road crossings, utility line backfill) as an expedited means for regulating discharges [39]. What are called wetlands constitute a vital natural resource in the U.S. and elsewhere. "Wetlands are areas where the frequent and prolonged presence of water at or near the soil surface drives a natural ecosystem, i.e., the kind of soils that form, the plants that grow, and the fish and wildlife that find habitat" [40]. Swamps, marshes, and bogs are common types of wetlands. The Everglades in Florida are perhaps the best known U.S. wetland.

Wetlands serve an important environmental health purpose, one in addition to serving as a habit for great numbers of birds, fish, mammals, plants, and trees. Wetlands are one of nature's water purifiers. Turbid surface waters that flow into wetlands drain off as freshwater. Regrettably, there has been a steady loss of wetlands acreage. For instance, the U.S. Fish and Wildlife Service estimates that between 1986 and 1997, a net of 644,000 acres of wetlands were lost, with 58,400 acres lost annually [41]. The principal causes of loss of wetlands were urban development, agriculture,

ENFORCEMENT EXAMPLE

Washington, D.C.—March 6, 2015: Coal producer Alpha Natural Resources Inc. agreed to $227.5 million in penalties and other costs to settle federal allegations that it illegally dumped large amounts of toxicants into waterways in Pennsylvania and four other states. The company will pay $27.5 million in penalties and spend $200 million upgrading its wastewater treatment systems to reduce illegal discharges. The Pennsylvania State Department of Environmental Protection will get $4.125 million from the fines to use in clean water programs. The fine is the largest assessed under federal clean water rules [42].

silviculture (i.e., the growing and culture of trees), and rural development [41]. As a matter of environmental policy, finding the best balance between the protection of wetlands and the need for land development is, and will remain in the future, a difficult calculus for policymakers.

In 1987, the CWAct was reauthorized and again focused on toxic substances. The amendments authorized citizen suit provisions and funded sewage treatment plants under a construction grants program. Prior to 1987, the CWAct only regulated pollutants discharged to surface waters from *point sources* (i.e., pipes, ditches, and similar conveyances of pollutants) unless a permit was obtained under provision in the Act. *Non-point sources* (e.g., stormwater runoff from agricultural lands) were covered by the Water Quality Act of 1987, which amended the CWAct. Pollution sources are required by the CWAct to treat their wastes to meet the more stringent of two sets of requirements, based on either technologic feasibility or attainment of desired levels of water quality [38]. From 1972 to 2003, the Nation has invested more than $300 billion (in constant dollars) to build and upgrade wastewater treatment systems [43].

Over the years, the sewage collection system in the U.S. has grown to more than a million miles of collection pipes. This system carries about 50 trillion gallons of raw sewage daily, delivered to approximately 20,000 sewage treatment plants [25]. This enormous system of sewage collection and treatment represents a vital public health resource to the U.S. public, preventing human exposure to the pathogens found in raw sewage. It also represents an indispensable global environmental contribution by reducing the pollution load deposited in the planet's oceans and seas.

3.6.1.3   The EPA Water Quality Trading Policy, 2003

On January 13, 2003, EPA announced a market-based "Water Quality Trading Policy" [44]. This is a variant of the cap-and-trade policy discussed in Chapter 2. The Trading Policy was written on the assumption that, if a total maximum daily load (TMDL) were in place, all trading partners would be covered by the TMDL. In this case, waste load allocations (WLAs) and load allocations (LAs) under the TMDL form the baseline for trading. In all cases, permits must be designed to meet water quality standards as required under CWAct section 301(b)(1)©. Inclusion of trading provisions in NPDES permits should facilitate meeting this requirement. The policy's aim is to authorize users of a waterbody to trade pollution credits among themselves in order to cost-effectively achieve the pollutant reductions mandated by the TMDL and other programs.

The 2003 Policy allows one source [of water pollution] to meet its regulatory obligations by using pollutant reductions created by another source. Entities that discharge into the same watershed may achieve increased flexibility by working together to reduce discharges of certain pollutants.

> Questions have arisen as to the extent of coverage of the CWAct. Specifically, what are the waters of the U.S. that are subject to the provisions of the act?

A study by Food and Water Watch, an NGO group, of this policy's efficacy yielded concern [45]. The NGO noted that water quality trading programs were underway in more than 20 states, covering the release of pollutants like nitrogen and phosphorus into U.S. waterways. Those nutrients are behind algae blooms that suck oxygen out of water supplies, killing fish and other wildlife and sometimes making people sick. But after reviewing more than 1,000 documents from pilot trading programs in Pennsylvania and Ohio, Food and Water Watch researchers came to the conclusion that the programs, though they sound reasonable on paper, operate very differently than predicted in the real world.

Specifically, according to researchers, with little state oversight, private contractors have been permitted to run pollution trading markets that offer highly regulated industrial polluters the chance to essentially swap places with farms, concentrated animal feeding operations (CAFOs), and feed lots whose runoff is not as tightly controlled under the CWAct. Researchers commented, "The big, big problem that we see on the credit generating side is that agriculture operations never have to monitor, sample, never have to verify that they actually generated the pollution that underlies these credits, [...] It's all based on modeling" [45].

3.6.1.4   Obama Administration's Waters of the United States Rule, 2015

Questions have arisen over the existence of the CWAct as to the extent of its coverage. Specifically, what are the waters of the U.S. that are subject to the provisions of the act? This rather basic question has arisen from state governments, industry, and individual property owners. The question has been litigated and resulted in U.S. Supreme Court decisions that left room for EPA's interpretation of what constituted "waters of the United States" under the CWAct's provisions. This resulted in EPA's issuance in 2015 of its Waters of the United States Rule. As characterized by EPA, "Protection for about 60% of the nation's streams and millions of acres of wetlands has been confusing and complex as the result of Supreme Court decisions in 2001 and 2006. The Waters of the United States Rule protects streams and wetlands that are scientifically shown to have the greatest impact on downstream water quality and form the foundation of our nation's water resources. EPA and the U.S. Army are ensuring that waters protected under the CWAct are more precisely defined, more predictable, easier for businesses and industry to understand, and consistent with the law and the latest science" [46].

"EPA and the U.S. Army Corps of Engineers finalized the Waters of the United States Rule to clearly protect the streams and wetlands that form the foundation of the nation's water resources. Protection for many of the nation's streams and wetlands has been confusing, complex, and time-consuming as a result of Supreme Court decisions in 2001 and 2006. The Waters of the United States Rule ensures that waters protected under the CWAct are more precisely defined, more predictably determined, and easier for businesses and industry to understand." Specifically, the Waters of the United States Rule:

- Clearly defines and protects tributaries that impact the health of downstream waters. The CWAct protects navigable waterways and their tributaries. The rule says that a tributary must show physical features of flowing water—a bed, bank, and ordinary high water mark— to warrant protection. The rule provides protection for headwaters that have these features and science shows can have a significant connection to downstream waters.

- Provides certainty in how far safeguards extend to nearby waters. The rule protects waters that are next to rivers and lakes and their tributaries because science shows that they impact downstream waters. The rule sets boundaries on covering nearby waters for the first time that are physical and measurable.
- Protects the nation's regional water treasures. Science shows that specific water features can function like a system and impact the health of downstream waters. The rule protects prairie potholes, Carolina and Delmarva bays, pocosins, western vernal pools in California, and Texas coastal prairie wetlands when they impact downstream waters.
- Focuses on streams, not ditches. The rule limits protection to ditches that are constructed out of streams or function like streams and can carry pollution downstream. Not covered are ditches that are not constructed in streams and that flow only when it rains.
- Maintains the status of waters within Municipal Separate Storm Sewer Systems. The rule does not change how those waters are treated and encourages the use of green infrastructure.

> The Obama administration's Clean Water Rule of 2015 was never implemented because of protracted litigation.

- Reduces the use of case-specific analysis of waters. Previously, almost any water could be put through a lengthy case-specific analysis, even if it would not be subject to the CWAct. The rule significantly limits the use of case-specific analysis by creating clarity and certainty on protected waters and limiting the number of similarly situated water features [47].

The Waters of the United States Rule was never implemented. Upon its release as an EPA rule, litigation ensued by several U.S. states and some industrial entities, including agricultural interests. A U.S. federal appeals court ordered the EPA in late 2015 to provide more background information on one of the Rule's specifications. With the incoming Trump administration came the decision to replace the Waters of the United States Rule with their following version of the Clean Water Act.

3.6.1.5 Trump Administration's Navigable Waters Protection Rule, 2020

As noted in the previous section, the EPA's Waters of the United States Rule was one of several key environmental regulations promulgated by the agency during the Obama administration. The Obama-era Waters of the United States rule was intended to construct a definition of which waters of the U.S. are subject to provisions of the CWAct. Along with another major regulation, the Clean Power Plan (Chapter 2), both EPA regulations touched on a common denominator, climate change. Concerning water, as ambient air temperatures rise, an effect on water resources will be of concern, as reported in the Fourth National Climate Assessment discussed in Chapter 2. Both the Clean Power Plan and the Waters of the United States Rule received vigorous litigious opposition from some U.S. states and various industrial and agricultural interests. Conversely, both regulations were generally embraced by environmental groups, some U.S. states, and some other non-government organizations.

Regarding the Waters of the United States Rule, on February 28, 2017, President Donald Trump (R-NY) instructed the EPA and the U.S. Army Corps of Engineers to review and reconsider the rule, stating his directive was, "paving the way for the elimination of this very destructive and horrible rule" that should have only applied to "navigable waters" affecting "interstate commerce" [48]. On January 24, 2020, the EPA Administrator and Assistant Secretary of the Army for Civil Works jointly announced the Trump administration's the Navigable Waters Protection Rule [49].

> The Trump administration's water rule of 2018 views the Clean Water Act as covering only major rivers, their primary tributaries, and wetlands along their banks.

The Navigable Waters Protection Rule rolls back key portions of the 2015 rule that had guaranteed protections under the 1972 Clean Water Act to certain wetlands and streams that run intermittently or run temporarily underground but also relieves landowners of the need to seek permits that the EPA had considered on a case-by-case basis before the Obama rule.

In the arid West, where the majority of streams flow only after rainfall or for part of the year, under this rule, entire watersheds would be left unprotected from pollution. In Arizona, for instance, as much as 94% of its waters could lose federal protection under the new definition, depending on how the agencies interpret key terms. Meanwhile, Arizona state law also prevents it from regulating waterways more stringently than the federal government requires. The proposed rule became subject to federal litigation in 2019 and its acceptability awaits court decisions [49].

3.6.1.6   Supreme Court Decision Regarding Groundwater, 2020

An important example of the role of the judiciary in shaping the execution of environmental policy is afforded by a Supreme Court 6-3 ruling of April 2020 [50]. In particular, the Supreme Court ruled on April 23, 2020, that the Clean Water Act applies to some pollutants that reach the sea and other protected waters indirectly through groundwater. The case concerned a wastewater treatment plant that used injection wells to dispose of some four million gallons of treated sewage each day by pumping it into groundwater about a half-mile from the Pacific Ocean. Some of the waste reached the ocean. The Trump administration filed a brief supporting the operator of the treatment plant.

3.6.1.7   Cost and Benefits of Waterbodies Pollution Control

In 2003, the White House's Office of Management and Budget (OMB) estimated that over the period 1992–2002 federal water pollution rules resulted in an annual benefit of $0.89–8.07 billion, with costs estimated at $2.4–2.9 billion. These figures have considerable uncertainty in the monetary benefits of water pollution control, with less uncertainty in the costs. The benefits/cost ratio, therefore, ranges from <1 to approximately 2.7 [51].

3.6.2 THE SAFE DRINKING WATER ACT, 1974

The SDWAct of 1974 is the second of the two major U.S. statutes on water quality policy. Whereas the CWAct of 1972, as amended, addresses the control of sources of water pollution, the Safe Drinking Water Act, as amended, focuses on the security of drinking water supplies in the U.S. In a sense, there is a duality of purpose between the two laws: both statutes apply to water in the U.S., but with different aims. Absent either statute, the nation's public health would be in jeopardy. Regarding safe drinking water, a discussion of the impacts of contaminated water on human and ecological health is discussed prior to a description of the Safe Drinking Water Act's public health policies.

3.6.2.1 Impacts of Non-potable Water on Human Health

Providing drinking water free of disease-causing agents, whether biological or chemical, is the primary goal of all water supply systems. During the first half of the 20th century, the causes for most waterborne disease outbreaks were bacteria, whereas beginning in the 1970s, protozoa and chemicals became the dominant causes [52]. Most outbreaks involve only a few individuals. However, failures in water treatment systems have occasionally led to instances of widespread waterborne disease. For example, more than 400,000 people were affected in 1993 when the Milwaukee, Wisconsin, water supply became contaminated with *Cryptosporidia* [53]. An example of a public water supply contaminated by a bacterium, *E. coli*, occurred in Walkerton, Ontario, Canada, in May 2000. *E. coli* is an intestinal bacterium that causes muscle cramps, fever, nausea, and severe diarrhea, and can cause kidney failure in extreme cases. The outbreak might have caused more than 2,000 cases of illness, including seven deaths [54].

Drinking water contaminated with microbial or chemical contaminants can cause human disease. Sources of drinking water (and percentage of waterborne disease outbreaks) comprise wells (70.5%), springs (5.9%), surface water (11.8%), and a combination of wells and springs (11.8%) [55]. Since 1971, CDC, EPA, and the Council of State and Territorial Epidemiologists have maintained a collaborative surveillance system for collecting and periodically reporting data related to occurrences and causes of waterborne-disease outbreaks (WBDOs). This surveillance system is the primary source of data concerning the scope and effects of waterborne disease outbreaks on persons in the U.S. [56].

> During the first half of the 20th century, the causes for most waterborne disease outbreaks were bacteria, whereas beginning in the 1970s, protozoa and chemicals became the dominant causes.

Public health agencies in the U.S. states and territories report information on waterborne disease outbreaks to the CDC Waterborne Disease and Outbreak Surveillance System. For an event to be defined as a waterborne disease outbreak, two or more persons must be linked epidemiologically by time, location of water exposure, and case illness characteristics; and the epidemiologic evidence must implicate water as the probable source of illness.

During 2013–2014, public health officials from 19 states reported 42 outbreaks associated with drinking water during the surveillance period [56]. These outbreaks resulted in at least 1,006 cases of illness, 124 hospitalizations (12% of cases), and 13 deaths. At least one etiologic agent was identified in 41 (98%) outbreaks. Counts of etiologic agents in this report include both confirmed and suspected etiologies, which differs from previous surveillance reports. Legionella was implicated in 24 (57%) outbreaks, 130 (13%) cases, 109 (88%) hospitalizations, and all 13 deaths. Eight outbreaks caused by two parasites resulted in 289 (29%) cases, among which 279 (97%) were caused by Cryptosporidium, and 10 (3%) were caused by *Giardia duodenalis*. Chemicals or toxins were implicated in four outbreaks involving 499 cases, with 13 hospitalizations, including the first reported outbreaks (two outbreaks) associated with algal toxins in drinking water.

> During 2013–2014, waterborne outbreaks in the U.S. resulted in at least 1,006 cases of illness, 124 hospitalizations (12% of cases), and 13 deaths.

Among the 1,006 cases attributed to drinking water-associated outbreaks, 50% of the reported cases were associated with chemical or toxin exposure, 29% were caused by parasitic infection (either Cryptosporidium or Giardia), and 13% were caused by Legionella bacterial infection [56]. Seventy-five percent of cases were linked to community water systems. Outbreaks in water systems supplied solely by surface water accounted for most cases (79%). Of the 1,006 cases, 86% originated from outbreaks in which the predominant illness was acute gastrointestinal illness. Three (7%) outbreaks in which treatment was not expected to remove the contaminant were associated with a chemical or toxin and resulted in 48% of all outbreak-associated cases [56].

The most commonly reported outbreak etiology was Legionella (57%), making acute respiratory illness the most common predominant illness type reported in outbreaks [56]. Thirty-five (83%) outbreaks were associated with public (i.e., regulated), community, or noncommunity water systems, and three (7%) were associated with unregulated, individual systems. Fourteen outbreaks occurred in drinking water systems with groundwater sources and an additional 14 occurred in drinking water systems with surface water sources. The most commonly cited deficiency, which led to 24 (57%) of the 42 drinking water-associated outbreaks, was the presence of Legionella in drinking water systems. In addition, 143 (14%) cases were associated with seven (17%) outbreak reports that had a deficiency classification indicating unknown or insufficient information [56].

In addition to drinking water-associated illnesses identified through surveillance systems, two investigations of the health consequences associated with contaminated drinking water have expanded the suite of potential adverse human health effects. In one study, babies born to mothers with high levels of perchlorate in drinking water during their first trimester are more likely to have lower IQs later in life [57]. The researchers analyzed perchlorate levels in the first trimester of 487 pregnant women in Cardiff, Wales, and Turin, Italy, who had iodine deficiency and thyroid dysfunction during pregnancy. Their children's IQ scores were evaluated at 3 years old. Children born to mothers with perchlorate levels in the highest 10% were more than three times as likely to have an IQ score in the lowest 10% of scores. It adds to evidence that the drinking water contaminant may disrupt thyroid hormones that are crucial for proper brain development. Perchlorate, which is both naturally occurring and manmade, is used in rocket fuel, fireworks, and fertilizers. It has been found in 4% of U.S. public water systems serving an estimated 5–17 million people, largely near military bases and defense contractors in the U.S. West [57].

In 2019, the EPA proposed setting the maximum contaminant level (MCL) of perchlorate at 56 parts per billion (ppb), significantly greater than the 15 ppb proposed under the Obama administration. Some U.S. states have also set their own standards for perchlorate, with 6 ppb in California and 2 ppb in Massachusetts. In June 2020, the EPA announced that it would not regulate perchlorate. The agency stated, "The chemical, perchlorate, does not meet the criteria for regulation as a drinking water contaminant under the Safe Drinking Water Act." The EPA further noted that up to 620,000 people might be consuming water that has a perchlorate concentration higher than "levels of concern," but commented in a draft final action that this number was too small to present a "meaningful opportunity for health risk reduction," and allowed U.S. states to continue managing any health risks associated with perchlorate in drinking water supplies [58]. The EPA also commented that the agency's decision was consistent with the Trump Administration's policy of reducing the number of environmental regulations.

Concerning a second water contaminant, a study compared 1,091 PCE-exposed pregnancies and 1,019 unexposed pregnancies among 1,766 women in Cape Cod, Massachusetts., where water was contaminated in the late 1960s to the early 1980s by the installation of vinyl-lined asbestos cement pipes, over time releasing tetrachloroethylene (PCE) into drinking water [59]. PCE exposure was estimated using water-distribution system modeling. Data on pregnancy complications were self-reported by mothers. Of the more than 2,000 pregnancies, 9% were complicated by pregnancy disorders associated with placental dysfunction. Pregnancies among women with high PCE exposure had 2.38 times the risk of stillbirth and 1.35 times the risk of placental abruption, compared to pregnancies not exposed to PCE.

Perspective: Public supplies of drinking water in the U.S. are generally safe to drink and use for other household purposes such as bathing and food preparation. Indeed, chlorination and ozonation of public water supplies are two of the 20th century major public health achievements. However, drinking water waterborne diseases can exist in low-income countries that lack the necessary resources to adequately manage water and sanitation services, a subject of concern to the WHO and other UN agencies.

3.6.2.2 Impacts of Non-potable Water on Ecosystem Health

The WHO says that poor access to sufficient quantities of water can be a key factor in water-related disease and is closely related to ecosystem conditions. The agency observes that about one-third of the world's population lives in countries with moderate to high water stress, and problems of water scarcity are increasing, partly due to ecosystem depletion and contamination. Two out of every three persons on the globe may be living in water-stressed conditions by the year 2025, if present global consumption patterns continue [56].

Further, WHO and others have observed that the sustainability of many water ecosystems has been impacted by development and land-use changes involving: elimination of marshes and wetlands; the diversion of surface water or alteration of flows; increased exploitation of underground aquifers; and contamination of water by waste and discharges from industry and transport, as well as from household and human waste [60]. The absolute quantity and the diversity of pollutants reaching freshwater systems have increased since the 1970s. These include not only biological contaminants, e.g. microorganisms responsible for traditional water-borne diseases, but also heavy metals and synthetic chemicals, including fertilizers and pesticides.

> The WHO notes that two-thirds of persons on the globe might be living in water-stressed conditions by the year 2025.

The WHO advises that water ecosystems should be valued for their protection of water supplies and suggests protection should include: (1) An "ecosystem approach" that recognizes and ascribes value, including economic value, to the "services" natural ecosystems provide in terms of water filtration and purification, and ensures their sustainability, through modern management regimes; (2) integrated water resource management; (3) protecting water from contamination from household to global level: careful disposal of waste and protection of health from contaminated water sources is a vital principle [60]. Given the public health importance of safe drinking water, a brief history of the Safe Drinking Water Act is instructive for an understanding of environmental health policymaking.

3.6.2.3   History of the Safe Drinking Water Act

The SDWAct of 1974 (SDWAct)[1] was enacted to protect the quality of drinking water in the U.S. This law focuses on all waters actually or potentially designed for drinking use, whether from surface or underground sources. Congress acted after a nationwide study of community water systems revealed widespread water quality and health risk problems resulting from poor operating procedures, inadequate facilities, and poor management of public water supplies in communities of all sizes. Further, the 1974 act was in response to congressional findings that chlorinated organic chemicals were contaminating major surface and underground water supplies, that widespread underground injection operations were a threat to aquifers, and that the infrastructures of public water supply systems were increasingly inadequate to protect the public health [61].

Perhaps the most important public health authority of the SDWAct, as amended in 1986 and 1996, is EPA's requirement to set drinking water standards. *Drinking water standards* are regulations that EPA sets to control the level of contaminants in the Nation's drinking water. These standards are part of the SDWAct's "multiple barrier" approach to drinking water protection, which includes assessing and protecting drinking water sources; protecting wells and collection systems; making sure water is treated by qualified operators; ensuring the integrity of distribution systems; and making information available to the public on the quality of

---

[1] The "Safe Drinking Water Act" consists of Title XIV of the Public Health Service Act (42 U.S.C. 300f-300j-D) as added by Public Law 93-523 (December 13, 1974) and subsequent amendments.

their drinking water. According to EPA, with the involvement of EPA, states, tribal nations, drinking water utilities, communities, and citizens, these multiple barriers ensure that tap water in the U.S. and territories is safe to drink. In most cases, EPA delegates responsibility for implementing drinking water standards to states and tribal nations.

There are two categories of drinking water standards [62]:

- A *National Primary Drinking Water Regulation* (NPDWR or primary standard) is a legally enforceable standard that applies to public water systems. Primary standards protect drinking water quality by limiting the levels of specific contaminants that can adversely affect public health and are known or anticipated to occur in water. They take the form of Maximum Contaminant Levels or Treatment Techniques, which are described below.

> The Safe Drinking Water Act establishes primary drinking water standards, regulates underground injection disposal practices, and establishes a groundwater control program.

- A *National Secondary Drinking Water Regulation* (NSDWR or secondary standard) is a non-enforceable guideline about contaminants that may cause cosmetic effects (such as skin or tooth discoloration) or aesthetic effects (such as taste, odor, or color) in drinking water. EPA recommends secondary standards to water systems but does not require systems to comply. However, states may choose to adopt them as enforceable standards.

The EPA classifies public water systems under its SDWAct authorities. The agency states that a public water system provides water for human consumption through pipes or other constructed conveyances to at least 15 service connections or serves an average of at least 25 people for at least 60 days a year. A public water system may be publicly or privately owned. There are more than 151,000 public water systems in the U.S. EPA classifies these water systems according to the number of people they serve, the source of their water, and whether they serve the same customers year-round or on an occasional basis.

> Drinking water standards are regulations that EPA sets to control the level of contaminants in the Nation's drinking water.

The EPA has defined three types of public water systems:

- **Community Water System (CWS):** A public water system that supplies water to the same population year-round.
- **Non-Transient Non-Community Water System (NTNCWS):** A public water system that regularly supplies water to at least 25 of the same people at least 6 months per year. Some examples are schools, factories, office buildings, and hospitals which have their own water systems.
- **Transient Non-Community Water System (TNCWS):** A public water system that provides water in a place such as a gas station or campground where people do not remain for long periods of time [63].

Community Water Systems constitute the vast majority of systems that supply water to the U.S. population. According to an EPA survey of CWSs, they provide water to more than 280 million persons in the U.S. The survey estimates that there are 49,133 community water systems in the 50 states and the District of Columbia. Nearly 75% of the nation's CWSs rely primarily on groundwater. Almost 9% rely primarily on surface water, while the remaining 18% purchases finished, partially treated, or untreated water [64]. Because of the large numbers of people serviced by these water systems, it is a sound environmental health policy to protect them from contamination in order to prevent waterborne illnesses.

3.6.2.4 Safe Drinking Water Act Amendments

As indicated in Table 3.4, the SDWAct has been amended several times since the original act of 1974. "The first major amendments, enacted in 1986, were largely intended to increase the pace at which EPA regulated contaminants. These amendments required EPA to (1) issue regulations for 83 specified contaminants by June 1989 and for 25 more contaminants every 3 years thereafter, (2) promulgate requirements for disinfection and filtration of public water supplies, (3) ban the use of lead pipes and lead solder in new drinking water systems, (4) establish an elective wellhead protection program around public wells, (5) establish a demonstration grant program for state and local authorities having designated sole-source aquifers to develop groundwater protection programs, and (6) issue rules for monitoring injection wells that inject wastes below a drinking water source. The amendments also increased EPA's enforcement authority" [65].

The Lead Contamination Control Act of 1988 added a new part F to the SDWAct. It was intended to reduce exposure to lead in drinking water by requiring the recall of lead-lined water coolers and required EPA to issue a guidance document and testing protocol to help schools and day care centers to identify and correct lead contamination in their drinking water [65]. The primary impetus for the act was a report to Congress that identified water coolers in schools as a potential source of children's exposure to lead in drinking water [66].

In 1996, Congress again made sweeping changes to the SDWAct. Originally, the SDWAct focused primarily on treatment as the means of providing safe drinking water at the tap. The 1996 amendments modified the existing law by recognizing source water protection, operator training, funding for water system improvements, and public information as important components of safe drinking water programs [67]. Implementation of the 1986 provisions had brought to the fore widespread dissatisfaction among states and communities. These concerns included inadequate regulatory flexibility and unfunded mandates. "As over-arching themes, the 1996 Amendments target resources to address the greatest health risks, increase regulatory and compliance flexibility under the Act, and provide funding for federal drinking water mandates.

> The 1996 SDWAct amendments recognized source water protection, operator training, funding for water system improvements, and public information.

Specific provisions revoked the requirement that EPA regulates 25 contaminants every 3 years, increased EPA's authority to consider costs when setting standards,

TABLE 3.4
Safe Drinking Water Act and Major Amendments [65]

Year	Act	Purpose
1974	Safe Drinking Water Act	
1977	Amendments	Authorized continuation of the agreement with the National Academy of Sciences to conduct a study of drinking water quality
1979	Amendments	Authorizes states regarding underground injection
1980	Amendments	control, extension/exemption of public water systems, permits grants to states for water filtration systems.
1986	Amendments	Creates a demonstration program to protect aquifers from pollutants, mandates state-developed critical wellhead protection programs, requires the development of drinking water standards for many contaminants now unregulated, imposes a ban on lead-content plumbing materials.
1988	Lead Contamination Control Act	Deals with the recall of lead-lined drinking water coolers.
1996	Amendments	Consumer confidence reports, cost-benefit analysis, drinking water state revolving fund, microbial contaminants and disinfection byproducts, operator certification, public information and consultation, small water systems
2002	Public Health Security and Bioterrorism Preparedness and Response Act	Large water system emergency response plan
2005	Amendments	2005 Energy Policy Act exempts hydraulic fracturing and oil and gas drilling from certain sections of the Safe Drinking Water Act of 1974 and the Clean Water Act of 1972.
2011	Reduction of Lead in Drinking Water Act (RLDWA)	Reduction of Lead in Drinking Water Act (RLDWA)
2013	Community Fire Safety Act	Evaluation of sources of lead in water distribution systems and alternate routing systems
2015	Amendments	Directs EPA to develop and submit to Congress a strategic plan for assessing and managing risks associated with algal toxins in drinking water provided by public water systems.

authorized EPA to consider overall risk reduction, established a state revolving loan program to help communities meet compliance costs, and expanded the Act's focus on pollution prevention through a new source water protection program" [67]. A cost-benefit analysis and a risk assessment are required before a standard can be set. The standards are initially based on health protection and the availability of technology.

They are called *Maximum Contaminant Levels*. The amendments required EPA to promulgate standards that maximize health risk reduction benefits at costs that are justified by the benefits.

The 1996 SDWAct amendments required that EPA establish criteria for a program to monitor unregulated contaminants found in drinking water supplies. Further, EPA must publish every 5 years a list of contaminants to be monitored in public drinking water supplies. One way to approach the requirement to regularly update a regulatory action is to establish a regulatory platform that first establishes criteria for updating—in this case a list of substances—then applying the criteria at specified intervals—in this instance, every 5 years. To develop such a platform is a policy decision.

In addition to the intent and authorizations of the original SDWAct of 1974, several other additions have been made to the law. In 2002, EPA promulgated The Lead and Copper Rule, 40 C.F.R., which was for the purpose of giving monitoring and reporting guidance for public water systems. Sections 141.80–141.91 require monitoring at consumer taps to identify levels of lead in drinking water that may result from corrosion of lead-bearing components in a public water system's distribution system or in household plumbing. These samples help assess the need for, or the effectiveness of, corrosion control treatment.

> The 1996 SDWAct amendments required that EPA establish criteria for a program to monitor unregulated contaminants found in drinking water supplies.

On December 22, 2020, EPA released its new Lead and Copper Rule. The new rule will speed notification to homeowners who are receiving lead-tainted water but does not force cities to act more quickly to replace the lead pipes that deliver the water [68]. The rule for the first time requires monitoring for lead at primary schools and child care centers and requires cities to notify residents of potential lead exposure within 24 hours. Cities will be required to replace just 3% of lead service lines each year rather than the previous 7%. EPA also will require cities to do the replacements for 2 years, rather than just one. The replacements are not required until a city detects high lead levels in 90% of the tested taps. The rule also creates a 10 ppb "trigger" level, where cities would be required to reevaluate their water treatment processes and possibly add corrosion-control chemicals to city water. But the rule keeps the current 15 ppb level that requires cities to begin replacing the nation's estimated 6 million lead service lines that connect homes to city water supplies—the underlying source of lead contamination.

3.6.2.5 Safe Drinking Water Compliance

The SDWAct of 1974 and its amendments establish the basic framework for protecting the drinking water used by public water systems in the U.S. This law contains requirements for ensuring the safety of the nation's public drinking water supplies. Public drinking water supplies include water systems that regularly serve 25 or more people per day or which have at least 15 service connections. The EPA sets national standards for drinking water to protect against health risks, considering available technology and cost. Each standard also includes monitoring and reporting requirements. The act allows U.S. states to take over the implementation of the program by obtaining "primacy."

The SDWAct's primary purpose is to ensure that tap water in the U.S. is safe to drink, with the EPA's promulgating water quality standards over the period of the law's existence and with states bearing the primary responsibilities for enforcement of water quality standards. Although generally successful, a substantial number of local water systems around the country have failed to meet these requirements. In a study of compliance with SDWAct standards, researchers found that, since 1982, between 3% and 10% of the country's water systems were in violation of SDWAct health standards each year. In 2015 alone, as many as 21 million Americans might have been exposed to unsafe drinking water [69].

> Since 1982, between 3% and 10% of the country's water systems were in violation of SDWAct health standards each year.

3.6.2.6 Bottled Drinking Water

Bottled water is big business globally. Rising concern for health and wellness, distrust of local drinking water sources, and evolution of new packaging initiatives are the major factors driving the global growth of the bottled water market. According to a market report, bottled water (neat, carbonated, flavored, and functional) was valued at $157.27 billion in 2013, which is expected to reach $279.65 billion by 2020, growing at a compound annual growth rate (CAGR) of 8.7% from 2014 to 2020. By volume, the global bottled water market is expected to grow at a CAGR of 8.3% during the forecast period from 2014 to 2020 to reach a market size of 465.12 billion liters (123 billion gals.) by 2020. In 2013, the volume of the market was 267.91 billion liters (70.8 billion gals.) [70].

According to an industry report, the volume of bottled water sold in the U.S. amounted to about 13.7 billion gallons in 2017, which made bottled water the top-selling liquid beverage in the U.S., exceeding sales of cola drinks [71].

In the U.S., tap water and bottled water are regulated by two different federal agencies, EPA and FDA. As described, EPA regulates tap water under its SDWAct authorities. But bottled drinking water is regulated as a food product, and as such, has been regulated since 1938 by FDA under the Food, Drug, and Cosmetic Act (FDCAct). FDA has established specific regulations for bottled water, including standard of identity regulations that define different types of bottled water, such as spring water and mineral water [72]. The agency has also established a standard of quality regulations that establish allowable levels for contaminants (chemical, physical, microbial, radiological) in bottled water.

Relevant to this chapter, section 305 of the SDWAct amendments of 1996 includes language that amends section 410 of the FDCAct as follows:

"(b)(1) Not later than 180 days before the effective date of a national primary drinking water regulation promulgated by the Administrator of the Environmental Protection Agency for a contaminant under section 1412 of the SDWAct, [t]he Secretary [of DHHS] shall promulgate a standard for that contaminant in bottled water or make a finding that such a regulation is not necessary to protect the public health because the contaminant is contained in water in public water systems [b]

ut not in water used for bottled drinking water." "(4)(A) If the Secretary does
not promulgate a regulation under this
subsection within the period described
in paragraph (1), the national primary
drinking water regulation referred to in
paragraph (1) shall be considered [a]s the

> According to industry sources, in
> 2014 the total volume of bottled
> water consumed in the U.S. was 11
> billion gallons, which translates into
> an average of 34 gallons per person.

regulation applicable under this subsection to bottled water." Stated more succinctly,
FDA must adopt EPA's MCLs if the contaminants appear in bottled water.

In addition to FDA, state and local governments also regulate bottled water. FDA
relies on state and local government agencies to approve water sources for safety and
sanitary quality, as specified in section 129.3 of the FDCAct. Additionally, states also
regulate the bottled water industry as well as the industry itself [71].

It is clear from the preceding language that congressional intent was to yoke
bottled drinking water quality with that of tap water as a means to ensure bottled
water quality. Has the intent been realized? According to one national environmental
group, the Natural Resources Defense Council (NRDC), "[b]bottled water sold in
the U.S. is not necessarily cleaner or safer than most tap water" [73]. This conclusion was predicated on findings from their study of 103 brands of bottled water.
Approximately one-third of the waters tested contained levels of contamination in
at least one sample that exceeded allowable limits under either state or bottled water
industry standards or guidelines. Moreover, the NRDC concluded that bottled water
regulations are inadequate because the FDA's rules exempt waters that are packaged
and sold within the same state, which is 60%–70% of all bottled water sales, and
approximately 20% of states don't regulate bottled water.

Setting aside the issue of source of bottled water, the product as a consumer item
has brought considerable environmental impacts predicated on the massive volume of
used plastic bottles that contained the water. This is a matter of plastic waste, which
will be discussed in Volume 2, that challenges waste disposal agencies. In order to
reduce the cost of waste disposal of plastic water bottles, some commercial entities
have discontinued selling bottled water. For example, in 2015, the Detroit Zoo ceased
selling water sold in plastic bottles and also installed 20 filtered water refill stations
[74]. Similarly, in 2011, the U.S. National Park Service announced a policy that permitted directors of the 408 national parks, monuments, and historical sites that they
could eliminate sales of disposable plastic water bottles, as long as refilling stations
and reusable bottles replaced them [75]. In April 2012, Concord, Massachusetts, residents voted to ban the sale of single-serving plastic water bottles and the measure
went into effect in January 2013. The ban was challenged by the water industry, but
a decision by the state's attorney general upheld Concord's ban [76].

3.6.3 IMPACTS OF UNREGULATED WATER SUPPLIES

Discussed to this point is the regulation of contaminants released into bodies of water
under the CWAct and the regulation of the quality of drinking water supplies under
the SDWAct. But there are unregulated supplies of water in the U.S. and other areas
of the globe. This is particularly the case in rural areas of the U.S. and similar locales

in other countries. To be discussed in this section are the human health implications of consumption of unregulated water supplies in the U.S.

Regarding public health in U.S. rural communities, the CDC has documented that cancer death rates have decreased nationwide. However, data from CDC show a slower reduction in cancer death rates in rural America (a decrease of 1.0% per year) compared with urban America (a decrease of 1.6% per year). This was driven in part by high death rates from lung, colorectal, prostate, and cervical cancers. Rates of new cases for lung cancer, colorectal cancer, and cervical cancer were also higher in rural counties. In contrast, rural counties were found to have lower rates of new cancers of the female breast and male prostate. This trend of disparity in cancer rates has been attributed to a number of factors, including heavy tobacco use and obesity compounded with generally inadequate public health structure and services [77]. While contaminated water supplies were not a factor evaluated in the CDC study, any consumption of impure water is inadvisable in any locale.

According to an investigative journalist, about one in seven U.S. residents relies on a private well for drinking water, as illustrated in Figure 3.3. Unlike the rest of the population served by the nation's many public water systems, these 44.5 million Americans are not protected by the SDWAct, which regulates 87 biological and chemical contaminants [78]. At best, private wells receive minimal oversight from

FIGURE 3.3   Rural U.S. private water well, 2021. (U.S. Geodetic Survey, 2021: https://www.usgs.gov/special-topic/water-science-school/science/groundwater-wells?qt-science_center_objects=0#qt-science_center_objects.)

local and state authorities, such as limited testing upon installation and, in some states, when properties change hands. Many wells escape regulation altogether, leaving the onus entirely on individuals to screen for pollutants and to mitigate them when they are discovered. Although rural areas generally have a high rate of domestic well use, the numbers vary. In New Hampshire, more than 46% of people rely on wells compared to 17% in West Virginia and just 6% in South Dakota, according to a national sampling of states by a newsmedia investigative team. In New England, 55% of people in Vermont and 57% in Maine have domestic wells [79].

> Unlike the rest of the population served by the nation's many public water systems, 44.5 million private well users in the U.S. are not protected by the SDWAct.

Many people who use private wells are left potentially exposed to harmful pollutants. In a 2013 study of nearly 4,000 private wells in rural Wisconsin, it was found that 47% exceeded at least one health-based water quality standard. And a study in North Carolina found that between 2007 and 2013, 99% of emergency department visits for acute gastrointestinal illness caused by microbial contamination of drinking water were associated with private wells [78]. These health data illustrate the potential public health impact of private wells left unevaluated for water potability. Local policymakers that support programs of public health should consider the merits of programs of private well water inspections, much as local health departments often inspect public swimming pools for sanitary conditions.

3.6.4 SOCIAL JUSTICE ISSUES

As with climate change and air pollution, disparities exist for persons of color and low income in the U.S. in regard to issues of water quality and security. In a 2019 study by the Natural Resources Defense Council (NRDC), nearly 40% of the U.S. population drinks water from unsafe systems, with communities of color facing an increased risk of exposure to unsafe water [80]. The NRDC identified 406 U.S. counties with the most severe health violations and water contamination—and also had the largest populations of low-income people, people of color, and nonnative English speakers. The percentage of water systems with violations for 12 consecutive quarters (i.e., systems in chronic noncompliance) was 40% greater in U.S. counties with the highest racial, ethnic, and language vulnerability compared to counties with the lowest racial, ethnic, and language vulnerability.

In a 2020 study of U.S. water disparities with social implications, researchers from King's College London, using census data from 2013 to 2017, found inequities in household water access [81]. Further, U.S. households headed by people of color were almost 35% more likely to live without piped water as compared to white households. In addition, plumbing poverty was also predicted by income inequality and precarious housing conditions such as living in rental accommodation and mobile homes. The researchers

> NRDC: U.S. households headed by people of color were almost 35% more likely to live without piped water as compared to white households.

assert that their findings add to a mounting body of evidence which reveal widespread inequities in access to clean, safe, affordable water in the U.S. in 2020.

3.7 GLOBAL WATER QUALITY POLICIES

Water quality and its security are global essentials and policies to protect water resources are cornerstones of environmental policies globally. The previous sections have presented and discussed the two key U.S. water policies.

3.7.1 GLOBAL WATER QUALITY AND HEALTH IMPACTS

While the quality of drinking water is generally good in the U.S. and other developed countries, that is not the case globally. In particular, developing countries struggle to build and maintain the environmental health infrastructure necessary to drastically reduce the public health toll of waterborne diseases. The diseases are caused by ingestion of water contaminated by human or animal feces or urine containing pathogenic bacteria or viruses. Waterborne diseases include cholera, typhoid, dysentery, and other diarrheal diseases. These diseases can be prevented through the disinfection of drinking water supplies and sanitary disposal of animal and human bodily wastes.

In 2016, the UNEP issued a warning concerning water pollution, estimating that 323 million people were at risk from life-threatening diseases caused by the pollution of rivers and lakes. The agency noted that cholera, typhoid, and other deadly pathogens are increasing in more than half of the rivers in Africa, Asia, and Latin America. Asia has been the worst hit, with up to 50% of all rivers affected by severe pathogen pollution caused by a mix of untreated wastewater disposal, agricultural pesticides run-off, and industrial pollution. The UNEP observed that among the groups most vulnerable to water quality deterioration in developing countries are women because of their frequent usage of surface water for household activities, children because of their play activities in local surface waters and because they often have the task of collecting water for the household, low-income rural people who consume fish as an important source of protein, and low-income fishers and fishery workers who rely on the freshwater fishery for their livelihood. Further, salinity levels have also risen in nearly a third of waterways, making them unsuitable for drinking water supply [82].

> In 2016, the UNEP estimated that 323 million people were at risk from life-threatening diseases caused by the pollution of rivers and lakes.

In 2017, the WHO and the UNICEF jointly released a study of global water security and noted that some 3 in 10 people worldwide, or 2.1 billion, lacked access to safe, readily available water at home, and 6 in 10, or 4.5 billion, lacked safely managed sanitation. Of the 2.1 billion people who did not have safely managed water, 844 million did not have even a basic drinking water service. This includes 263 million people who had to spend more than 30 minutes per trip collecting water from sources outside the home and 159 million who still drank untreated water from surface water sources, such as streams or lakes. Two out of three people with safely managed drinking water and three out of five people with safely managed sanitation services live in urban areas. Of the 161 million people using untreated surface water (from lakes, rivers, or irrigation channels), 150 million resided in rural areas. Regarding sanitation, the

> The WHO and UNICEF jointly estimated that some 3 in 10 people worldwide, or 2.1 billion, lacked access to safe, readily available water at home.

report commented "Of the 4.5 billion people who do not have safely managed sanitation, 2.3 billion still do not have basic sanitation services. This includes 600 million people who share a toilet or latrine with other households, and 892 million people—mostly in rural areas—who defecate in the open" [83].

Waterborne diseases have accompanied human development perhaps from the time of using surface water as a source of drinking water. Of these diseases, perhaps cholera has been the most damaging to global human health, as described in Volume 2, Chapter 1, where wastewater-borne infectious diseases are discussed. Of the remaining waterborne diseases, diarrhea is a disease of particular concern because of its cause of premature mortality in young children. The disease is especially important for persons in lower-income countries where water sanitation resources might be limited or absent. The WHO characterizes diarrhea as a major killer. In 1998, diarrhea was estimated to have killed 2.2 million people, most of whom were less than 5 years of age. Each year there are approximately 4 billion cases of diarrhea worldwide, according to the WHO. Diarrhea is fatal to 2,195 children every day—more than AIDS, malaria, and measles combined. Diarrheal diseases account for 1 in 9 child deaths worldwide, making diarrhea the second leading cause of death among children under the age of 5 years old. The WHO notes that diarrhea is a symptom of infection caused by a host of bacterial, viral, and parasitic organisms most of which can be spread by contaminated water. It is more common when there is a shortage of clean water for drinking, cooking, and cleaning and basic hygiene is important in prevention [84].

> Each year there are approximately 4 billion cases of diarrhea worldwide, according to the WHO. Diarrhea is fatal to 2,195 children every day.

An emerging water quality concern is the impact of personal care products and pharmaceuticals, such as birth control pills, analgesics, and antibiotics, on aquatic ecosystems. Little is known about their long-term human or ecosystem impacts, although some are believed to mimic natural hormones in humans and other species [85]. A 2019 study, the first ever such study, reinforced the concern of the presence of pharmaceuticals in global water supplies [86]. Researchers looked for 14 commonly used antibiotics in rivers of 72 countries across six continents and found antibiotics at 65% of the sites monitored. Concentrations of antibiotics found in some of the world's rivers exceeded "safe" levels by up to 300 times. The most prevalent antibiotic was trimethoprim, which was detected at 307 of the 711 sites tested and is primarily used to treat urinary tract infections.

Poor water quality has a direct impact on water quantity in a number of ways. Polluted water that cannot be used for drinking, bathing, industry, or agriculture effectively reduces the amount of useable water within a given area. Major water pollutants include microbes, nutrients, heavy metals, organic chemicals, oil, and sediments; heat, which raises the temperature of the receiving water, can also be a pollutant. Pollutants are typically the cause of major water quality degradation around the world.

Every day, 2 million tons of sewage and other effluents drain into the world's waters. Every year, more people die from unsafe water than from all forms of violence, including war. The most significant sources of water pollution are lack of inadequate treatment of human wastes and inadequately managed and treated industrial and agricultural wastes. The health impact has been estimated by the United Nations Environment Programme (UNEP) to put 323 million people are at risk from life-threatening diseases caused by the pollution of rivers and lakes [82]. According to the UNEP report, cholera, typhoid, and other deadly pathogens are increasing in more than half of the rivers in Africa, Asia, and Latin America. Salinity levels have also risen in nearly a third of waterways. Further, Asia has been the worst hit, with up to 50% of all rivers now affected by severe pathogen pollution caused by a cocktail of untreated wastewater disposal, agricultural pesticides run-off, and industrial pollution [82].

> Every day, 2 million tons of sewage and other effluents drain into the world's waters. Annually, more people die from unsafe water than from all forms of violence.

In general, the quality of water necessary for each human use varies, as do the criteria used to assess water quality. For example, the highest standards of purity are required for drinking water, whereas it is acceptable for water used in some industrial processes to be of less quality [87]. Poor water quality and human and ecological health are interrelated. As characterized by the WHO [88]:

- Lacking safe drinking water: almost 1 billion people lack access to an improved supply,
- Diarrheal disease: 2 million annual deaths attributable to unsafe water, sanitation, and hygiene,
- Cholera: more than 50 countries still report cholera to WHO,
- Cancer and tooth/skeletal damage: millions exposed to unsafe levels of naturally occurring arsenic and fluoride,
- Schistosomiasis[2]: an estimated 260 million infected,
- Emerging challenges: increasing the use of wastewater in agriculture is important for livelihood opportunities, but also associated with serious public health risks.

In recognition of this dimension of adverse effects, WHO offers guidance on their prevention:

- 4% of the global disease burden could be prevented by improving water supply, sanitation, and hygiene,

---

[2] "Schistosomiasis is a disease of poverty that leads to chronic ill-health. Infection is acquired when people come into contact with fresh water infested with the larval forms (cercariae) of parasitic blood flukes, known as schistosomes" [89].

- A growing evidence base is needed on how to target water quality improvements in order to maximize health benefits,

> WHO: 4% of the global disease burden could be prevented by improving water supply, sanitation, and hygiene,

- Better tools and procedures are needed to improve and protect drinking-water quality at the community and urban level, for example, through water safety plans,
- There is a need to make available simple and inexpensive approaches to treat and safely store water at the household level [88].

Perspective: As a matter of international environmental health policy, should the developed countries provide the resources necessary to prevent the global loss of life and illnesses attributable to waterborne diseases? The public health answer is obviously, yes. Altruistic reasons include children's welfare, improved quality of life, decreased disabilities, and increased longevity. Prevention of waterborne diseases would also improve national economic development by increased workforces, lower health care costs, improved social stability, and larger tax bases from employed workers and from the products they would produce.

3.7.2   UN AND US POLICIES TO PROTECT OCEAN WATERS

In addition to the subject of water quality and security of groundwater supplies and policies to protect human and ecological health, ocean waters have been and remain a vital resource for humankind. In fact, about 71% of the Earth's surface is water-covered, and the oceans hold about 96.5% of all Earth's water. Among many benefits of oceans, they are a source of marine life that is harvested for human food and oceans serve as avenues of marine commerce. But as discussed elsewhere in this book, contemporary factors are making serious, deleterious impacts on the globe's oceans and the life therein. Climate change, as noted in Chapter 1, has increased waters' temperature and caused further acidification, while, as noted in Volume 2, Chapter 2, plastics and other wastes dumped into ocean waters are adversely affecting marine life. As described in the following sections, both the UN and the US have adopted policies to protect ocean waters and their marine life.

3.7.2.1   UN Ocean Protection Policies

Turning to environmental policy, global policies have been implemented as means to both use and protect the Earth's oceans. As commented by the UN, "The oceans had long been subject to the freedom-of-the-seas doctrine—a principle put forth in the 17th century, essentially limiting national rights and jurisdiction over the oceans to a narrow belt of sea surrounding a nation's coastline. The remainder of the seas was proclaimed to be free to all and belonging to none. While this situation prevailed into the twentieth century, by mid-century there was an impetus to extend national claims over offshore resources" [90].

These considerations by the UN led to adoption of the 1982 Law of the Sea Convention. According to the UN, the Convention has resolved a number of important issues related to ocean usage and sovereignty, such as: established freedom-of-navigation rights, set territorial sea boundaries 12 miles offshore, set exclusive economic zones up to 200 miles offshore, set rules for extending continen-

> The UN's 1982 Law of the Sea Convention has resolved a number of important issues related to ocean usage and sovereignty.

tal shelf rights up to 350 miles offshore, created the International Seabed Authority, and created other conflict-resolution mechanisms (e.g., the UN Commission on the Limits of the Continental Shelf) [89].

Additionally, after more than 5 years of negotiations, UN members are poised to agree to draw up a new Law of the Sea rulebook by 2020, which could establish conservation areas, catch quotas, and scientific monitoring. The motion is supported by 140 nations, which is more than the two-thirds needed for passage. Delegates are drafting in 2019 a legally binding instrument on the conservation and sustainable use of marine biological resources [90].

3.7.2.2 U.S. Marine Debris Act, 2006

The NOAA Marine Debris Program is authorized by Congress to work on marine debris through the Marine Debris Act, signed into law in 2006 and amended in 2012 and 2018. The Act requires the program to "identify, determine sources of, assess, prevent, reduce, and remove marine debris and address the adverse impacts of marine debris on the economy of the U.S., marine environment, and navigation safety" [91].

In October 2018, President Donald Trump (R-NY) signed the Save our Seas Act of 2018. This law amends and reauthorizes the Marine Debris Act for 4 years; promotes international action to reduce marine debris in our ocean; authorizes cleanup and response actions needed as a result of severe marine debris events, such as hurricanes or tsunamis; and updates the membership of the Interagency Marine Debris Coordinating Committee. Additionally, the Act authorizes and requires NOAA to work with other Federal agencies to develop additional outreach and education strategies to address sources of marine debris [91].

ENFORCEMENT EXAMPLE

Miami, June 3, 2019: Carnival Corp. reached a settlement Monday with federal prosecutors in which the world's largest cruise line agreed to pay a $20 million penalty because its ships continued to pollute the oceans despite a previous criminal conviction aimed at curbing similar conduct [92].

3.8 WATER SECURITY: AVAILABILITY

Water security is essentially a two-prong subject. One prong is the quality of water; contaminated water is insecure in the sense of its presenting a potential health hazard. The second prong of water security is availability, for it is a matter of reality that the purest water in the world is useless to a thirsty population if the water is simply unavailable. So the subject of water availability comprises the content of this section of the chapter.

3.8.1   U.S. State of Water Availability

The U.S. has immense amounts of water. The country has an estimated 4.3% of the world's population yet contains more than 7% of global renewable freshwater resources. It is home to the largest freshwater lake system in the world, the Great Lakes, which holds 6 quadrillion gallons of water. And the Mississippi River flows at 4.4 million gallons per second at its mouth in New Orleans, which supplies water to about 15 million people along its way [93]. As of 2015, the U.S. uses 322 billion gallons of water per day (Bgal/day). The three largest water-use categories were irrigation (118 Bgal/day), thermoelectric power (133 Bgal/day), and public supply (39 Bgal/day), cumulatively accounting for 90% of the national total [94].

> Nationwide, the amount of water that is lost each year in the U.S. is estimated to top 2 trillion gallons.

Unfortunately, especially in a time of climate change and increasing human population, much of the U.S. water availability is lost to breaks in an aging water delivery infrastructure. Nationwide, the amount of water that is lost each year is estimated to top 2 trillion gallons, according to the American Water Works Association. That's about 14%–18% (or one-sixth) of the water the nation treats for drinking water and industrial purposes [95]. Where does it go? Much of it just leaks away because of aging and leaky pipes, broken water mains, and faulty meters. The loss of water translates into a need for replacement. While U.S. freshwater supplies remain plentiful, drilling for groundwater supplies has been shown by researchers to require deeper drilling, suggesting lesser available water in aquifers [96]. This observation reinforces the importance of water conservation.

> As of 2015, the U.S. uses 322 billion gallons of water per day (Bgal/day). The three largest water-use categories were irrigation, thermoelectric power, and public supply.

Fixing the U.S. water delivery infrastructure would be enormously costly. The American Water Works Association estimates a cost of $1 trillion, about half of that amount would be to replace existing infrastructure. The other half will be putting into the ground new infrastructure to serve population growth and areas that currently aren't receiving water [95]. To put this figure in perspective, the Congressional Budget Office estimated the federal government would spend $4.109 trillion in the fiscal year 2018 to fund all federal departments and programs and social support commitments.

3.8.2   Global State of Water Availability

The WHO observes that global freshwater consumption rose six-fold between 1900 and 1995—at more than twice the rate of population growth, and that for many of the world's poor, one of the greatest environmental threats to health is lack of access to safe water and sanitation. The agency estimates that more than 1 billion people globally lack access to safe drinking-water supplies, while 2.6 billion lack adequate sanitation; diseases related to unsafe water, sanitation, and hygiene result in an estimated

1.7 million deaths every year [85]. A second UN agency, the Food and Agriculture Organization (FAO), estimated in 2020 that more than 3 billion people live in agricultural areas with high to very high levels of water shortages and scarcity [97]. The agency asserts that "improved water management, supported by effective governance and strong institutions—including secure water tenure and rights, underpinned by sound water accounting and auditing—will be essential to ensure global food security and nutrition."

Further, according to the UN, declining water quality has become a global issue of concern as human populations grow, industrial and agricultural activities expand, and climate change threatens to cause major alterations to the hydrological cycle [85]. Globally, the most prevalent water quality problem is eutrophication,[3] a result of high-nutrient loads (mainly phosphorus and nitrogen), which substantially impairs beneficial uses of water, i.e., its availability. Major nutrient sources include agricultural runoff, domestic sewage (also a source of microbial pollution), industrial effluents, and atmospheric inputs from fossil fuel burning and bush fires. Lakes and reservoirs are particularly susceptible to the negative impacts of eutrophication because of their complex dynamics, relatively longer water residence times, and their role as an integrating sink for pollutants from their drainage basins.

In 2019 the World Resources Institute published a report on global water availability [98]. According to their analysis, countries that are home to one-fourth of Earth's population face an increasingly urgent risk: the prospect of running out of water. From India to Iran to Botswana, 17 countries around the world are currently under extremely high water stress, meaning they are using almost all the water they possess. Many are arid countries to begin with; some are squandering what water they have. Several are relying too heavily on groundwater, which instead they should be replenishing and saving for times of drought, according to the Institute. In those countries are several big, thirsty cities that have faced acute shortages in 2018–2019, including São Paulo, Brazil; Chennai, India; and Cape Town.

> World Resources Institute: Countries that are home to one-fourth of Earth's population face an increasingly urgent risk: the prospect of running out of water.

The Institute further observed that climate change heightens the risk of water shortage. As rainfall becomes more erratic, the water supply becomes less reliable. At the same time, as the days grow hotter, more water evaporates from reservoirs just as demand for water increases [98].

Relatively unrecognized as a matter of water security is the plight of some of the world's largest lakes [99]. Warming climates, drought, and overuse are draining crucial water sources, threatening habitats and cultures. In particular, climate change is warming many lakes faster than it is warming the oceans and the ambient air. This heat accelerates water evaporation, along with other anthropogenic factors, to intensify water shortages, pollution, and loss of habitat for birds and fish. As an example, East Africa's Lake Tanganyika has warmed so much that fish catches that

---

[3] The process by which a body of water becomes enriched in dissolved nutrients (such as phosphates) that stimulate the growth of aquatic plant life usually resulting in the depletion of dissolved oxygen.

feed millions of poor people in four surrounding countries are at risk. Of all the challenges lakes face in a warming world, the starkest examples are in closed drainage basins where waters flow into lakes but don't exit into rivers or a sea [99]. Lake Chad in Africa has shrunk to a shell of its former self since the 1960s, heightening shortages of fish and irrigation water. As global air temperatures continue to climb absent global efforts to mitigate climate change, loss of lakes will continue to increase, with the consequent impacts on human and ecological health.

3.8.3 THIRSTY CITIES AND WATER INSECURITY

The UN has observed that in 2016, an estimated 54.5% of the world's population lived in urban settlements. By 2030, urban areas were projected to house 60% of people globally and one in every three people will live in cities with at least half a million inhabitants. Urban living presents many benefits to its residents as well as consequential challengers to both residents and urban planners and authorities. Not the least of urban challenges is the provision of potable water and wastewater management. Regarding the provision of water to urban dwellers, the UN notes that humans use about 4,600 km^3 of water every year, of which 70% goes to agriculture, 20% to industry, and 10% to households [100]. The UN water analysis concluded that more than 5 billion people could suffer water shortages by 2050 due to climate change, increased demand, and polluted supplies. This troubling prediction has in 2019 already begun to reveal itself in water shortages in a number of the world's cities, as illustrated in the following.

> The UN notes that humans use about 4,600 km^3 of water every year, of which 70% goes to agriculture, 20% to industry, and 10% to households.

- **Beijing, China:** After depleting groundwater reserves, China's capital is becoming increasingly reliant on water pumped from the country's flood-prone South. Currently, China's South-to-North water diversion project is providing the bulk of Beijing's water supply. The project, which is two-thirds complete, comprises three massive aqueducts [101].
- **Cape Town, South Africa:** In 2015, Cape Town began experiencing a 3-year drought, the worst in a century, necessitating the imposition of residents' water restrictions. Residents were asked to curb the daily amount of municipal water to just 50 L. The city upgraded its water systems and implemented desalination and water-recycling projects during and subsequent to the drought [102].
- **Delhi, India:** A report in June 2019 from a government-chartered think tank warned that Delhi, along with 20 other Indian cities, could reach "zero groundwater levels" by 2020. Collectively, water users in the region are extracting groundwater faster than the natural recharge rate. The city has begun water conservation policies [103].

The water availability problems of these three cities are illustrative of similar problems in other cities. However, some cities have successfully implemented policies and practices that promote sustainable water security. As observed by the World Bank, a number of cities and states, such as Amman, Malta, Windhoek, Los Angeles, Singapore, and Perth, have had substantial success tackling physical water scarcity by (1) diversifying water resources and supply strategies to hedge against risks associated with their depletion or pollution, and/or (2) closing the urban water cycle to increase its resilience to external climate shocks, for example by managing treated wastewater or seawater as alternative resources, capturing rainwater for onsite use or stormwater to replenish aquifers, and using or treating brackish groundwater for specific uses [104].

3.8.4 DROUGHT AND WATER INSECURITY

Drought is the second environmental threat to global water security. A drought is an extended period of dry weather caused by a lack of rain or snow. As temperatures rise due to global climate change, more moisture evaporates from land and water, leaving less water behind. Climate change affects a variety of factors associated with drought. According to the Union of Concerned Scientists, drought ranked second in terms of national weather-related economic impacts in 2015, with annual losses nearing $9 billion per year in the U.S. [105]. Beyond direct economic impacts, drought can threaten drinking water supplies and ecosystems, and can even contribute to increased food prices. Within the last decade, drought conditions have hit the Southeastern U.S., the Midwest, and the Western U.S. In 2011, Texas had the driest year since 1895. In 2013, California had the driest year on record [105].

In July 2016, the U.S. drought monitor (Drought.gov) estimated that 22% of the contiguous U.S. was experiencing moderate to extreme drought.

There are different types of drought. One form is hydrological drought, or how decreased precipitation affects stream flow, soil moisture, reservoir, and lake levels, and groundwater recharge. Farmers are most concerned with agricultural drought, which occurs when available water supplies are not able to meet crop water demands. Agricultural droughts can occur for a variety of reasons, including low precipitation, the timing of water availability, or decreased access to water supplies. For instance, earlier snowmelt might not change the total quantity of water available but can lead to earlier runoff that is out of phase with peak water demand in the summer. Thus, it is possible to suffer an agricultural drought in the absence of a meteorological drought [105]. Illustrated in Figure 3.4 is an area of land that has deteriorated due to drought. Notable is the relative absence of plants and the presence of cracked, bare soil.

3.8.5 STATE OF CALIFORNIA'S DROUGHT POLICIES

California has experienced a severe drought for several years. In response, the government of California implemented policies that were intended to lessen the burden of water shortage. In April 2015, following 4 years of drought conditions, California's governor issued an executive order that cut urban potable water use by

FIGURE 3.4   Example of a drought effect on soil. (CDC [77].)

25% statewide. Implementing the executive order became the responsibility of the State Water Resources Control Board. In the board's first proposal, a community was placed in one of four tiers based on how much water it used in September 2014, which was later based on how much water communities used during July, August, and September 2014. Further, the original set of four tiers, with cuts ranging from 10% to 35%, was expanded to nine tiers, with cuts ranging from 4% to 36%.

Failure to meet water conservation goals would incur a penalty of up to $100,000 per day, although most water agencies were reluctant to impose fines, preferring educational outreach and warnings to their customers [106]. In October 2015, California state officials commenced imposing fines on water systems that failed to meet water conservation goals. While announcing that the state overall met its monthly conservation goals in September 2014, officials said Beverly Hills, Indio, Redlands, and the Coachella Valley Water District missed their mandates by wide margins. Each was fined $61,000 [107].

In November 2015, California's governor extended his executive order that mandated a 25% reduction in water usage across the state. The original order, issued in April, could now be extended until October 2016 if the drought persisted through January 2016. Although the 25% state goal was not achieved, some water conservation did occur during 2015. The state's experience revealed that some policy elements were more effective than others. For example, asking homeowners to voluntarily conserve water was ineffective, but one of the state's most successful water-reduction efforts was rebates to encourage homeowners to tear out their grass lawns in favor of artificial turf or desert-friendly native plants [108].

Due to a welcome wet winter in California, the state's governor revised his water conservation executive order. On May 9, 2016, he ordered state water regulators to extend some drought protections, such as a prohibition on irrigating lawns and landscapes so intensely that water runs down the sidewalk or into the street. He also demanded a new plan for making conservation a way of life over the long term. But the governor's order did not include an extension of the mandatory 25% water use cutback he had ordered in 2015 [109,110]. In early 2016, a strong series of storms left parts of Northern California rehydrated, with reservoirs full. But for Southern California, no drought relief arrived. To deal with the dichotomy, the State Water

> In November 2015, California's governor extended his executive order that mandated a 25% reduction in water usage across the state.

Resources Control Board ruled on May 18, 2016, that local water districts were allowed to set their own savings targets based on water supply and demand forecasts tailored to their areas [111].

3.8.6 Climate Change's Impact on Water Security

Freshwater is expected to become increasingly scarce in the future, and this is partly due to climate change. The distribution of water on the planet is approximately 98% salty and only 2% is fresh. Of that 2%, almost 70% is snow and ice, 30% is groundwater, less than 0.5% is surface water (lakes, rivers, etc.), and less than 0.05% is in the atmosphere [112].

Climate change has several effects on these proportions on a global scale. The main one is that warming causes polar ice to melt into the sea, which turns freshwater into seawater, although this has a little direct effect on water supply. Another effect of warming is to increase the amount of water that the atmosphere can hold, which in turn can lead to more and heavier rainfall when the air cools. Although more rainfall can add to freshwater resources, heavier rainfall leads to more rapid movement of water from the atmosphere back to the oceans, reducing our ability to store and use it. Warmer air also means that snowfall is replaced by rainfall and evaporation rates tend to increase.

> The IPCC special report on climate change adaptation estimates that around 1 billion people in dry regions might face increasing water scarcity.

Yet another impact of higher temperatures is the melting of inland glaciers. This will increase water supply to rivers and lakes in the short to medium term, but this will cease once these glaciers have melted. In the subtropics, climate change is likely to lead to reduced rainfall in what are already dry regions. The overall effect is an intensification of the water cycle that causes more extreme floods and droughts globally [112].

Global climate change affects a variety of factors associated with drought. There is high confidence that increased temperatures will lead to more precipitation falling as rain rather than snow, earlier snowmelt, and increased evaporation and transpiration. Thus, the risk of hydrological and agricultural drought increases as temperatures rise. Much of the U.S. Mountain West has experienced declines in spring snowpack, especially since mid-century. These declines are related to a reduction in precipitation falling as snow (with more falling as rain) and a shift in the timing of snowmelt. Earlier snowmelt, associated with warmer temperatures, can lead to water supply being increasingly out of phase with water demands. While there is some variability in the models for western North America as a whole, climate models unanimously project increased drought in the U.S. Southwest. The Southwest is considered one of the more sensitive regions in the world for the increased risk of drought caused by climate change [105].

In July 2016, the U.S. drought monitor (Drought.gov) estimated that 22% of the contiguous U.S. was experiencing moderate to extreme drought, with the U.S. Southwest experiencing the most extreme drought conditions. Global data on the prevalence of drought are scarce. But one example will demonstrate one country's drought struggles. In January 2014, the Government of Kenya declared an impending drought with an estimated 1.6 million people affected. After a poor performance of the long rains between March and May 2014 in the arid and semiarid zones, the drought situation continued to affect both pastoral and marginal agriculture livelihood zones with an impact on households' food availability as well as livestock productivity. These conditions in Kenya have not eased, causing food shortages and inadequate access to potable water supplies [113].

The Intergovernmental Panel on Climate Change (IPCC) cautions that the global picture is less important than the effect of warming on freshwater availability in individual regions and in individual seasons. An IPCC technical report on climate change and water concludes that, despite global increases in rainfall, many dry regions, including the Mediterranean and southern Africa, will suffer badly from reduced rainfall and increased evaporation [112]. As a result, the IPCC special report on climate change adaptation estimates that around 1 billion people in dry regions might face increasing water scarcity. Of special import, a report from UNICEF in 2017 forecast that one in four of the world's children will be living in areas with extremely limited water resources by 2040. The agency estimates that 600 million children will reside in regions enduring extreme water stress as a consequence of climate change [114].

> UNICEF in 2017 forecast that one in four of the world's children will be living in areas with extremely limited water resources by 2040.

Knowing that humankind and many other animals with which we share the planet all rely on freshwater for existence, it becomes important to forecast the potential impact of climate change on freshwater supplies [115]. In that regard, scientists from the UK, Australia, Canada, Germany, France, and the U.S. as a team used modeling and hydrological data sets to estimate the impact of climate change on aquifers [116]. Of Earth's 7.5 billion people in 2019, 2 billion rely on groundwater stored in underground deposits called aquifers. The researchers found that climate-related changes to rainfall in the next century will make it harder for 44% of the world's aquifers to recharge, particularly the shallower ones, which are currently relied on to fill up faster. That means within the next 100 years, nearly half the world's groundwater supply will become less reliable [116].

In addition to the implications of climate change on water security, there are implications of climate change on water quality. For example, one study examined the effect of temperature increase and rainfall on water quality [117]. Investigators from the Carnegie Institution for Science spent several years studying the effects of nitrogen runoff and how expected changes in precipitation patterns due to climate change could lead to greater risks to water quality. Nitrogen from agriculture and other human activities washes into waterways, which, in excess, creates a dangerous phenomenon called eutrophication. They found that climate was a key factor in how much nitrogen ended up in the water system, with warming springtime temperatures and accompanying storms having a direct impact on nitrogen runoff. The researchers observed that simply decreasing the amount of nitrogen released by industry and farming will not be enough. The decreased usage will have to be enough to offset greater risk due to increased precipitation, too.

3.8.7 WATER SECURITY AND PRODUCED WATER

Increased human population, climate change, and economic development are among several factors that have resulted in strained supplies of water. This need translates into using some nontraditional sources of water. One such source is produced water, which is water trapped in underground formations that are brought to the surface during oil and gas exploration and production. Because the water has been in contact with the hydrocarbon-bearing formation for centuries, it has some of the chemical characteristics of the formation and the hydrocarbon itself. It may include water from the reservoir, water injected into the formation, and any chemicals added during the drilling, production, and treatment processes. Produced water can also be called "brine," "saltwater," or "formation water."

The physical and chemical properties of produced water vary considerably depending on the geographic location of the field, the geological formation from which it comes, and the type of hydrocarbon product being produced. Produced water properties and volume can even vary throughout the lifetime of a reservoir. The major constituents of interest in produced water are salt, oil and grease, various inorganic and organic chemicals, and naturally occurring radioactive material. Generally, the radiation levels in produced water are very low and pose no health risk. Most produced water needs some form of treatment before it can be used. The levels of specific constituents found in a particular produced water sample and the desired type of reuse will determine the types of treatment that are necessary.

Produced water is by far the largest volume byproduct stream associated with oil and gas exploration and production. Approximately 21 billion barrels (1 bbl = 42 U.S. gallons) of produced water are generated annually in the U.S. from about 900,000 wells. This is equivalent to a volume of 2.4 billion gallons per day. For perspective, the Denver Water agency, which supplies drinking water to 1.3 million customers, has a combined total capacity of approximately 745 million gallons per day. Several western U.S. states are using treated produced water for beneficial purposes. Some of these uses include domestic, livestock watering, industrial, and commercial, agriculture irrigation, mining, fire protection, and dust suppression. The determination of a specific beneficial use depends on federal and state jurisdiction, and the circumstances of each case. As an example, five states (Colorado, Montana, New Mexico, Utah, Wyoming) use treated produced water for various domestic purposes such as irrigation of lawns and crops [118].

3.8.8 Water Security and Graywater

The reuse of graywater is an emerging environmental policy issue. Graywater is generally defined as all wastewater generated from household activities except that produced from toilets, which is called "blackwater." Graywater includes water from dishwashers, clothes washers, household washbasins, showers, and bathtubs. Such "waste" can be collected and used for outdoor watering of plants, trees, lawns, and irrigation of crops. The average amount of graywater produced in the U.S. is 40 gallons per day per person, which equates to about 65% of a household's daily water consumption [119]. This is a significant amount of water that is potentially available for recycling. Graywater is important because several U.S. states are considering its use as a component of water conservation programs. These programs are largely nascent and are in response to shortages of water supplies needed to meet the needs of households, industry, and agriculture. The causes of the shortages vary, but factors include increased human populations, fragile groundwater supplies, greater water demand by industry, and drought conditions. Making maximum use of existing water supplies is an environmental policy that will become increasingly important in many countries, including the U.S., as climate changes due to greenhouse gas emissions continue to appear.

> The average amount of graywater produced in the U.S. is 40 gallons per day per person, which equates to about 65% of a household's daily water consumption.

One state, Colorado, with a pending water shortfall is considering graywater as part of a state plan for water security. More specifically, Colorado water providers, facing a shortfall of 163 billion gallons, are developing the first statewide water plan to sustain population and industrial growth. State water planners hypothesize that if even the worst sewage could be cleaned to the point where it is potable—filtered through super-fine membranes or constructed wetlands, treated with chemicals, irradiated with ultraviolet rays—then the state's dwindling aquifers and rivers could be saved [120].

Similar to Colorado, California water managers are incorporating recycled water into their water security plans. In the last couple of decades, some coastal water managers have attempted to recycle some of this water for human use. So-called purple pipe systems take sewage that has been filtered and cleansed and use it to irrigate crops, parks, and golf courses. This water, however, is not currently used as drinking water. The Metropolitan Water District of Southern California is considering developing what could be one of the largest recycled water programs in the world [121].

> For areas short of water, the use of recycled water (graywater) will predictably become more common.

For areas short of water, the use of recycled water (graywater) will predictably become more common, even given the $1 billion cost of wastewater treatment plants adequate to produce potable water [121].

3.8.9   WATER SECURITY AND DESALINATION

Desalination of ocean water for conversion into drinking water supplies is another method for water security in some geographic areas where freshwater is limited. According to an MIT report, already, some 700 million people worldwide suffer from water scarcity, but that number is expected to swell to 1.8 billion in just 10 years. Some countries, like Israel, already rely heavily on desalination; more will follow suit [122]. The current costs of constructing a desalinization plant are enormous, as is the huge amount of energy used for desalination. For example, in 2015 the Western Hemisphere's largest ocean desalination plant opened in Carlsbad, San Diego County, California [123]. The plant is capable of producing 50 million gallons of freshwater daily, about 10% of the county's total water use. The plant's cost was approximately $1 billion, with annual operating costs to be determined. The State of California is considering the construction of an additional 15 desalination plants, given the state's water shortage and dependence. The costs of construction and operation of desalination plants are forecast to decrease as improved technology and more energy-efficient equipment are developed.

As noted by the United Nations Educational, Scientific and Cultural Organization (UNESCO) in 2014, there were more than 14,000 desalination plants in more than 150 countries—exemplifying the growing reliance on this technology. About 50% of this capacity exists in the West Asia Gulf region, while North America has about 17%, Asia (apart from the Gulf) about 10%, and North Africa and Europe about 8% and 7%, respectively [124].

On a much smaller scale than desalination plants, research is ongoing for the purpose of developing less expensive methods for producing potable water. For example, researchers at Alexandria University in Egypt have developed a procedure called pervaporation to remove the salt from seawater and make it potable. Specially made synthetic membranes are used to filter out large salt particles and impurities and then the rest is heated, vaporized, and condensed back into clean water. Crucially, the membranes are easily fabricated from inexpensive materials that are available locally, and the vaporization part of the process can rely on solar power. This means the new method is both inexpensive and suitable for areas without a regular power supply—both factors that are very important for low-income countries [125].

> UNESCO: In 2014, there were more than 14,000 desalination plants in more than 150 countries. About 50% of this capacity exists in the West Asia Gulf region [121].

Perspective: Inadequate water security has become a reality for areas afflicted because of climate change and other factors, including unwise use of freshwater sources, population increase, and irrigation of agriculture. Water security policies will need to be implemented for both afflicted and nonimpacted areas, as a matter of current and future urgency. For example, water restrictions in areas of protracted drought, development of recycled water treatment resources, use of produced water, and construction of desalination plants are examples of policies being contemplated or in actual practice.

3.9   WATER CONSERVATION AND INNOVATIVE TECHNOLOGY

As data in this chapter establish, supplies of freshwater are not meeting human and ecological requirements. Climate change, population grown, agricultural irrigation, and wasteful individual behaviors (e.g., failure to fix leaky faucets) are factors contributing to this shortage.

The current shortfall of freshwater will worsen with the effects of climate change and population growth.

3.9.1   WATER CONSERVATION

Wasting of water by individual households is a major factor contributing to shortfalls of available water. For example, EPA estimates that the U.S. population needlessly wastes 1 trillion gallons of water annually. The waste is due to the leaky kitchen and bathroom faucets, malfunctioning toilets, errant sprinkler systems, and such. The loss equals the annual household water use of more than 11 million U.S. homes, according to the EPA [126].

The EPA further notes that the U.S. population has doubled over the past 50 years, while the country's thirst for water has tripled. With at least 40 states anticipating water shortages by 2024, the need to conserve water is critical [127]. As an example of successful water conservation policies and outcomes, the state of California implemented mandated water conservation policies during a statewide drought that occurred in 2012–2016. That episode of drought was discussed in a previous section of this chapter. But as a subject of water conservation here, one can ask if the state's water conservation policies and outcomes have been sustained following the end of the drought. As observed by one reviewer of the state's draught policies, it was noted that the historic dry spell in California had prompted many state residents to reduce their water consumption, as did strict regulations imposed by state agencies and individual water districts [128]. Whether they wanted to or not, urban Californians reduced their use of the state's most precious resource by about 25%.

> EPA estimates that the U.S. population needlessly wastes 1 trillion gallons of water annually.

But with the end of the drought and the end of mandated water conservation rules, did water conservation continue? After mandatory water conservation targets were lifted in April 2017 following a very wet winter, many Californians continued using less water than they were prior to the drought. In Sacramento, Los Angeles, most of the San Francisco Bay Area, and Orange County, urban residential water use was down between 20% and 26% since 2013, often used by water agencies as the benchmark year for pre-drought water consumption, according to the State Water Resources Control Board. However, a report from the State Water Resources Control Board shows statewide savings on urban water use for June 2017 totaled 17%, a figure that suggests a rebound in water use. California water conservation officials forecast that water conservation will continue due to a public better educated in water use efficiency that many Californians had installed more efficient appliances during the drought and replaced water-hogging lawns with drought-tolerant California native plants—actions that should translate into permanent water conservation savings [128].

A second example of a successful water conservation policy comes from Kansas [129]. As background, in 2013, a band of farmers in northwest Kansas decided that pumping prodigious volumes of water from the Ogallala Aquifer was a path to ruin. The vast Ogallala, an underground reserve stretching from South Dakota to Texas, was shrinking. If Sheridan County farmers kept pumping, their piece of the aquifer might effectively be tapped before their heirs had a chance to work land that families revered. So the farmers decided to use less water. Taking advantage of a new state law, they agreed to cut water withdrawals by 20% per year through 2017. They saw self-restraint as a test of their farming skills.

> Five years of water conservation in Sheridan County, Kansas, cut the rate of decline in groundwater levels by two-thirds.

The Kansas Geological Survey found that the 5 years of conservation in Sheridan County had cut the rate of decline in groundwater levels by two-thirds. If the aquifer had been falling by 1 foot per year, it was now losing only 4 inches. Further, studies by Kansas State University compared farm production within the water conservation area with production in a three-mile radius outside its boundaries. The control group, which was of a similar geographic size to the water conservation area, was not restricted in its water use. Findings showed that corn producers within the conservation area used 23% less groundwater than their nonrestricted neighbors. Further, corn yields for farmers within the conservation area were only 1.2% less than those outside the conservation boundary while cash flow was 4.3% greater. The Sheridan farmers had produced a little less corn but used a lot less water and had significantly fewer water input costs.

3.9.2 U.S. ENERGY POLICY ACT, 1992

U.S. households use a large amount of water daily. Water used for flushing toilets and showers can amount to significant volumes of utilized water. Low-flow toilet and shower fixtures are one means for conserving water. In 1992, Congress overwhelmingly passed—and President George H.W. Bush (R-TX) signed the Energy Policy Act (EPAct), which among other energy-related policies, set national water efficiency standards for toilets, faucets, and showerheads, as established by EPA regulations. Those standards went into effect in 1994. Since then, new toilets sold in the U.S. must not exceed 1.6 gallons per flush (gpf); faucets are capped at 2.5 gallons per minute (gpm), and showerheads may not spray more than 2.5 gpm. Toilets installed prior to 1994 used an average of 3.5 gallons per flush, more than double that of toilets currently allowed. As to water conservation, the EPAct's standards have conserved more than 18 trillion gallons of water through 2014 due to more efficient toilets alone, according to an estimate by the Alliance for Water Efficiency [130]. The same source estimates that the total cost savings from federal water efficiency standards—including savings from reducing the amount of energy to heat water—is more than $100 per household every year.

In 2010, the U.S. Department of Energy waived federal preemption of the national standards for toilets, faucets, and showerheads so that states could set stronger standards. Several have. For example, Georgia legislators responded in 2012 to the state's low water resources, in part from a drought in 2007. The General Assembly enacted—and then-Governor Sonny Perdue signed—a law in response that included standards more stringent than EPA's for toilets (1.28 gpf) and building code requirements for faucets (1.5 gpm for bathrooms, and 2 gpm for kitchens) [130].

Perspective: It is interesting to note in both a conservative political policy and a public health context the responses to the EPAct's water conservation standards. In a public health context, water conservation is vital during periods of drought and other episodes of water insecurity. In a conservative policy context, federal government regulations on toilets, showerheads, and faucets are deemed outside the province of the government's responsibility.

3.9.3 INNOVATIVE WATER TECHNOLOGY

The state of global water insecurity has been painted here as a dire consequence of climate change, population growth, and other factors. And as discussed in Chapter 1, absent global commitment to mitigate climate change, water insecurity will be but one of several global threats to human well-being. Having said what some might consider hyperbole, there are some bright lights in the development of technology that is intended to aid in better water management and security. The following examples will illustrate some of this technology.

- An Ohio company has patented a perpetual use water system that recycles wastewater into potable water [131]. The system pumps wastewater through a filtration system that removes solids and biological organisms. Then ultraviolet light and reverse-osmosis membranes filter and sterilize the water, resulting in drinking water of ultra-purity, then chlorinated and held for everyday use.
- Smoothing out the rough patches of a material widely used to filter saltwater could make producing freshwater more affordable. Researchers have made a super-smooth version without the divots that trap troublesome particles. That could cut costs for producing freshwater, making desalination more broadly accessible [132].
- A university research team is testing ferrate, an ion of iron, as a replacement for several water treatment steps [108]. First, ferrate kills bacteria in the water. Next, it breaks down carbon-based chemical contaminants into smaller, less harmful molecules. Finally, it makes ions like manganese less soluble in water so they are easier to filter out. With its multifaceted effects, ferrate could potentially streamline the drinking water treatment process or reduce the use of chemicals, such as chlorine.

Scientists have improved how a crop uses water by 25% without compromising yield by altering the expression of one gene.

- By draping black, carbon-dipped paper in a triangular shape and using it to both absorb and vaporize water, university researchers have developed a method for using sunlight to generate clean water with near-perfect efficiency. The low-cost technology could provide drinking water in regions where resources are scarce, or where natural disasters have struck [133].
- Scientists have improved how a crop uses water by 25% without compromising yield by altering the expression of one gene that is found in all plants. The team improved the plant's water-use-efficiency—the ratio of CO_2 entering the plant to water escaping—by 25% without significantly sacrificing photosynthesis or yield in real-world field trials. This result could translate into the need for less freshwater used in agriculture [134].

Perspective: These examples of successful management of water resources illustrate the feasibility of water conservation policies. While the methods of water conservation surely varied between EPA's water management in office building and laboratories and the Kansas farmers' management of water used for irrigation and other farming activities, there seems a necessary common factor in all three examples of successful water conservation: education, to wit: Individuals and organizations were educated about the need and reasons for water conservation, so they accepted the adoption of water conservations policies and methods. State and local health departments can and should play a role in providing education about the human health implications of water insecurity.

The examples of new technology that address aspects of water insecurity illustrate how engineering and science can aid in achieving improved water security. But some caution must be expressed. Technology alone will not abate the consequences of climate change; only global policies and actions to reduce greenhouse gases will suffice. The abatement actions will surely require visionary technology such as renewable sources of energy and new methods of transportation, but this role of technology requires sustained support of climate change mitigation policies.

3.9.4 BEST WATER PRACTICES POLICIES

The EPA has implemented its own best water practices that apply to its programs. It is noteworthy that about two-thirds of water used in the agency's office buildings is for sanitary and heating/cooling purposes. Following are the top 10 water best management practices that EPA has implemented to reduce water use at its own office buildings and laboratories: meter/measure/manage, optimize cooling towers, replace restroom fixtures, eliminate single-pass cooling, use water-smart landscaping and irrigation, reduce steam sterilizer tempering water use, reuse culture water, control reverse osmosis system operation, recover rainwater, and recover air handler condensate [127]. In fiscal year (FY) 2014, EPA reduced its water use by 40.4% compared to an FY 2007 baseline, exceeding the water intensity reductions required by Executive Order (EO) 13514. This is an example of water conservation achieved by a large user of water resources.

Another example of a federal policy on water conservation is found in President Obama's Executive Order (EO) 13514, issued in 2009, that sets sustainability goals for U.S. federal agencies and focuses on making improvements in their environmental, energy, and economic performance. One goal is a 26% improvement in water efficiency by 2020 [135]. This goal was set aside by the Trump administration.

3.9.5   EXAMPLES OF SUCCESSFUL WATER MANAGEMENT

Although the image of water insecurity is grim, there are other images of successful outcomes in water management and security. For example, agriculture has emerged as the biggest threat to water quality in many parts of the U.S. The nutrients phosphorus and nitrogen from manure and synthetic fertilizers are causing problems in the Midwest, and elsewhere. A Midwest farm association is promoting new conservation strategies, such as building artificial wetlands and underground "bioreactors" to capture nutrients in drainage systems. One farmer uses rapeseed and rye plants in his cornfields to absorb nutrients that might otherwise find their way into a local river, a runoff that would contribute to algal contamination in adjacent lakes. These and similar agriculture strategies will lessen water pollution attributable to agricultural runoff of fertilizers and nutrients that in turn contribute to water pollution [136].

> A farmer uses rapeseed and rye plants in his cornfields to absorb nutrients that might otherwise find their way into a local river.

Another example illustrates that polluted waterways can be renewed as secure water sources. The Harbor Raritan Estuary between New York and New Jersey is the most densely developed urban estuary in the U.S. Since 2009, more than $1 billion of federal, state, and local funds have been invested in projects of conservation, restoration, and development of publicly accessible waterfront spaces. To date, this effort has resulted in more than 80 habitat restoration and land conservation projects. The improvements through 2015 include the restoration of more than 200 acres of wetlands and the creation or enhancement of more than 500 acres of new parks and public spaces [137].

In a further example, Medford, Oregon's wastewater treatment plant was discharging warm water into the Rogue River in violation of EPA regulations. But instead of spending millions on expensive machinery to cool the water to federal standards, the city is planting trees. Shade trees cool rivers, and the end goal is 10–15 miles of new native vegetation along the Rogue River [138].

These examples suggest that policies of water quality protection and/or restoration can help in reducing the magnitude of water insecurity, but require time and support from the public and policymakers alike.

3.9.6   THE "ONE WATER" MOVEMENT

A growing movement worldwide—called "One Water" in the U.S.—intends to manage water more holistically. The idea represents a shift from the standard separate management of drinking water, wastewater, stormwater, and water for the environment [139]. Advocates proclaim that One Water is about managing water to achieve multiple benefits, citing myriad ways people are experimenting with managing water in an integrated way to meet multiple needs simultaneously—such as providing water for ecosystems, flood control, drinking, and irrigation. As an example, some wastewater utilities using the One Water frame of reference are seeing new value in their effluent. Some are cleaning it to standards safe enough for some kinds of reuse, such as watering landscaping. And they are beginning to recover valuable products from sewage, such as biosolids for fertilizer. As demands for freshwater continue to increase due to climate change and other forces, the One Water movement could become a social policy adopted by concerned policymakers.

> Advocates proclaim that the One Water movement is about managing water to achieve multiple benefits.

3.9.7 CALIFORNIA WATER FUTURES MARKETING

Beginning in December 2020, water joined gold, oil, and other commodities traded on the Wall Street stock exchange, highlighting worries that the life-sustaining natural resource may become scarce across more of the world. Farmers, hedge funds, and municipalities alike are now able to hedge against—or bet on—future water availability in California, the biggest U.S. agriculture market and world's fifth-largest economy. CME Group Inc.'s January 2021 contract, linked to California's $1.1 billion spot water market, traded on December 6, 2020, at 496 index points, equal to $496 per acre-foot [140]. Making water a traded commodity like oil will promote water conservation and development of new supplies.

3.10 IMPACT OF THE COVID-19 PANDEMIC

While restaurants, gyms, schools, and other buildings are closed indefinitely to prevent the spread of COVID-19, water left sitting in pipes could change in quality. It's possible that water left sitting for long periods of time could contain excessive amounts of heavy metals and pathogens concentrated in pipes nationwide, say researchers who have begun a field study on the impact of a pandemic shutdown on buildings. Reopening building will necessitate assessing water quality prior to use by occupants.

3.11 HAZARD INTERVENTIONS

As described in this chapter, water quality and security are threatened in many parts of the globe, and unabated climate change and continued population growth will exacerbate the problem. Interventions to reduce or interdict the hazard of inadequate water quality will be required. These include global, regional, national, local, and individual policies for the protection of water security. Environmental health water policies such as those described for the U.S. and the EU are examples. Other examples are water policies being set in place in countries with emerging national economies, such as China and India. But any of these water protective policies are only as effective as their implementation. Strong policies require strong implementation and follow-through.

Water conservation policies and use of unconventional sources of water, for example, produced or recycled water, must be made part of the water security calculus. More efficient uses of water in agriculture and industrial operations must also be part of the calculus, as exemplified by some examples cited in this chapter. An interesting source of unconventional water is the invention of hydropanels that generate water from the air and solar power, even when humidity is as low as 10%. The developer asserts the product uses solar thermal energy to heat the surrounding air, and using solar-powered fans to move the heated air over a hygroscopic material inside the hydropanel, which only attracts pure water molecules, thence collected in a reservoir. The product has been used to produce water in an insecure water village in Belize [141].

Individuals must assume their responsibilities for personal policies of water quality protection. Persons should know and keep track of their water expenditures. In industrialized countries, how much water is used for household purposes, for example, laundry, dishes, baths, and lawn care should be subject to water conservation. Eschewing water service at restaurants is another policy. And most importantly, controlling the amount of water used for toilet flushing can be achieved by reducing the frequency of flushes. In geographic areas where water security is an extant problem, individuals should be educated on the importance of water conservation and the application of some innovative methods (e.g., rain collection devices) for water conservation.

3.12   SUMMARY

Water is the fluid of life. This chapter has presented the global consequences of inadequate water quality and security. As a reminder, the global state of water security has been characterized by the UN as follows:

- Worldwide, infectious diseases such as waterborne diseases are the number one killer of children less than 5 years old and more people die from unsafe water annually than from all forms of violence, including war.
- Unsafe water causes 4 billion cases of diarrhea annually and results in 2.2 million deaths, mostly of children less than 5 years old.
- Two-thirds of the global population (4.0 billion people) are estimated to live under conditions of severe water scarcity for at least 1 month of the year. Nearly half live in India and China. Half a billion people in the world face severe water scarcity all year round.
- Worldwide, 2.5 billion people live without improved sanitation. More than 70% of these people who lack sanitation, or 1.8 billion people, live in Asia.
- Every day, 2 million tons of sewage and industrial and agricultural waste are discharged into the world's water.
- Reflection on U.S. water quality data provides a troubling characterization, given that 44% of U.S. rivers, 64% of lakes, and 30% of estuaries are assessed as "impaired." The leading causes of impairment are viruses and

bacteria. However, public drinking water supplies in the U.S. are generally safe, given available water treatment resources.

Policies to protect water quality and security were described. For the U.S. just as the CA Act controls the emissions in the U.S. of contaminants released into outdoor air in the U.S., the CWAct, as amended, controls the discharges of contaminants into U.S. bodies of water. As with the CA Act, the CWAct contains a number of policies of importance to public health. One of the earliest provisions, continuing today, is the awarding of grants to states for the construction of sewage treatment plants, an example of federalism. The public health benefits of treating raw sewage in order to achieve sanitary and healthful conditions are obvious. As another important policy, the CWAct requires U.S. states, or EPA where states choose to defer to EPA, to issue permits that limit the amount of contaminants discharged into bodies of water. These emission standards are for the purpose of meeting water quality standards established under the SDWAct. The CWAct also adopts the policy that the regulated community (i.e., those entities that release contaminants into water) must use the Best Available Technology in their waste management operations. This policy leads to updates and improvements in waste management as technology changes. The policy also moves the regulated community toward a uniform technology.

The SDWAct, as amended, is the complement of the CWAct. The act establishes drinking water quality standards. The SDWAct contains policies of import to public health practice. The act creates the policy of dual drinking water regulations. Primary regulations are intended to protect human health and are enforceable under law. Secondary regulations, which are voluntary unless adopted as law by individual states, pertain to welfare considerations such as odor, appearance, and taste of water. The SDWAct also contains a second set of dual standards. Specifically, EPA must establish Maximum Contaminant Levels (MCLs) for individual water contaminants, which are legally enforceable by states, unless individual states have promulgated more stringent standards. MCLs, as policy, must consider the availability of technology necessary to achieve desired contaminant levels. Where technology is lacking, Maximum Contaminant Level Goals (MCLGs) are established. The policy of having both MCLs and MCLGs marries the present to the future. Public health's benefits when water quality standards shift when new or improved water treatment technologies are adopted.

Also described in this chapter is EU legislation that provides for measures against chemical pollution of surface waters. There are two components: the selection and regulation of substances of EU-wide concern (the priority substances) and the selection by Member States of substances of national or local concern (river basin specific pollutants) for control at the relevant level. The first component constitutes the major part of the Union's strategy against the chemical pollution of surface waters. It is set out in Article 16 of the Water Framework Directive 2000/60/EC, which is the primary EU policy on water quality and security.

As the world's human population continues to expand, with corresponding demands for food and water security, evolving policies on water conservation and water reinforcement will become increasingly important.

REFERENCES

1. Spector, D. 2016. How long a person can survive without water? *Independent*, February 14.
2. NOAA (National Oceanic and Atmospheric Administration). 2016. Harmful Algal Bloom and Hypoxia Research and Control Act. https://coastalscience.noaa.gov/research/habs/habhrca.
3. USGS. Water use in the United States. 2015. Washington, D.C.: National Water Use Science Project.
4. Reubold, T. 2019. America uses 322 billion gallons of water each day. Here's where it goes. *Ensia*, February 19.
5. UN (United Nations). 2015. Fresh water. The United Nations inter-agency mechanism on all water related issues. New York: Office of Secretary General, Media Centre.
6. United Nations. 2019. Water. https://www.un.org/en/sections/issues-depth/water/.
7. Mekonnen, M. M. and A. Y. Hoekstra. 2016. Four billion people facing severe water scarcity. *Sci. Adv.* 2:e1500323.
8. Iowa Water Center. 2015. USGS: Insecticides similar to nicotine found in about half of sampled streams across the United States. Ames: Iowa State University, School of Veterinary Medicine.
9. Bradley, P. M., et al. 2016. Metformin and other pharmaceuticals widespread in wadeable streams of the southeastern United States. *Environ. Sci. Technol. Lett.* 3:243–249.
10. EPA (Environmental Protection Agency). 2016. Proposed rule: Management standards for hazardous waste pharmaceuticals. Washington, D.C.: Office of Water.
11. EPA (Environmental Protection Agency). 2019. Management standards for hazardous waste pharmaceuticals and amendment to the p075 listing for nicotine. *Federal Register* 84 FR 5816. February 22.
12. *Reuters*. 2020. US drinking water contamination with 'forever chemicals' far worse than scientists thought. *The Guardian*, January 22.
13. Beitsch, R. 2020. EPA will regulate 'forever chemicals' in drinking water. *The Hill*, February 20.
14. EPA (Environmental Protection Agency). 2016. Harmful algal blooms. Washington, D.C.: Office of Water.
15. NOAA (National Oceanic and Atmospheric Administration). 2007. Distribution of HABs throughout the world. https://www.whoi.edu/redtide/regions/world-distribution.
16. Philpot, T. 2019. This year's wild, wet spring is feeding massive blobs of toxic algae. *Mother Jones*, July 10.
17. Gustin, G. 2019. Toxic algae blooms occurring more often, may be caught in climate change feedback loop. *Inside Climate News*, May 15.
18. Staff. 2018. Algae control methods to prevent algal blooms. LG Sonic. Zoetermeer: The Netherlands. https://www.lgsonic.com/blogs/algae-control/.
19. McMyn, N. 2020. A remedy for harmful algal blooms? Scientist thinks he's found one. *Environmental Health News*, February 3.
20. Vanderhart, D. 2018. As climate warms, algae blooms in drinking water supplies. *NPR*, September 3.
21. DHHS (U.S. Department of Health and Human Services). 2010. Healthy People 2010, vol. 1, 8-1. Washington, D.C.: Office of the Secretary.
22. Blackburn, B. G., et al., 2004. Surveillance for waterborne-disease outbreaks associated with drinking water – United States, 2001–2002. *Mort. Morb. Weekly Report* 53(SS08):23–45.
23. Hlavsa, M. C., et al. 2018. Outbreaks associated with treated recreational water — United States, 2000–2014. *MMWR* 67(19):547–51.

24. EPA (Environmental Protection Agency). 2017. National Water Quality Inventory: Report to Congress. Washington, D.C.: Office of Water.

25. NRDC (Natural Resources Defense Council). 2004. Swimming in sewage. Washington, D.C.: Office of the Director.

26. EPA (Environmental Protection Agency). 2017. National water quality inventory: Report to Congress. EPA 841-R-16-011. Washington, D.C.: Office of Water.

27. UNEP (United Nations Environment Programme). 2010. The Nairobi scientific communique on water quality challenges and responses. Nairobi: Office of Director-General, Media Centre.

28. Arnold, K., et al. 2014. Assessing risks and impacts of pharmaceuticals in the environment on wildlife and ecosystems. *Philos. Trans. R. Soc. B* 371(1701–10).

29. Kidd, K. A. 2014. Direct and indirect responses of a freshwater food web to a potent synthetic oestrogen. *Philos. Trans. R. Soc. B* 369(1656):20130578.

30. Weise, J. 2003. Historic drinking water facts. Anchorage: Drinking Water and Wastewater Program, ADEC Division of Environmental Health.

31. Ohio DEP. 2003. Understanding Ohio's surface water quality standards. Columbus: Director, Ohio Department of Environmental Protection.

32. Copeland, C. 1999. Clean Water Act: Summaries of environmental laws administered by EPA. Washington, D.C.: Congressional Research Service.

33. EPA (Environmental Protection Agency). 2016. Toxic and priority pollutants under the Clean Water Act. Washington, D.C.: Office of Water.

34. EWG. 2000. Clean water report card: How the regulators are keeping up with keeping our water clean. Washington, D.C.: Environmental Working Group.

35. EPA (Environmental Protection Agency). 2005. Efforts to manage backlog of water discharge permits need to be accomplished by greater program integration, report 2005-P-00018. Washington, D.C.: EPA Inspector General.

36. USPIG (U.S. Public Interest Research Group). 2006. Troubled waters: An analysis of Clean Water Act compliance. Washington, D.C., U.S. Public Interest Research Group Education Fund.

37. Copeland, C. 2002. Clean Water Act and total maximum daily loads (TMDLs) of pollutants. Washington, D.C.: Congressional Research Service.

38. Miller, J. 1992. Environmental law and the science of ecology. In *Environmental and Occupational Medicine*, 2nd edition, p. 1307. Rom, W., (ed.). Boston: Little, Brown and Co.

39. EPA (Environmental Protection Agency). 2003. Section 404 of the Clean Water Act: An overview. http://www.epa.gov/owow/wetlands/facts/fact10.html.

40. EPA (Environmental Protection Agency). 2003. Section 404 of the Clean Water Act: How wetlands are defined and identified. http://www.epa.gov/owow/wetlands/facts/fact11.html.

41. Dahl, T. E. 2000. National wetlands inventory. Summary findings. http://wetlands.fws.gov/bha/SandT/SandTSummaryFindings.html.

42. Puko, T. 2014. Alpha Natural Resources to pay record $227.5M in water pollution settlement. *TribLive*. http://triblive.com/business/headlines/5710782-74/alpha-coal-federal.

43. EPA (Environmental Protection Agency). 2003. 2002–2003: The year of clean water. Washington, D.C.: Office of Water.

44. Ross, D. R. 2019. Memorandum, February 6: Updating the EPA's Water Quality Trading Policy. Washington, D.C.: Office of Water.

45. Kelly, S. 2015. Water pollution trading programs under fire as report finds lax oversight, "shell games" put waterways at risk. DESMOG. http://www.desmogblog.com/2015/11/19/water-quality-trading-programs-under-fire-report-finds-lax-oversight-shell-games-put-waterways-risk.

46. EPA (Environmental Protection Agency). 2015. Clean water rule protects streams and wetlands critical to public health, communities, and economy. News release, May 27. Washington, D.C.: Office of Water.
47. EPA (Environmental Protection Agency). 2015. What the clean water rule does. https://www.epa.gov/cleanwaterrule/what-clean-water-rule-does.
48. Eilperin, J. and A. Phillip. 2017. Trump directs rollback of Obama-era water rule he calls 'destructive and horrible.' The Washington Post, February, 28.
49. EPA (Environmental Protection Agency). 2020. More widespread support for EPA and Army's Navigable Waters Protection Rule – a new definition of WOTUS. Washington, D.C.: EPA Press Office, January 24.
50. Liptak, A. 2020. Clean Water Act covers groundwater discharges, Supreme Court says. The New York Times, April 23.
51. OMB (White House Office of Management and Budget). 2003. Informing regulatory decisions: 2003 report to Congress on the costs and benefits of federal regulations and unfunded mandates on state, local, and tribal entities. Washington, D.C.: Office of Information and Regulatory Affairs.
52. Craun, G. 1986. Statistics of waterborne outbreaks in the U.S. (1920–1980). In: Waterborne Disease Outbreaks in the United States. Craun, G. F., (ed.). Boca Raton, FL: CRC Press.
53. Mac Kenzie, W. R, N. J. Hoxie, M. E. Proctor, et al. 1994. A massive outbreak in Milwaukee of cryptosporidium infection transmitted through the public water supply. NEJM 331(3):161–67.
54. McLaughlin, T. 2000. Walkerton e. coli outbreak declared over. The Globe and Mail, June 4.
55. Barwick, R. S., et al. 2000. Outbreaks-United States. 1997–1998. Mort. Morb. Weekly Rep. 49(SS04):1–35.
56. Benedict, K. M., et al. 2017. Surveillance for waterborne disease outbreaks associated with drinking water — United States, 2013–2014. MMWR 66:1216–21.
57. Bienkowski, B. 2014. Lower IQ in children linked to chemical in water. Scientific American, September 30.
58. Frazin, R. 2020. EPA declines to regulate rocket chemical tied to developmental damage. The Hill, Jun 18.
59. Boston University. 2014. Contaminated water linked to pregnancy complications, study finds. News release, September 30. Boston: School of Public Health.
60. WHO (World Health Organization). 2017. Water, health and ecosystems. http://www.who.int/heli/risks/water/water/en/.
61. Randle, R. V. 1991. Safe drinking water act. In Environmental Law Handbook, p. 149. Arbuckle, J. G., et al., (ed.). Rockville: Government Institutes, Inc.
62. EPA (Environmental Protection Agency). 2005. Setting standards for safe drinking water. http://www.epa.gov/OGWD/standard/setting.html.
63. EPA (Environmental Protection Agency). 2017. Drinking water requirements for states and public water systems. Washington, D.C.: Office of Water.
64. EPA (Environmental Protection Agency). 2006. Community water system survey. Washington, D.C.: Office of Water.
65. Tiemann, M. 1999. Safe Drinking Water Act. Summaries of environmental laws administered by the EPA. Washington, D.C.: Congressional Research Service.
66. ATSDR (Agency for Toxic Substances and Disease Registry). 1988. The nature and extent of lead poisoning in children in the United States: A report to Congress. Atlanta: U.S. Department of Health and Human Services. Public Health Service, Centers for Disease Control and Prevention.
67. EPA (Environmental Protection Agency). 1999. Understanding the Safe Drinking Water Act, EPA 810-F-99-008. Washington, D.C.: Office of Public Information.

68. Beitsch, R. 2020. EPA revises lead rule, sidestepping calls for more stringent standards. *The Hill*, December 22.

69. Allaire, M., H. Wu, U. Lall. 2018. National trends in drinking water quality violations. *Proc. Nat. Acad. Sci.* 115(9):2078–83.

70. Transparency Market Research. 2015. Global bottled water market is expected to reach USd 279.65 billion in 2020. By volume, global bottled water market is expected to reach 465.12 billion liters in 2020. Albany: Office of Research.

71. Conway, J. 2019. Volume of bottled water in the U.S. 2010–2017. Statista, August 29.

72. FDA (Food and Drug Administration). 2002. Bottled water regulation and the FDA. Washington, D.C.: Center for Food Safety and Applied Nutrition.

73. NRDC (Natural Resources Defense Council). 2000. Bottled water: Pure drink or pure hype? www.nrdc.org/water/drinking/nbw.asp.

74. Gitau, B. 2015. Refill stations: Why the Detroit Zoo no longer sells bottled water. *The Christian Science Monitor*, October 3.

75. Rein, L. 2015. Park Service to big water: No federal funding for bottled water bans? We'll find our own money, thanks. *The Washington Post*, July 30.

76. Lefferts, J. F. 2012. Concord bottled water ban OK'd. *The Boston Globe*, September 6.

77. CDC (Centers for Disease Control and Prevention). 2019. Rural health. https://www.cdc.gov/ruralhealth/cancer.html.

78. Seltenrich, N. 2017. Unwell: The public health implications of unregulated drinking water. *Environ. Health Perspect.* 125(11):114001. Published online Nov 1. doi:10.1289/EHP2470.

79. Caswell, B. and F. O'Leary. 2017. Troubled water: Wells aren't regulated and septic tanks aren't inspected. Troubled Water project produced by the *Carnegie-Knight News 21*, September 6.

80. Fedinick, K. P., S. Taylor, M. Roberts. 2019. Water Downed Justice. Natural Resource Defense Council. https://www.nrdc.org/sites/default/files/watered-down-justice-report.pdf.

81. Kakhani, N. 2020. People of color more likely to live without piped water in richest US cities. *The Guardian*, November 2.

82. Neslen, A. 2016. River pollution puts 323m at risk from life-threatening diseases, says UN. *The Guardian*, September 22.

83. WHO (World Health Organization). 2017. 2.1 billion people lack safe drinking water at home, more than twice as many lack safe sanitation. News release. Geneva: Office of Communications.

84. WHO (World Health Organization). 2019. Water-related diseases. https://www.who.int/water_sanitation_health/diseases-risks/diseases/diarrhoea/en/.

85. WHO (World Health Organization). 2016. The health and environment linkages initiative: Water, health and ecosystems. Geneva: Office of Director-General, Media Centre.

86. University of York. 2019. Antibiotics found in some of the world's rivers exceed 'safe' levels, global study finds. *Science Daily*, May 27.

87. UN (United Nations). 2014. International decade for water for life: Decade for action 2005–2015. New York: Department of Economic and Social Affairs.

88. WHO (World Health Organization). 2016. Water sanitation health. Geneva: Office of Director-General, Media Centre.

89. WHO (World Health Organization). 2013. Schistosomiasis: A major public health problem. Geneva: Office of Director-General, Media Centre.

90. Watts, J. 2017. UN poised to move ahead with landmark treaty to protect high seas. *The Guardian*, December 22.

91. NOAA (National Oceanic and Atmospheric Administration). 2019. The Marine Debris Act. Washington, D.C.: Office of Response and Restoration Program.

92. Anderson, C. 2019. Carnival will pay $20m over pollution from its cruise ships. *ABC News*, June 3.

93. Water Footprint. 2018. How the United States uses water. https://www.watercalculator.org/footprints/how-united-states-uses-water/.

94. USGS. How much water is used by people in the United States? https://www.usgs.gov/faqs/how-much-water-used-people-united-states?qt-news_science_products=0#qt-news_science_products.

95. Schaper, D. 2014. As infrastructure crumbles, trillions of gallons of water lost. *NPR News*, October 29.

96. Perrone, D. and S. Jasechko. 2019. Deeper well drilling an unsustainable stopgap to groundwater depletion. *Nat. Sustain.* doi:10.1038/s41893-019-0325-z.

97. FAO (Food and Agriculture Organization). 2020. How to overcome water challenges in agriculture. Rome: Media Relations Office, November 26.

98. Sengupta, S. and W. Cai. 2019. A quarter of humanity faces looming water crisis. *The New York Times*, August 6.

99. Weiss, K. R. 2019. Some of the world's biggest lakes are drying up. Here's why. *National Geographic*, March issue.

100. Watts, J. 2018. Water shortages could affect 5bn people by 2050, UN report warns. *The Guardian*, March 19.

101. Ritter, K. 2018. Water-stressed Beijing exhausts its options. *Circle of Blue*, March 14.

102. Said-Moorhouse, L. and G. Mezzofiore. 2018. Cape Town cuts limit on water use by nearly half as 'Day Zero' looms. *CNN News*, February 1.

103. Ritter, K. 2019. Groundwater plummets in Delhi, city of 29 million. *Circle of Blue*, January 10.

104. World Bank Group. 2018. Water scarce cities initiative. http://pubdocs.worldbank.org/en/588881494274482854/Water-Scarce-Cities-Initiative.pdf.

105. Union of Concerned Scientists. 2015. Causes of drought: What's the climate connection? http://www.ucsusa.org/global_warming/science_and_impacts/impacts/causes-of-drought-climate-change-connection.html#.V5vPkrgrLy0.

106. Nagourney, A. 2015. California releases revised water consumption rules. *The New York Times*, April 18.

107. Reese, P., D. Kasler, R. Sabalow. 2015. Beverly Hills, other cities hit with fines for failure to meet water conservation targets. *The Sacramento Bee*, October 30.

108. Hamers, L. 2018. Engineers are plugging holes in drinking water treatment. *Science News*, November 25.

109. Jackson, M. 2015. Gov. Brown extends California's water limits: We are in a new era. *The Christian Science Monitor*, November 15.

110. Bernstein, S. 2016. California to lift severe mandatory water conservation rules. *Reuters*, May 9.

111. Stevens, M. 2016. California board allows water districts to set their own conservation targets. *Los Angeles Times*, May 18.

112. Grantham Institute, Imperial College London. 2012. How will climate change impact on fresh water security? *The Guardian*, December 21.

113. *ReliefWeb*. 2016. Kenya: Drought – 2014–2016. http://reliefweb.int/disaster/dr-2014-000131-ken.

114. Quinn, B. and S. K. Dehghan. 2017. World Water Day: One in four children will live with water scarcity by 2040. *The Guardian*, March 21.

115. NECASC. 2016. Climate impacts on freshwater resources and ecosystems. Amherst: University of Massachusetts, Northeast Climate Adaptation Science Center.

116. Cuthbert, M. O., et al. 2019. Global patterns and dynamics of climate–groundwater interactions. *Nat. Clim. Change* 9:137–41.

117. Ballard, T. C., E. Sinha, A. M. Michalak. 2019. Long-term changes in precipitation and temperature have already impacted nitrogen loading. *Environ. Sci. Tech.* doi:10.1021/acs.est.8b06898.

118. Colorado School of Mines. 2016. Produced Water Treatment and Beneficial Use Information Center. http://aqwatec.mines.edu/produced_water/index.htm.
119. Noah, M. 2002. Graywater use still a gray area. *J. Environ. Health* 64(10):22–5.
120. Finley, B. 2014. Colorado weighs taking "waste" out of wastewater to fix shortfall. *The Denver Post*, November 23.
121. Stevens, M. 2015. California seeks to build one of world's largest recycled water programs. *Los Angeles Times*, September 22.
122. Talbot, D. 2014. *Desalination Out of Desperation*. Cambridge: MIT Technology Review.
123. Fikes, B. J. 2015. $1-billion desalination plant, hailed as model for state, opens in Carlsbad. *Los Angeles Times*, December 14.
124. UNESCO (UN Educational, Scientific, and Cultural Organization). 2014. Intergovernmental Oceanographic Commission. Addressing the impacts of harmful algal blooms on water security. New York: Office of Director-General, Media Centre.
125. Nield, D. 2015. This new technology converts sea water into drinking water in minutes, Science Alert, September 16.
126. Mooney, C. 2015. The incredibly stupid way that Americans waste 1 trillion gallons of water each year. *The Washington Post*, March 17.
127. EPA (Environmental Protection Agency). 2016. Water conservation at EPA. Washington, D.C.: Office of Water.
128. Bland, A. 2017. With drought restrictions long gone, California keeps conserving water. *Water Deeply*, September 6.
129. Walton, B. 2018. Kansas farmers cut Ogallala water use – and still make money. *Circle of Blue*, March 6.
130. Scerbo, M. 2019. Congress set toilet standards in 1992.Here's the data showing they're saving water and energy. *Alliance to Save Energy*, December 11.
131. McCarty, J. E. 2018. Newly patented wastewater recycling system turns toilet water into purified drinking water. *The Plain Dealer*, January 7.
132. Temming, M. 2018. A filter that turns saltwater into freshwater just got an upgrade. *Science News*, August 16.
133. Staff. 2018. Engineers upgrade ancient, sun-powered tech to purify water with near-perfect efficiency. *Science Daily*, May 3.
134. Staff. 2018. Scientists engineer crops to conserve water, resist drought. *Science Daily*, March 6.
135. White House. 2009. Executive Order 13514–Focused on federal leadership in environmental, energy, and economic performance. Washington, D.C.: Office of the Press Secretary.
136. Mertens, R. 2015. Agriculture is big threat to water quality. These farmers are doing something about it. *The Christian Science Monitor*, December 27.
137. Sullivan, S. P. 2015. New York-New Jersey harbors see 'substantial progress' after decades of pollution, group says. http://www.nj.com/news/index.ssf/2015/02/new_york-new_jersey_harbors_see_substantial_progre.html.
138. Bienkowski, B. 2015. Forget the fancy machinery: Oregon wastewater plant's warm water discharges offset by restoring riverside. But not everyone is on board. *Environmental Health News*, July 30.
139. Gies, E. 2018. How the growing "one water" movement is not only helping the environment but also saving millions of dollars. *Ensia*, May 8.
140. Chipman, K. 2020. California water futures begin trading amid fear of scarcity. *Bloomberg Green*, December 6.
141. Osborne, H. 2020. Remote Belize village with no access to safe drinking water gets solar-powered device to suck it out of the air. *Newsweek*, April 28.

4 Food Safety and Security

4.1 INTRODUCTION

This chapter is about the third of humankind's three basic needs, food. Prior chapters have presented the other two survival needs: air quality and water quality/security. This chapter addresses food safety and food security as companion challenges to the well-being of global populations, for both are required for human survival. A definition is in order. The World Food Summit of 1996 defined food security as existing "when all people at all times have access to sufficient, safe, nutritious food to maintain a healthy and active life" [1]. One can observe that safe food is integral to this definition, suggesting that the definition of food security could embody safe food. However, this chapter will consider safe food and food security as separate concepts, since existing policies have for historical reasons made this kind of distinction. Therefore, for the purposes of this chapter, safe food will refer to food that when consumed will cause no ill effects to the consumer. And food security will refer to the availability of safe food. In support of this distinction, one can have safe food within a food supply, but if it is not accessible (i.e., insecure), one would experience hunger.

The effects of unsafe food and insecure food supplies on human and ecological health are presented, as well as hazard interventions to prevent the development of unsafe food supplies. The global consequences and implications of food insecurity are described, along with several factors that contribute to food security.

As a reminder, we humans have evolved from ancestors who hunted feral animals for food, later learning to cultivate grains, vegetables, and fruits. Food grown and prepared before the invention of refrigeration became commonplace in America was generally consumed soon

> Until the early 20th century, food safety in the U.S. was deemed to be the responsibility of individual consumers and therefore not a matter for government intervention.

after its acquisition. This was in order to avoid contact with pathogens found in deteriorating food. Decay of food is caused by microorganisms that parasitize dead plant or animal tissue and thereby render food unsafe for human consumption, as illustrated in Figure 4.1.

To preserve food, meat was salted or exposed to dense smoke from hardwood fires. Vegetables and fruits were canned or stored in root cellars. With refrigeration came the ability to chill or freeze food and store it safely for short to long periods of time. In modern times, irradiation has been used to preserve food, but this method has not gained widespread acceptance due to concerns that irradiation might allegedly cause harmful changes in food, which then could affect consumers of irradiated food. The problem of pathogens in deteriorating food was theoretically overcome through refrigeration. However, the equally important health consequences of how food is prepared remain a problem. Fresh, or properly preserved, food that is

DOI: 10.1201/9781003253358-4

FIGURE 4.1 Decaying moldy food unsafe for human consumption. (Food Mold. 2021. Pet Poison Hotline, https://www.petpoisonhelpline.com/poison/mycotoxin/2021.)

contaminated with pathogens from human contact during growing, transporting, preparing, or serving food has the potential to cause human illnesses.

Government's involvement in protecting the public against adulterated or impure food is a relatively recent occurrence in the U.S. Until the early 20th century, food safety was deemed to be the responsibility of individual consumers and therefore not a matter for government intervention. As will be described in the following sections, U.S. states and territories had the primary authority over issues of food safety—an arrangement that still prevails. However, federal government gradually assumed a strong role in food safety as a matter of interstate commerce. Federal involvement in food safety brought together the public health triad of federal, state, and local governments directed to a common purpose—in this case, prevention of foodborne illnesses.

4.2 FOOD SAFETY AND HEALTH

Foodborne illnesses are a serious environmental health problem, although the U.S. food supply is relatively safe overall. However, how food is grown, transported, prepared, and served can introduce pathogens and other hazards of potential harm to human health. In the U.S., there are three federal statutes that concern food safety. To be described are the Food, Drug, and Cosmetic Act (FDCAct), the Federal Meat Inspection Act (FMIAct) (and analogous laws for poultry and eggs), and the FDA Food Modernization Act.

4.2.1 UNSAFE FOOD AND HUMAN HEALTH

Some estimates of the global public health burden of foodborne disease are persuasive as to the gravity of unsafe

An estimated 600 million—almost one in ten people in the world—fall ill after eating contaminated food and 420,000 die every year.

food as an environmental hazard. In 2015, the WHO released its first report on the global prevalence of foodborne diseases. The agency's 2019 update acknowledged that access to sufficient amounts of safe and nutritious food is key to sustaining life and promoting good health. Additionally, the agency cited the following sobering health statistics [2]:

- Unsafe food containing harmful bacteria, viruses, parasites, or chemical substances causes more than 200 diseases—ranging from diarrhea to cancers.
- An estimated 600 million—almost one in ten people in the world—fall ill after eating contaminated food and 420,000 die every year, resulting in the loss of 33 million healthy life years (DALYs).
- Children <5 years of age carry 40% of the foodborne disease burden, with 125,000 deaths every year.
- Diarrheal diseases are the most common illnesses resulting from the consumption of contaminated food, causing 550 million people to fall ill and 230,000 deaths every year.
- Foodborne diseases impede socioeconomic development by straining health care systems and harming national economies, tourism, and trade.

Data for foodborne illness in the U.S. come from CDC, which reported, "Food safety is an important public health priority. Foodborne illness (sometimes called 'foodborne disease', 'foodborne infection', or 'food poisoning') is a common, costly—yet preventable—public health problem. CDC estimates that annually about 1 in 6 Americans (or 48 million people) get sick, 128,000 are hospitalized, and 3,000 die of foodborne diseases" [3].

> In 2018 the CDC's FoodNet identified 25,606 infections, 5,893 hospitalizations, and 120 deaths due to foodborne illnesses.

The CDC collects foodborne illness reports from U.S. states, territories, and other sources in order to assess etiologic agents. To evaluate progress toward the prevention of enteric and foodborne illnesses in the U.S, the Foodborne Diseases Active Surveillance Network (FoodNet) monitors the incidence of laboratory-confirmed infections caused by nine pathogens transmitted commonly through food in ten U.S. sites. Foodborne diseases represent a major health problem in the U.S. The 2018 report summarizes preliminary 2018 data and changes since 2015. "During 2018, FoodNet identified 25,606 infections, 5,893 hospitalizations, and 120 deaths. The incidence of most infections is increasing, including those caused by Campylobacter and Salmonella, which might be partially attributable to the increased use of culture-independent diagnostic tests (CIDTs). The incidence of Cyclospora infections increased markedly compared with 2015–2017, in part related to large outbreaks associated with produce (1). More targeted prevention measures are

> In 2019, the American Academy of Pediatrics, together with the American Heart Association, endorsed a range of strategies designed to curb children's consumption of sugary drinks.

needed on produce farms, food animal farms, and in meat and poultry processing establishments to make food safer and decrease human illness" [4].

The issue of sugary beverages consumed by children and adolescents has also become in the U.S. a matter of food safety, according to pediatricians [5]. Pediatricians have expressed concerns about the risks of consuming too many sugary drinks and the association with type 2 diabetes and obesity. In 2019 the American Academy of Pediatrics, together with the American Heart Association, endorsed a range of strategies designed to curb children's consumption of sugary drinks. The groups' strategies included taxes on sugary drinks, limits on marketing sugary drinks to children, and financial incentives to encourage more healthful beverage choices. Exemplifying the pediatricians' recommendations, a sugar tax was established in 2016 by the UK government, taking effect in 2018. The tax applies to beverages with total sugar content of more than 5 g per 100 mL. This is estimated to raise around £520 million ($655 million) a year, which will be spent on increasing funding sports activities in primary schools [6].

Relating to the public health issue of sugary drinks, a 2019 population-based cohort study of 451,743 individuals from ten countries in Europe found greater consumption of total, sugar-sweetened, and artificially sweetened soft drinks were associated with a higher risk of all-cause mortality [7]. Consumption of artificially sweetened soft drinks was associated with deaths from circulatory diseases, and sugar-sweetened soft drinks were associated with deaths from digestive diseases.

A line is drawn here between "unsafe food" and "unhealthful food". The former term was discussed in the previous sections. Unhealthful food is discussed in the following sections. One source notes that unhealthful eating and physical inactivity are leading causes of death in the U.S. Unhealthful diet contributes to ~678,000 deaths each year in the U.S., due to nutrition- and obesity-related diseases, such as heart disease, cancer, and type 2 diabetes. In the last 30 years, obesity rates have doubled in adults, tripled in children, and quadrupled in adolescents [8]. A regular diet of "fast food" can be too high in calories, saturated fat, sodium, and added sugars, and does not have enough fruits, vegetables, whole grains, calcium, and fiber, conditions that can contribute to poor health.

4.2.2 UNSAFE MEAT AND HUMAN HEALTH

Episodes of meat-associated foodborne disease can occur because of failures in meat production, preparation, or delivery to and preparation by consumers. A significant provision of the FMIAct gives the U.S. Department of Agriculture (USDA) authority to take action against producers of meat or meat products when found to be unsafe for human consumption. Specifically, Section 673 of the act states, "...[any carcass, part of a carcass, meat or meat product... is capable of use as human food and is adulterated or misbranded... shall be liable to be proceeded against and seized and condemned, at any time, on a libel of information in any U.S. district court or other proper court..."

States with their own food safety statutes can also suspend operations or close facilities found to be producing meat or meat products contaminated with pathogens.

Preventing contaminated meat from reaching consumers is, in public health terms, an act of primary prevention, i.e., hazard interdiction. Two examples will suffice to illustrate the interdiction of contaminated meat from reaching consumers. In October 2002, USDA recalled 27.4 million pounds of poultry found contaminated with Listeria, made into delicatessen products, produced at a Pennsylvania processing plant, the largest recall at that time in U.S. history. In April 2021, USDA issued a public health alert that 211,406 pounds of ground turkey were potentially linked to *Salmonella hadar* illness [9]. The alert warns consumers of the risk and vendors to remove the product from sales.

4.2.3 Unsafe Food and Ecosystem Health

Data on the impact of unsafe food on ecosystem health are lacking. However, an important aspect of food security is food wastage, which does impact ecosystem health. The seminal study on this subject was conducted in 2013 by the UN's Food and Agriculture Organization (FAO) [10]. The FAO estimated that each year, approximately one-third of all food produced for human consumption in the world is lost or wasted. This food wastage represents a missed opportunity to improve global food security but also mitigate environmental impacts.

The FAO study provided a global account of the environmental footprint of food wastage (i.e., both food loss and food waste) along the food supply chain, focusing on impacts on climate, water, land, and biodiversity. A model was developed to answer two key questions: What is the magnitude of food wastage impacts on the environment? And what are the main sources of these impacts, in terms of regions, commodities, and phases of the food supply chain involved—with a view to identify "environmental hotspots" related to food wastage?

> The FAO estimates that approximately one-third of all food produced for human consumption in the world is lost or wasted.

The FAO's study estimated the global volume of food wastage to be 1.6 Gtonnes of "primary product equivalents," while the total wastage for the edible part of food was 1.3 Gtonnes. This amount can be weighed against total agricultural production (for food and non-food uses), which is about 6 Gtonnes.

FAO also related food wastage with the impact on climate change. Without accounting for greenhouse gas (GHG) emissions from land use change, the carbon footprint of food produced and not eaten was estimated to be 3.3 Gtonnes of CO_2 equivalent: as such, food wastage ranks as the third top emitter of GHG after the U.S. and China. Globally, the blue water footprint (i.e., the consumption of surface and groundwater resources) of food wastage was about 250 km^3, which is equivalent to the annual water discharge of the Volga River, or three times the volume of Lake Geneva. Finally, produced but uneaten food occupies almost 1.4 billion hectares of land, which represents about 30% of the world's agricultural land area. While it was difficult for FAO to estimate impacts on biodiversity at a global level, food wastage unduly compounds the negative externalities that mono-cropping and agriculture expansion into wild areas create on biodiversity loss, including mammals, birds, fish, and amphibians.

The FAO observed that the loss of land, water, and biodiversity, as well as the negative impacts of climate change, represent huge costs to society that are yet to be quantified. The direct economic cost of food wastage of agricultural products (excluding fish and seafood), based on producer prices only, is about \$750 billion, equivalent to the GDP of Switzerland in 2013.

The FAO study highlighted global environmental hotspots related to food wastage at regional and subsectoral levels, noting for consideration by policymakers engaged in waste reduction the following observations:

- Wastage of cereals in Asia emerges as a significant problem for the environment, with major impacts on carbon, blue water, and arable land. Rice represents a significant share of these impacts, given the high carbon intensity of rice production methods (e.g., paddies are major emitters of methane), combined with high quantities of rice wastage.
- Wastage of meat, even though wastage volumes in all regions are comparatively low, generates a substantial impact on the environment in terms of land occupation and carbon footprint, especially in high-income regions (that waste about 67% of meat) and Latin America.
- Fruit wastage emerges as a blue water hotspot in Asia, Latin America, and Europe because of food wastage volumes.
- Vegetable wastage in industrialized Asia, Europe, and South and South-East Asia constitutes a high carbon footprint, mainly due to large wastage volumes.

The FAO opined that by highlighting the magnitude of the environmental footprint of food wastage, the results of their study—by regions, commodities, or phases of the food supply chain—allow prioritizing actions and defining opportunities for various actors' contributions to resolving this global challenge [10].

4.2.4 UNSAFE MEAT AND ECOSYSTEM HEALTH

Although unsafe meat that is discarded into environmental media can potentially harm ecosystems as water pollutants (pathogens of tissue decay) or air emissions (fetid odors), the primary ecosystem health problem derives from the farming of livestock. In particular, waste from large commercial livestock operations can be a hazard to ecosystem health if not properly managed by farmers. As commented by the USDA, "Animal production has the potential to negatively affect surface water quality (from pathogens, phosphorus, ammonia, and organic matter); groundwater quality (from nitrate); soil quality (from soluble salts, copper, arsenic, and zinc); and air quality (from odors, dust, pests, and aerial pathogens). Manure and other byproducts of animal production, if not carefully managed, will have a significant negative impact on the environment. Agricultural production has been identified by the Environmental Protection Agency (EPA) as the largest single contributor to water quality impairment for rivers and lakes" [11].

As a potential contributor to water pollution, EPA has regulations specific to large animal production facilities. In particular, "Animal Feeding Operations (AFOs) are agricultural operations where animals are kept and raised in confined situations. An AFO is a lot or facility (other than an aquatic animal production facility) where the following conditions are met: animals have been, are, or will be stabled or confined and fed or maintained for a total of 45 days or more in any 12-month period, [...]. AFOs that meet the regulatory definition of a concentrated animal feeding operation (CAFO) are regulated under EPA's National Pollutant Discharge Elimination System (NPDES) permitting program. The NPDES program regulates the discharge of pollutants from point sources to waters of the U.S. CAFOs are point sources, as defined by the Clean Water Act [Section 502(14)]). To be considered a CAFO, a facility must first be defined as an AFO, and meet the criteria established in the CAFO regulation" [12].

In addition, AFOs that qualify are subject to EPA's regulations on emissions of greenhouse gases (GHG). Specifically, in response, in 2008 EPA issued the Mandatory Reporting of Greenhouse Gases Rule, which requires reporting of GHG data and other relevant information from large sources and suppliers in the U.S. In general, the Rule is referred to as 40 CFR Part 98 (Part 98). Implementation of Part 98 is referred to as the Greenhouse Gas Reporting Program (GHGRP).

4.3   U.S. FOOD SAFETY POLICIES

Many national governments and one regional government, the EU, have implemented food safety and security policies. These will be summarized in the following sections. In the U.S. there exist three primary federal statutes bearing on food safety, as administered by the U.S. FDA and the USDA. Food security policies lie primarily with the USDA, as will be highlighted herein.

The three principal federal food safety statutes are the FDCAct, the FMIAct, and the FDA Food Modernization Act (FSMA). As will be related, two of these U.S. federal statutes date from the early years of the 20th century and have been amended over the years. The third statute, the FSMA of 2011, is a more recent statute. Regarding food security in the U.S., the USDA has significant responsibilities and authorities that will be summarized in this section.

4.3.1   U.S. FOOD, DRUG, AND COSMETIC ACT, 1906

This act is a powerful statement by the U.S. federal government of the value of protecting the nation's food security, therapeutic drugs, cosmetics, and medical devices. The FDA is the principal federal agency with regulatory authorities to administer the provisions of the act. Given the purpose of this chapter, the act's provisions specific to food safety are presented. Other provisions (e.g., drugs and medical devices) while important lie outside the focus of this chapter.

In the 19th and early 20th centuries, any government control of food and drugs was the responsibility of states.

4.3.1.1   History

Americans of the 21st century expect not to be harmed by the food and medicinal drugs and therapeutic devices with which they come into contact. The expectation is the product of personal experience (e.g., a few of us have had protracted illnesses from eating impure food), and there is general trust in public health systems (e.g., restaurant inspections). While episodes of illness occur as the result of impure food (e.g., undercooked meat in hamburgers), the current situation is vastly different from that of our ancestors.

> During the 1870s, the grassroots Pure Food Movement arose and soon became the principal source of political support for federal food and drugs legislation.

In the 19th and early 20th centuries, any government control of food and drugs was the responsibility of states. State laws, if enacted, varied greatly between states. In that era, the use of chemical preservatives and toxic colors added to food was virtually uncontrolled [13]. Instances of morbidity surely occurred, given current bacteriological and toxicological knowledge, but no health reporting system was in place then to record the extent of morbidity. As public concern grew in the late 19th century about unsanitary conditions in the meatpacking industry, a similar concern arose about the harm caused by drugs, medications, and concoctions sold for alleged medicinal purposes. "Medicines" containing opium, morphine, heroin, and cocaine were sold without any restriction [13]. Moreover, labels gave no indication about the ingredients of "over-the-counter" drugs and medications. The policy of "buyer beware" prevailed during this period.

> Remarkably, Congress enacted on the same day, June 30, 1906, both the Pure Food and Drug Act and the FMIAct.

During the 1870s, the grassroots Pure Food Movement arose and soon became the principal source of political support for federal food and drugs legislation [13]. In 1903, Dr. Harvey W. Wiley became the director of the U.S. Department of Agriculture's Division of Chemistry and soon thereafter aroused public opinion against impure consumer products that his staff had identified. In a sense, Dr. Wiley was serving as a surrogate surgeon general, informing the public and advocating for public health legislation. Strenuous opposition to Wiley's campaign for a federal food and drug law came from whiskey distillers and the patent medicine firms, many of which thought they would be put out of business by federal authorities and regulation of their industries. Supporting the need for federal legislation were agricultural organizations, some food processors, public health professionals, and state food and drug officials. The political scale was tipped toward legislative action through the intercession of club women who rallied behind the pure food cause [13]. Remarkably, Congress enacted on the same day, June 30, 1906, both the Pure Food and Drug Act and the FMIAct.

The Food and Drugs Act of 1906 prohibited the manufacture and interstate shipment of adulterated and mislabeled foods and drugs. The law enabled the federal government to initiate litigation against alleged illegal products but lacked affirmative requirements to guide compliance with the law. The 1906 law also lacked key provisions necessary to make it effective in identifying harmful food and drug products. For example, food adulteration continued to flourish because judges could find no authority in the law for any standards of purity and content established by FDA [14]. In time, the 1906 law became obsolete because FDA lacked enforcement authorities and due to technological changes in how food and drugs were produced.

TABLE 4.1
Chapters of the Food, Drug, and Cosmetic Act [118]

Chapter	Title
I & II	Short title and definitions
III	Prohibited acts and penalties
IV	Food
V	Drugs and devices
VI	Cosmetics
VII	General authority
VIII	Imports and exports
IX	Tobacco products
X	Miscellaneous

The provisions of the 1906 Food and Drugs Act simply were not sufficiently robust to keep up with technology changes in the food and drug industries. Thirty-two years were to pass before the act was updated.

In June 1938, President Franklin D. Roosevelt (D-NY) signed into law the Federal FDCAct, which replaced the Food and Drug Act of 1906. The 1938 law contained many significant changes, including those shown in Table 4.1 [14]. Even the 1938 law was found to be in need of further improvements. For instance, the 1938 law prohibited poisonous substances but required no evidence that food ingredients were safe for human consumption.

In 1949, Congress began lengthy hearings on the FDCAct, resulting in three substantive amendments: the Pesticide Amendment of 1954, the Food Additives Amendment of 1958, and the Color Additive Amendments of 1960 [14]. These amendments effectuated an environmental health policy that no substance can legally be introduced into the U.S. food supply unless there has been a prior determination that it is safe. Moreover, these amendments required manufacturers to conduct the research necessary to establish their products' safety. FDA became a reviewer of manufacturers' data, with the authority to reject products or to request more data from manufacturers.

PRODUCT CATEGORIES REGULATED BY THE FDA

Biologics
Cosmetics
Drugs
Foods
Medical Devices
Radiating Devices
Tobacco Products
Veterinary Products

The FDCAct has been amended several times since 1960. In particular, Congress added products to the list of items regulated by the FDA. The key product amendments are listed in Table 4.2. Noteworthy is that FDA was given regulatory coverage over medical devices, dietary food supplements, and devices that radiate energy.

TABLE 4.2

Congressional Amendments Adding to FDA's Regulatory Authority [118]

Year	Title	Public Law
1968	Radiation Control for Safety and Health Act	PL 90-602, 82 Stat 1173
1976	Medical Device Regulation Act	PL 94-295, 90 Stat 539
1990	Nutrition Labeling and Education Act	PL 101-535, 104 Stat 2353
1990	Safe Medical Device Amendments	PL 101-629, 104 Stat 4511
1994	Dietary Supplement Health and Education Act	PL 103-417, 108 Stat 4332
1997	Food and Drug Administration Modernization Act	PL 105-115, 111 Stat 2296

4.3.1.2   Key Provisions of the FDCAct Relevant to Public Health

The scope of FDA's regulatory authority is very broad. The agency's responsibilities are closely related to those of several other government agencies. The following is a list of traditionally recognized product categories that fall under FDA's regulatory jurisdiction; however, this is not an exhaustive list. In general, FDA regulates the following [15]:

- **Foods, including**: dietary supplements, bottled water, food additives, infant formulas, other food products (although the USDA plays a lead role in regulating aspects of some meat, poultry, and egg products),
- **Drugs, including**: prescription drugs (both brand name and generic), non-prescription (over-the-counter) drugs,
- **Biologics, including**: vaccines, blood and blood products, cellular and gene therapy products, tissue and tissue products, allergenics,
- **Medical devices, including**: simple items like tongue depressors and bedpans, complex technologies such as heart pacemakers, dental devices, surgical implants, and prosthetics,
- **Electronic products that give off radiation, including**: microwave ovens, X-ray equipment, laser products, ultrasonic therapy equipment, mercury vapor lamps, sunlamps,
- **Cosmetics, including**: color additives found in makeup and other personal care products, skin moisturizers and cleansers, nail polish and perfume,
- **Veterinary products, including**: livestock feeds, pet foods, veterinary drugs and devices,
- **Tobacco products, including**: cigarettes, cigarette tobacco, roll-your-own tobacco, smokeless tobacco.

ENFORCEMENT EXAMPLE

Washington, D.C., May 7, 2012, Abbott Laboratories Inc. plead guilty and agreed to pay $1.5 billion to resolve its criminal and civil liability arising from the company's unlawful promotion of the prescription drug Depakote for uses not approved as safe and effective by FDA. The resolution is the second-largest payment by a drug company [16].

4.3.1.3 Public Health Implications of the FDCAct

The FDCAct, with its amendments, has resulted in the removal of unsafe food additives from the U.S. food supply, tighter pesticide regulations on levels of in food sources, and review of therapeutic drugs intended for medical use [17]. These actions benefit the public's health and represent primary prevention measures.

Of note for the prevention of foodborne illness is FDA's *Food Code*, which is a set of guidelines that represent best practices for the retail and food service industries [18]. The *Food Code* was first issued by FDA in 1993 and is currently updated every 4 years. According to FDA, more than 1 million retail and food service establishments use the *Food Code's* provisions as a model to develop or update their own food safety rules. While the *Food Code* is not mandated of states, a survey found that 48 of the 56 U.S. states and territories, which cover 79% of the U.S. population, have adopted food safety codes modeled after the *Food Code* [19]. As environmental health policy, the widespread voluntary adoption of the *Food Code* by states, territories, and the food industry is an example of how prevention of foodborne illness can be reduced by the adoption of a common set of food safety practices.

The removal or reduction of antibiotics in the American food chain addresses a long-standing concern of medical and public health agencies that excessive use of antibiotics is a problem. In particular, reliance too often on antibiotics has led to pathogens' developing drug resistance. FDA had initiated work in 1977 under its FDCAct authorities on how to limit antibiotics in livestock food. But the agency's efforts languished until litigated by environmental and public health groups. The plaintiffs argued that using common antibiotics in livestock feed has contributed to the rapid growth of antibiotic-resistant bacteria in both animals and humans. Further, plaintiffs cited data that indicated antibiotic-resistant infections cost Americans more than $20 billion annually. The litigation led to a federal judge on March 22, 2012, to order FDA to start proceedings to withdraw approval for the use of common antibiotics in animal feed, unless makers of the drugs could produce evidence that their use is safe. The court cited concerns that overuse is endangering human health by creating antibiotic-resistant "superbugs" [20].

ENFORCEMENT EXAMPLE

Washington, D.C., October 1, 2015

Three former officials of the Peanut Corporation of America were sentenced to prison for their roles in shipping Salmonella-positive peanut products. On September 21, 2015, PCA's former president received 28 years in federal prison, the largest criminal sentence ever given in a food safety case. The tainted food caused more than 700 reported cases of Salmonella poisoning in 46 states, including nine deaths [21].

In response to the aforementioned court decision, the agency chose a strategy to work with industry to protect public health by providing documents to help phase out the use of medically important antimicrobials in food animals for production purposes (e.g., to enhance growth or improve feed efficiency), and to bring the therapeutic uses of such drugs (to treat, control, or prevent specific diseases) under the oversight of licensed veterinarians.

In particular, the 2012 document, *New Animal Drugs and New Animal Drug Combination Products Administered in or on Medicated Feed or Drinking Water of Food-Producing Animals: Recommendations for Drug Sponsors for Voluntarily Aligning Product Use Conditions with GFI #209* (Guidance #213), provides guidance for drug companies to voluntarily revise the FDA-approved labeled use conditions to (1) remove the use of antimicrobial drugs for production purposes; (2) add, where appropriate, scientifically supported disease treatment, control, or prevention uses; and (3) change the marketing status from over-the-counter to Veterinary Feed Directive for drugs administered through feed or to prescription status for drugs administered through water in order to provide for veterinary oversight or consultation [22]. In brief, FDA chose to work with the drug and livestock industries for their voluntary phase-out of some antibiotics used in livestock production [23].

This FDA strategy of voluntary action by industrial sources has seemed ineffective. For example, FDA announced in 2015 that U.S. sales of medically important antibiotics approved for use in livestock rose by 23% between 2009 and 2014, again raising concerns about risks to humans from antibiotic-resistant bacteria [24].

In contrast to the federal strategy, California has enacted the strictest law on antibiotic use in farms. The state enacted a law to sharply limit the use of antibiotics in farm animals, making it the first state to ban the routine use of the drugs in animal agriculture. The law bans the use of medically important antibiotics to promote growth in cows, chickens, pigs, and other animals raised for profit. Meat producers will only be able to administer the drugs with the approval of a veterinarian when animals are sick, or to prevent infections when there is an "elevated risk" [25].

In departure from FDA's stance on antibiotics in livestock, the agency has taken action on removal of trans fats from the U.S. food supply and, separately, acted to prohibit three chemicals used in food packaging. Regarding the former, in June 2015 FDA finalized its determination that partially hydrogenated oils (PHOs), the primary dietary source of artificial trans fat in processed foods, are not "generally recognized as safe" or GRAS for use in human food. Food manufacturers had 3 years to remove PHOs from products [26]. Trans fat is associated with the clogging of arteries and its removal contributes to improved public health.

In another action, in January 2016, FDA announced that it would withdraw its approval for three chemicals used to make grease, stain, and water-repelling food packaging. The banned chemicals are all perfluorinated compounds (PFCs), a class of chemicals used to coat products like pizza boxes, pastry wrappers, take-out food containers, paper plates, and non-stick cookware. In lab studies, PFCs have been linked to adverse effects on hormones, reproductive, developmental, neurological, and immune systems, and to certain cancers [27]. In 2016, FDA also announced that it had begun reviewing certain food additives for possible restrictions.

Regarding food additives, some environmental groups, supplemented by social media campaigns, have questioned the safety of some food additives. In response, in advance of any FDA regulatory action, several food companies have begun to voluntarily remove some food additives. In particular, Taco Bell and Pizza Hut committed in 2015 to remove artificial flavorings and coloring from their food products [28]. Similarly, cereal manufacturers Kellogg and General Mills both plan to stop using artificial colors and flavors in their cereal and other food products by the end of 2018 [29]. Further, Kraft is removing artificial preservatives from its most popular individually wrapped cheese slices [30]. Some of this voluntary action by food manufacturers might be influenced by their anticipation of food labeling revisions by FDA, perhaps leading to closer scrutiny by consumers of food products.

4.3.2 U.S. FEDERAL MEAT INSPECTION ACT, 1906

The U.S. diet has always contained animal-derived protein from hunted feral animals or farmed livestock. Meat as a food source is the subject of the FMIAct of 1906. And as will be described, the act's provisions of meat inspection by government or private sources and removal of contaminated meat from the food chain have brought public health benefits.

4.3.2.1 History

Consumption of meat and meat products has long been part of the human diet, although debate continues about the ethics of raising animals as a food source. Our ancestors, whether indigenous people or colonists, hunted the forests, plains, and bodies of water for birds, mammals, fish, and shellfish as food sources. With the passage of time, rural Americans grew their own food in gardens and processed domesticated animals into meat and meat products. Farmers slaughtered animals in the fall and winter when temperatures were cool so as to diminish the deterioration of meat products. Salt rubbed into the meat and smoke from wood fires was used to preserve meats so that consumption of the meat could occur during warmer seasons. A family's meat quality and personal health protection were therefore at the mercy of a farmer's skill and resources in food preservation. Government had no role to play in what were essentially personal matters of family diet and health.

> Until 1906, U.S. states had the primary responsibility for protecting the public against impure food, including meat products.

As villages and cities grew in numbers and population, meat was supplied by butchers and sold in butcher shops. In the 19th century, cities such as Chicago and Cincinnati became renowned as centers of the meatpacking industry. As matters of public health, how animals were slaughtered and under what conditions were matters of indifference to the U.S. public. During this period, until 1906, states had the primary responsibility for protecting the public against impure food, including meat products. Needless to say, food inspection programs varied considerably between the states.

The public's ignorance of the conditions in the meatpacking industry began to change in the early years of the 20th century. In particular, a major influence on public opinion was Upton Sinclair's 1906 book, *The Jungle*, which graphically described unsanitary conditions and inhumane slaughter of animals in the Chicago meatpacking industry. Sinclair's book, much like Rachael Carson's book *Silent Spring* 56 years later, served to turn a spotlight on a major environmental health problem. On June 30, 1906, Congress enacted the Federal Meat Inspection Act (FMIAct). The act was substantially amended by the Wholesome Meat Act of 1967. The FMIAct was the first federal government involvement in the food safety of meat and meat products.

The primary goals of the Meat Inspection Act, as amended, are to prevent adulterated or misbranded livestock and products from being sold as food and to ensure that meat and meat products are slaughtered and processed under humane and sanitary conditions. These requirements apply to animals and their products produced and sold within U.S. states and territories as well as to imports, which must be inspected under equivalent foreign standards [31]. Excerpted key provisions of the Meat Inspection Act follow. The "Secretary" refers to the Secretary of the U.S. Department of Agriculture [32]:

Section 602, Congressional Statement of Findings: "Meat and meat food products are an important source of the Nation's total supply of food. They are consumed throughout the Nation and the major portion thereof moves in interstate or foreign commerce. It is essential in the public interest that the health and welfare of consumers be protected by assuring that meat and meat food products distributed to them are wholesome, not adulterated, and properly marked, labeled, and packaged [...]."

The primary goals of the FMIAct, as amended, are to prevent adulterated or misbranded livestock and products from being sold as food and to ensure that meat and meat products are slaughtered and processed under humane and sanitary conditions.

ENFORCEMENT EXAMPLE

Omaha, Nebraska, November 3, 2015

An Omaha, Neb., company recalled nearly 84 tons of ground beef after USDA inspectors traced *E. coli* to the company. *E. coli* is deadly and can cause dehydration, bloody diarrhea, and abdominal cramps. All 167,427 pounds were produced on October 16, 2015, for sale in 60- and 80-pound boxes shipped to retailers nationwide [33].

Section 603: Inspection of Meat and Meat Food Products: (1) Examination of animals before slaughtering; diseased animals slaughtered separately and carcasses examined. "For the purpose of preventing the use in commerce of meat and meat food products which are adulterated, the Secretary shall cause to be made, by inspectors appointed for that purpose, an examination and inspection of all cattle, sheep, swine, goats, horses, mules, and other equines before they shall be allowed to enter into any slaughtering, packing, meat-canning, rendering, or similar establishment, in which they are to be slaughtered and the meat and meat food products thereof are to be used in commerce [...]."

(2) Humane Methods of Slaughter: "For the purpose of preventing the inhumane slaughtering of livestock, the Secretary shall cause to be made, by inspectors appointed for that purpose, an examination and inspection of the method by which cattle, sheep, swine, goats, horses, mules, and other equines are slaughtered and handled in connection with slaughter in the slaughtering establishments inspected under this chapter [...]."

Section 604: Post Mortem Examination of Carcasses and Marking or Labeling; Destruction of Carcasses Condemned; Reinspection: "For the purposes hereinbefore set forth the Secretary shall cause to be made by inspectors appointed for that purpose a post mortem examination and [...] found to be not adulterated shall be marked, stamped, tagged, or labeled as 'Inspected and passed'; and said inspectors shall label, mark, stamp, or tag as 'Inspected and condemned' all carcasses and parts thereof of animals found to be adulterated; and all carcasses and parts thereof thus inspected and condemned shall be destroyed for food purposes by the said establishment in the presence of an inspector [...]."

Section 606: Inspectors of Meat Food Products; Marks of Inspection; Destruction of Condemned Products; Products for Export: "For the purposes hereinbefore set forth the Secretary shall cause to be made, by inspectors appointed for that purpose, an examination and inspection of all meat food products prepared for commerce in any slaughtering, meat-canning, salting, packing, rendering, or similar establishment, and for the purposes of any examination and inspection and inspectors shall have access at all times, by day or night, whether the establishment be operated or not, to every part of said establishment; [...]."

In summary, the FMIAct, as amended, provides a framework to inspect, label, and enforce standards of meat and meat products. Inspections include both visual examination of carcasses as well as tests for microbacterial contamination. Federal meat inspectors bear the public health responsibility for approving or condemning meat or meat products intended for human consumption. As an issue of environmental health policy, some of the inspection work has been delegated to company inspectors, with USDA meat inspectors performing related, but different duties. This change in policy will be described subsequently.

4.3.2.2 Public Health Implications of the FMIAct

Pathogens in meat and meat products have the potential to cause human illness. There, however, are few data on the incidence of foodborne illnesses specific to meat and meat products. That does not mean that foodborne illnesses from meat and other foods are inconsequential. Quite to the contrary. One source examined 3,500 foodborne food outbreaks, representing 115,700 individual illnesses between the years 1990 and 2003, and found that beef and beef dishes were associated with 338 outbreaks and 10,795 cases, which represented about 9% of the total cases [34]. Moreover, it is known that contaminated meat has been associated with individual outbreaks of illness. For example, in 1996, CDC investigated an outbreak of *Salmonella* serotype Thompson infections that were associated with a restaurant in Sioux Falls, South Dakota. Fifty-two infections were found in persons who had eaten food prepared by the restaurant. Results of the investigation revealed that cooking times and storage temperatures for roast beef were inadequate to prevent *Salmonella* proliferation [35].

In 2002, the USDA announced more stringent regulations that were intended to reduce *E. coli* contamination of meat and meat products, particularly ground beef. In support of the revised regulations, USDA noted that 43% of animal carcasses were contaminated with *E. coli* [36]. Moreover, the department referenced CDC data showing foodborne transmission of *E. coli* annually causes more than 62,000 illnesses and 52 deaths. Under the proposed regulations, no slaughter plants would be exempt from random *E. coli* testing (some small production meat processors were previously exempt). Further, the new regulations would require meat processors to add microbiological testing to actions required of them.

Under the FMIAct, the USDA must regulate the operations of meatpackers for purposes of preventing contaminants in meat from reaching consumers. However, how

USDA develops and enforces its meat inspection authorities have historically been subject to policy challenges. The meatpacking industry has argued that federal meat inspectors should have lesser authority to inspect meat and meat products, asserting that such inspections impede a plant's productivity. The industry preferred an inspection system whereby meat inspections would be conducted by a plant's personnel, but overseen by federal meat inspectors. To date, the industry's proposal has not been fully implemented.

In 2014, USDA revised its chicken slaughter inspection rule to allow greater involvement by company inspectors. Specifically, USDA states, "the new rule establishes a New Poultry Inspection System (NPIS) for young chicken and all turkey slaughter establishments. The NPIS does not replace the current Streamlined Inspection System (SIS), the New Line Speed Inspection System (NELS), or the New Turkey Inspection System (NTIS). As such, young chicken and turkey slaughter establishments may choose to operate under the NPIS or might continue to operate under their current inspection system, i.e., SIS, NELS, NTIS, or Traditional Inspection, as modified by this final rule. The NPIS is designed to facilitate pathogen reduction in poultry products by shifting Agency resources to allow Food Safety and Inspection Service (FSIS) inspectors to perform more offline inspection activities that are more effective in ensuring food safety, while providing for a more efficient and effective online carcass-by-carcass inspection [...]."

Key elements of the NPIS include: "(1) Requiring that establishment personnel sort carcasses and remove unacceptable carcasses and parts before the birds are presented to the FSIS carcass inspector; (2) shifting Agency resources to conduct more offline inspection activities that are more effective in ensuring food safety, which will allow for one offline verification inspector per line per shift and will reduce the number of online inspectors to one; (3) replacing the Finished Product Standards (FPS), which will apply to establishments that continue operating under SIS, NELS, and NTIS, with a requirement that establishments that operate under the NPIS maintain records to document that the products resulting from their slaughter operations meet the definition of ready-to-cook (RTC) poultry; and (4) authorizing young chicken slaughter establishments to operate at a maximum line speed of 140 birds per minute (bpm), provided that they maintain process control" [37].

Reaction to this revised rule was mixed. Some consumer groups disavowed the increased role of company inspectors, which lessened the inspection of USDA food inspectors. In contrast, some public health groups supported USDA's increased attention to assays of poultry products for the presence of pathogens.

In addition to the FMIAct, analogous acts pertain to poultry and egg products [38]. The Poultry Products Inspection Act and the Egg Products Inspection Act mandate inspections of producers of those products and authorize the USDA to take actions similar to those in the FMIAct in order to prevent contaminated poultry and eggs from causing foodborne illnesses. The Food Safety and Inspection Service of the USDA is the administrative unit that bears the responsibility for enforcing the provisions of the FMIAct, the Poultry Products Inspection Act, and the Egg Products Inspection Act. All three acts require states to cooperate with the USDA and require the states to establish their own statutes that comply with the three federal statutes. This is another example of federalism in action, a characteristic of the main body of federal environmental health legislation.

4.3.3.3 Organic Animal Welfare Rules, 2017

A group of farmers and advocates pushed for many years to see new rules, called the Organic Livestock and Poultry Practices (OLPP), put in place. Developed by the Obama administration the rule was published in the *Federal Register* on January 19, 2017, by the USDA's Agricultural Marketing Service and was scheduled to go into effect on May 14, 2018. The OLPP specified a set of standards for organic livestock and poultry designed to minimize animals' stress, facilitate natural behaviors, and promote well-being. Championed throughout the organic industry—from farmers to consumer groups to retailers and animal-welfare advocates—the OLPP was intended as a course correction for the $43 billion food industry, formulated to bring organic dairy, eggs, and meat production into line with consumers' expectations of higher animal welfare. However, on March 12, 2018, the Trump administration's USDA announced that it would withdraw the OLPP rules, ending the possibility that they would be implemented. The USDA stated that the OLPP rule exceeded the department's statutory authority [39]. However, on April 14, 2020, the U.S. District Court for the District of Maryland vacated the USDA rule, concluding that it violated the Administrative Procedure Act because the 2018 rule differed significantly from the administration's 2017 interim rule setting up the final standards [40].

4.3.3 NUTRITION LABELING AND EDUCATION ACT, 1990

For several years, Members of Congress were aware of consumer and industry interest in nutrition labeling, leading to various legislative acts that addressed some of the issues of nutrition labeling. These preliminary Congressional

> The Nutrition Labeling and Education Act (NLEA) of 1990 was the most significant food labeling legislation in 50 years.

efforts culminated in November 1990 with the passage of the Nutrition Labeling and Education Act (NLEA) of 1990. As described by the Nutrition Labeling Committee of the Institute of Medicine, this was the most significant food labeling legislation in 50 years [41]. The NLEA amended the FDCAct to give FDA explicit authority to require nutrition labeling on most food packages and specified the nutrients to be listed on the nutrition label.

The NLEA also required that nutrients be presented in the context of a consumer's daily diet; specified that serving sizes should represent "an amount customarily consumed and which is expressed in a common household measure that is appropriate to the food"; and provided for a voluntary nutrition labeling program for raw fruits, vegetables, and fish. The NLEA also required standard definitions to be developed by FDA that characterized the level of nutrients and required that FDA provide for approved health claims.

The NLEA's requirements for the content of the nutrition label were very similar to those in FDA's 1990 proposal except that the NLEA included complex carbohydrates and sugars in the list of required nutrients. It also permitted the agency to add or delete nutrients based on a determination that such a change would "assist consumers in maintaining healthy dietary practices." General principles for nutrient content claims and the definition of terms for claims to be allowed were also proposed, as were general principles for health claims.

The NLEA pertains only to those labels of food products regulated by FDA, which has label authority over the majority of foods. However, meat and poultry product labels are under the authority of the USDA, and alcoholic beverage product labels are under the authority of the Alcohol and Tobacco Tax and Trade Bureau of the Department of the Treasury [41].

4.3.4 FDA AMENDMENTS ACT OF 2007

The Food and Drug Administration Amendments Act of 2007 (FDAAA) was signed into law on September 27, 2007, by U.S. President George W. Bush (R-TX). This law reviewed, expanded, and reaffirmed several existing pieces of legislation regulating the FDA. These changes allowed the FDA to perform more comprehensive reviews of potential new drugs and devices. It was passed unanimously by the Senate. The FDAAA extended the authority of FDA to levy fees to companies applying for approval of drugs, expanded clinical trial guidelines for pediatric drugs, and created the priority review voucher program, among other items.

4.3.5 HEALTHY, HUNGER-FREE KIDS ACT, 2010

Improving child nutrition is the focal point of the Healthy, Hunger-Free Kids Act of 2010. The legislation authorized funding and set policy for USDA's core child nutrition programs: the National School Lunch Program, the School Breakfast Program, the Special Supplemental Nutrition Program for Special Supplemental Nutrition Program for Women, Infants, and Children, the Summer Food Service Program, and the Child and Adult Care Food Program. The Act allowed USDA, for the first time in more than 30 years, the opportunity to make reforms to the school lunch and breakfast programs by improving the critical nutrition and hunger safety net for millions of children. The Act strengthened nutrition standards for meals and beverages provided through the National School Lunch, Breakfast, and Smart Snacks Programs, affecting 50 million children daily at 99,000 U.S. schools.

> For children in poverty, the risk of obesity declined substantially each year after the act's implementation.

Public health researchers at Harvard tested whether the legislation was associated with reductions in child obesity risk over time using an interrupted time series design for years 2003–2018 among 173,013 youth in the National Survey of Children's Health. The researchers reported, "We found no significant association between the legislation and childhood obesity trends overall. For children in poverty, however, the risk of obesity declined substantially each year after the act's implementation, such that obesity prevalence would have been 47 percent higher in 2018 if there had been no legislation. These results suggest that the Healthy, Hunger-Free Kids Act's science-based nutritional standards should be maintained to support healthy growth, especially among children living in poverty" [42].

4.3.6 FDA FOOD MODERNIZATION ACT, 2011

The FDA Food Modernization Act (FSMAct) was signed into law by President Barack Obama (D-IL) on January 4, 2011 [43]. This law is the most sweeping reform of U.S. food safety laws in more than 70 years. The FSMAct aims to ensure the U.S. food supply is safe by shifting the focus from responding to contamination to preventing it. Prevention of disease and disability is the cornerstone of public health and is the focus of the FSMAct.

The FSMAct provides FDA with new enforcement authorities designed to achieve higher rates of compliance with prevention and risk-based food safety standards and to better respond to and contain problems when they do occur. The law also gives FDA important new tools to hold imported foods to the same standards as domestic foods and directs FDA to build an integrated national food safety system in partnership with U.S. state, territorial, and local authorities.

The following are among FDA's key new authorities and mandates. Specific implementation dates specified in the law are noted in parentheses [43]:

Prevention: For the first time, FDA will have a legislative mandate to require comprehensive, science-based preventive controls across the food supply. This mandate includes:

- **Mandatory preventive controls for food facilities**: Food facilities are required to implement a written preventive controls plan. This involves: (1) evaluating the hazards that could affect food safety; (2) specifying what preventive steps, or controls, will be put in place to significantly minimize or prevent the hazards; (3) specifying how the facility will monitor these controls to ensure they are working; (4) maintaining routine records of the monitoring; and (5) specifying what actions the facility will take to correct problems that arise.
- **Mandatory produce safety standards**: FDA must establish science-based, minimum standards for the safe production and harvesting of fruits and vegetables. Those standards must consider naturally occurring hazards, as well as those that may be introduced either unintentionally or intentionally, and must address soil amendments (materials added to the soil such as compost), hygiene, packaging, temperature controls, and animals in the growing area and water.
- **Authority to prevent intentional contamination**: FDA must issue regulations to protect against the intentional adulteration of food, including the establishment of science-based mitigation strategies to prepare and protect the food supply chain at specific vulnerable points.

> The FSMA of 2011 is the most sweeping reform of U.S. food safety laws in more than 70 years. The act aims to ensure the U.S. food supply is safe by shifting the focus from responding to food contamination to preventing it.

Inspection and compliance: The FSMAct recognizes that preventive control standards improve food safety only to the extent that producers and processors comply with them. Therefore, it will be necessary for FDA to provide oversight, ensure compliance with requirements, and respond effectively when problems emerge. The FSMAct provides FDA with important new tools for inspection and compliance, including:

Mandated inspection frequency: The FSMAct establishes a mandated inspection frequency, based on risk, for food facilities and requires the frequency of inspection to increase immediately. All high-risk domestic facilities must be inspected within 5 years of enactment and no less than every 3 years thereafter. Within 1 year of enactment, the law directs FDA to inspect at least 600 foreign facilities and double those inspections every year for the next 5 years.

Records access: FDA will have access to records, including industry food safety plans, and the records firms will be required to keep documenting the implementation of their plans.

Testing by accredited laboratories: The FSMAct requires certain food testing to be performed by accredited laboratories and directs FDA to establish a program for laboratory accreditation to ensure that U.S. food testing laboratories meet high-quality standards (establishment of accreditation program due 2 years after enactment).

Response: The FSMAct recognizes that FDA must have the tools to respond effectively when problems emerge despite preventive controls. New authorities include:

Mandatory recall: The FSMAct provides FDA with the authority to issue a mandatory recall when a company fails to voluntarily recall unsafe food after being asked to by FDA.

Expanded administrative detention: The FSMAct provides FDA with a more flexible standard for administratively detaining products that are potentially in violation of the law (administrative detention is the procedure FDA uses to keep suspect food from being moved).

Suspension of registration: FDA can suspend the registration of a facility if it determines that the food poses a reasonable probability of serious adverse health consequences or death. A facility that is under suspension is prohibited from distributing food (effective 6 months after enactment).

Enhanced product tracing abilities: FDA is directed to establish a system that will enhance its ability to track and trace both domestic and imported foods. In addition, FDA is directed to establish pilot projects to explore and evaluate methods to rapidly and effectively identify recipients of food to prevent or control a foodborne illness outbreak (implementation of pilots due 9 months after enactment).

Additional recordkeeping for high-risk foods: FDA is directed to issue proposed rulemaking to establish recordkeeping requirements for facilities that manufacture, process, pack, or hold foods that the Secretary designates as high-risk foods (implementation due 2 years after enactment).

Imports: The FSMA gives FDA unprecedented authority to better ensure that imported products meet U.S. standards and are safe for U.S. consumers. New authorities include:

Importer accountability: For the first time, importers have an explicit responsibility to verify that their foreign suppliers have adequate preventive controls in place to ensure that the food they produce is safe (final regulation and guidance due 1 year following enactment).

Third-party certification: The FSMAct establishes a program through which qualified third parties can certify that foreign food facilities comply with U.S. food safety standards. This certification may be used to facilitate the entry of imports (establishment of a system for FDA to recognize accreditation bodies due 2 years after enactment.)

Certification for high-risk foods: FDA has the authority to require that high-risk imported foods be accompanied by a credible third-party certification or other assurance of compliance as a condition of entry into the U.S.

Voluntary qualified importer program: FDA must establish a voluntary program for importers that provides for expedited review and entry of foods from participating importers. Eligibility is limited to, among other things, importers offering food from certified facilities (implementation due 18 months after enactment).

Authority to deny entry: FDA can refuse entry into the U.S. of food from a foreign facility if FDA is denied access by the facility or the country in which the facility is located.

Enhanced partnerships: The FSMAct builds a formal system of collaboration with other government agencies, both domestic and foreign. In doing so, the statute explicitly recognizes that all food safety agencies need to work together in an integrated way to achieve our public health goals. The following are examples of enhanced collaboration:

State and local capacity building: FDA must develop and implement strategies to leverage and enhance the food safety and defense capacities of State and local agencies. The FSMAct provides FDA with a new multiyear grant mechanism to facilitate investment in State capacity to more efficiently achieve national food safety goals.

Foreign capacity building: The law directs FDA to develop a comprehensive plan to expand the capacity of foreign governments and their industries. One component of the plan is to address training of foreign governments and food producers on U.S. food safety requirements.

Reliance on inspections by other agencies: FDA is explicitly authorized to rely on inspections of other federal, state, and local agencies to meet its increased inspection mandate for domestic facilities. The FSMAct also allows FDA to enter into interagency agreements to leverage resources with respect to the inspection of seafood facilities, both domestic and foreign, as well as seafood imports.

Under the provision of the FSMAct, in 2016 FDA updated the requirements for food labeling under the terms of the Nutrition Labeling and Education Act of 1990. The new Nutrition Facts label will include the following [44]:

- An updated design to highlight "calories" and "servings," two important elements in making informed food choices.
- Requirements for serving sizes that more closely reflect the amounts of food that people currently eat. By law, the Nutrition Labeling and Education Act requires that serving sizes be based on what people actually eat.
- Declaration of grams and a percent daily value (%DV) for "added sugars."
- "Dual column" labels to indicate both "per serving" and "per package" calorie and nutrition information for certain multiserving food products that could be consumed in one sitting or multiple sittings.

- Updated daily values for nutrients like sodium, dietary fiber, and vitamin D, consistent with Institute of Medicine recommendations and the 2015–2020 Dietary Guidelines for Americans.
- Declaration of Vitamin D and potassium that will include the actual gram amount, in addition to the %DV. The %DV for calcium and iron will continue to be required, along with the actual gram amount.
- "Calories from Fat" will be removed because research shows the type of fat is more important than the amount. "Total Fat," "Saturated Fat," and "Trans Fat" will continue to be required.
- An abbreviated footnote to better explain the %DV.

The FDA asserts that it is also making minor changes to the Supplement Facts label found on dietary supplements to make it consistent with the Nutrition Facts label. Most food manufacturers will be required to use the new label by July 26, 2018. Manufacturers with <$10 million in annual food sales will have an additional year to comply with the new rules. The FDA plans to conduct outreach and education efforts on the new requirements.

In an additional action under the FSMAct, FDA announced in 2015 a final rule to add selenium to the list of required nutrients for infant formula and to establish both minimum and maximum levels of Se in infant formula. By amending regulations to add Se to the list of required nutrients for infant formula and establish a safe range for this use, the FDA is able to require manufacturers currently marketing infant formula in the U.S. to add Se within this safe range and to require any manufacturer newly entering the U.S. market to adopt this practice as well [45].

4.3.7 USDA FOOD SECURITY POLICIES

The U.S. Department of Agriculture (USDA) has responsibilities under various federal statutes to provide food assistance to both domestic and global groups in need of improved food security. Domestic programs of agriculture involve conducting research and providing services to U.S. agriculture programs and businesses. Additionally, the department has responsibilities for providing the following food assistance programs via its Food and Nutrition Service [46]:

- Supplemental Nutrition Assistance Program (SNAP): benefits for eligible low-income families,
- Special Supplemental Nutrition Program for Women, Infants, and Children (WIC) low-income women, infants, and children up to age 5 through nutritious foods, information on healthy eating, and healthcare referrals,
- Child Nutrition Programs: nutrition assistance for children,
- Nutrition Programs for Seniors: programs that focus on the needs of older Americans,
- USDA commodity distribution programs, including Commodity Supplemental Food Program for low-income elderly people at least 60 years.

In addition to these domestic programs of food assistance, USDA supports global food security through in-country capacity building, basic and applied research, and support for improved market information, statistics, and analysis. Since 2010, USDA has aligned appropriate programs to Feed the Future plans to support agriculture development in target countries and regions: Ghana, Kenya, East Africa, Bangladesh, Haiti, Guatemala, and Central America. USDA international food aid programs benefited about 34 million individuals globally with assistance valued at nearly $1.6 billion through the year 2010 [47].

4.3.8 U.S. State, Territorial, and Local Food Safety Policies

U.S. state, territorial, and local governments have the primary responsibility for enforcing food safety regulations. The authorities of states vary, as do the degrees of local government involvement. States typically establish standards for the transportation, storage, preparation, and serving of food by food service establishments. Local health departments conduct inspections of restaurants and commercial food processors, typically under authorities in state laws. Issuance of permits to food service establishments is at the heart of state food safety laws. Without approved permits, food service establishments cannot legally operate.

4.3.8.1 Example of a U.S. State's Food Safety Policies

As an illustration of one state's approach to food safety, consider the state of Georgia. In Georgia, two state agencies have the primary responsibilities for protecting the public against foodborne illness. One agency, the Division of Public Health, has the state's primary authority for illness attributable to food services. The other state agency, the Georgia Department of Agriculture, has the authority to regulate and inspect food supplies.

> U.S. state and local governments have the primary responsibility for enforcing food safety regulations.

Excerpts from the key public health food safety provisions administered by the Georgia Division of Public Health include the following [48]:

Section II 290-5-4-02: "Selected Provisions: (1)(a): It shall be unlawful for any person to operate a food service establishment, or mobile food unit, a temporary food service operation or a restricted food service operation without having first obtained a valid food service permit from the health authority pursuant to this chapter. [...] (d): The permit shall be prominently displayed at all times, [...]."

Section III 290-5-14-03: "Selected Provisions: Food Care: (1) Food Supplies: (a): Food shall be in sound condition, free from spoilage, filth, or other contamination and shall be safe for human consumption. (b) Food shall be obtained from approved sources that comply with all laws relating to food processing and shall have no information on the label that is false or misleading. [...] (d) Fluid milk and fluid milk products used or served shall be pasteurized and shall meet the Grade A quality standards as established by law. [...] (g) Only clean whole eggs, with shell intact and without cracks or checks, or pasteurized liquid, frozen, or dry eggs or pasteurized dry egg products shall be used [...]. (2) Food Protection: (a) At all times, including while being stored, prepared, displayed, served, or transported, food shall be protected from potential contamination, [...] (3) Food Storage: (a) Food, whether raw or prepared, if removed from the container or package in which it was obtained, shall be stored in an approved, clean, and covered container [...] (g) Enough conveniently

located refrigeration facilities or effectively insulated facilities shall be provided [...].
(4) Food Preparation: (a) Food shall be prepared with the least possible manual contact with suitable utensils, and on surfaces that prior to use have been cleaned, rinsed
and sanitized to prevent cross-contamination. (b) Raw fruits and vegetables shall be
thoroughly washed with potable water under pressure before being cooked or served.
[...] (5) Food Display and Service: (g) Food on display shall be protected from consumer contamination [...]. (6) Food Transportation: (a) During transportation, food
and food utensils shall be kept in covered containers or completely wrapped or packaged so as to be protected from contamination and spoilage."

Section XI 290-5-14-11: "Selected Provisions: Compliance Procedures: (1)
Permits: (a) Issuance: Permits shall be issued by the health authority. Such permits
shall be valid until suspended or revoked. (2) Inspections: (a) Inspection Frequency:
An inspection of a food service establish-
ment shall be performed at least twice
annually. [...] (b) Access: Representatives
of the health authority, after proper identification, shall be permitted to enter any
food service establishment or operation
[...]."

Without a permit from the state's
public health department (or county
health department, if delegated by
the state), no food service operations
are allowed to operate.

A moment of reflection on Georgia's food safety law and regulations shows a
program centered on permits issued to food service establishments. Without a permit
from the state's public health department (or county health department, if delegated
by the state), no food service operations are allowed to operate. Moreover, the state
can revoke a permit if sufficient unsanitary conditions are found by local health
department inspectors. Of particular note are *critical violations* found by health
inspectors, as distinguished from *minor violations*. Critical violations are those findings that have direct implications for the public's health, e.g., service personnel not
wearing protective gloves or food stored at temperatures that permit the growth of
bacteria.

The state's regulations provide detailed specifications on food transportation,
storage, preparation, and service. While the regulations and public health systems
of inspections and reporting are generally impressive, they are only as effective as
available budgets and personnel permit.

The Georgia Department of Agriculture's food safety authorities complement
those of the Georgia Division of Public Health. The department's primary food safety
authorities derive from several Georgia State laws and include the following [49]:

- Enforce state laws, rules, and regulations by conducting sanitation inspection of retail food stores, salvage food operations, mobile meat trucks, and
  rolling stores.
- Inspect food storage warehouses, wholesale bakeries, bottled water, and flavored drink processors, seafood processors, and wholesale fish dealers, and
  sanitation in establishments where food is handled and manufactured.
- Enforce federally mandated programs of inspection and sampling of dairy
  farms and dairy processing plants. This authority extends to the inspection
  of out-of-state mile products shipped to Georgia, along with the authority to

inspect tanker trucks, route trucks, and warehouses that are used to transport or store dairy products.

- Respond to consumers' inquiries about sanitary conditions relative to food and foodborne illness.

A comparison of the food safety authorities administered by the two Georgia State agencies shows both similarities and differences. As to similarities, both the Division of Public Health and the Department of Agriculture derive their food safety authorities from state laws. Without authorizing statutes, the agencies would have no specific food safety authority. Also, the prevention of foodborne illness is at the heart of both agencies' authorities and programs. This prevention focus is achieved primarily by requiring food supplies and food service establishments to be registered under state control and, second, to conduct inspections of food producers, transporters of food products, storage facilities where food products are stored, and food service establishments.

Regarding differences between the two agencies' food safety authorities, the Division of Public Health focuses on the **registration and inspection** of food service (emphasis added) establishments; whereas the Department of Agriculture focuses on registration and inspection of food producers, transporters, and those who store food, such as warehouse operators. As a policy observation, this kind of sharing of public health responsibility for food safety is much like the duality of responsibility found throughout environmental health. For example, on matters of toxic substances, EPA has primacy in controlling the release into the environment of substances that can harm human and ecological health; whereas the U.S. Public Health Service agencies conduct research on the toxicity and human health implications of toxic substances and work with states to collect surveillance health data and exposure data that can be used to help determine regulatory standards developed by EPA or other regulatory agencies.

4.3.8.2 Example of a County's Food Safety Policies

In addition to the State of Georgia's responsibilities in food protection, county and city health departments play a critical role as well. For example, Georgia's DeKalb County Department of Health's food protection unit reviews and approves plans for new food service establishments, issues permits, and conducts ongoing inspections. Approximately 1800 food establishments and services are inspected by the county's health department each year. The results from restaurant inspections are made available to the public by: (1) posting a copy of the inspection report in a prominent place in each restaurant inspected, and (2) placing the inspection reports on Dekalb County's website. As environmental health policy, providing the public with information with which to make personal health decisions is a matter of right-to-know.

Several environmental health policy issues pertain to food safety. Federalism is one such issue. The entry in 1906 of the federal government into the areas of meat inspection, food, drugs, and cosmetics, somewhat diminished the food safety role of free enterprise in the food industry. Heretofore, food safety was largely a matter of "let the buyer beware," supported by state food safety laws. Outbreaks of foodborne illnesses were considered then as a matter of personal health and consumer choice.

While individual consumer choice and an informed public remain essential for preventing foodborne illnesses, stronger federal and state laws, girded with local health departments' inspections, are essential for food safety.

4.3.8.3 Food Service Inspection Scores

Another policy issue is how to inform the public about food service establishments that fail to meet standards of food safety. Some local health departments place on their websites the results, current and past, of restaurant inspections. How these are presented to the public is a challenge. The inspection report must be factually accurate but should not create unrealistic fears in the public. This difficult balance in health communication has led some food safety authorities to suggest that internet posting of individual food service scores is inappropriate, possibly raising unreasonable fears in the public.

There are several arguments against posting food establishment's inspection reports on the internet or given to local newsmedia. Some inspectors have expressed concern that the public could be misled by unabridged inspection scores, citing problems in inspection procedures that do not clearly distinguish between

> A study of foodborne-disease hospitalizations found that restaurant hygiene grading with public posting of results was an effective means for reducing the incidence of foodborne disease.

critical and noncritical findings [50]. They note, depending on the kind of inspection system used, that a restaurant with a score of 95, based on a critical health finding like prepared food left unrefrigerated, would be seen as preferable to a restaurant with a score of 88, based primarily on administrative failures, such as inappropriate placement of the food inspection score within the food service establishment. Further, some food inspectors have expressed their concern that their professional relationship with food service managers can be hindered when inspection scores are made available to the public [50].

On the other hand, in support of communicating food service inspection reports to the public is the acceptance by many health departments that posted reports help improves food safety. In a study of foodborne-disease hospitalizations in Los Angeles County, California, it was found that restaurant hygiene grading with public posting of results was an effective means for reducing the incidence of foodborne disease [51]. Investigators reported a decrease of 13% in the number of foodborne diseases in the year following the implementation of a public posting program for restaurant inspections. As this study suggests, public perception can be a powerful motivator for change. Much like how the Toxics Release Inventory data have led to voluntary reductions of emission from industrial facilities, food establishments fear a poor rating of their services. Some, therefore, argue that public availability of inspection scores helps reinforce food quality standards and practices [51].

4.3.8.4 Food Sovereignty Laws

The food advocacy group La Vía Campesina defined "food sovereignty is the right of peoples to healthy and culturally appropriate food produced through ecologically sound and sustainable methods, and their right to define their own food and agriculture systems. It puts those who produce, distribute, and consume food at the heart of food systems and policies rather than the demands of markets and corporations."

The group asserts that it is a coalition of 182 organizations in 81 countries, advocating family-farm-based sustainable agriculture and was the group that coined the term "food sovereignty" [52].

Maine's first-in-the-nation food sovereignty law, An Act to Recognize Local Control Regarding Food Systems, allows local governments in the state to pass ordinances that exempt many direct-to-consumer food sales within city limits from burdensome state licensing and inspection requirements [53]. As of 2019, 45 Maine cities and towns have now passed food safety ordinances (FSOs). Local ordinances are popular with cities because they can factor local agricultural products into local food safety policies. However, Maine's food sovereignty law must comport with USDA food safety regulations.

A second kind of food policy established by three U.S. states is called food freedom acts. Wyoming was the first U.S. state to enact such a law. The general purpose of the Wyoming Food Freedom Act is to allow for the sale and consumption of homemade foods. Local food entrepreneurs can develop and sell to the public their food products without state review and approval. North Dakota enacted its own food freedom law in 2017, followed by Utah 1 year later. Aspiring entrepreneurs as well as existing farmers and ranchers are alleged to be beneficiaries of U.S. state food freedom acts. Some public health groups have raised concerns regarding potential food safety issues, but Wyoming's foodborne illness records do not reflect a problem [54].

Perspective: Regardless of which side one takes on the argument about the public's access to food inspection reports, the trend seems clear. The U.S. public will continue to want access to government information that has health and safety relevance to them. This trend has been accelerated because of the rapid growth of social media and the public's access to it. Moreover, the well-publicized newsmedia reports of occasional food poisonings have compounded the public's concerns and personal interests. The challenge is therefore not whether to report food establishment ratings, but how to do it in a responsible manner.

4.4   FOOD SECURITY IN THE U.S.

Even though the U.S. is an affluent country in many respects, including food production, the country's disparities in income and cultural structure have manifested in food insecurity for a portion of the country's population. Put simply, hunger exists in the U.S. The USDA conducts surveys of food patterns in the U.S. From these surveys, USDA reports that most U.S. households have consistent, dependable access to enough food for active, healthy living—they are food secure. But a minority of American households experience food insecurity at times during the year, meaning that their access to adequate food is limited by a lack of income and other resources. USDA's food and nutrition assistance programs increase food security by providing low-income households access to food, a healthful diet, and nutrition education.

The USDA also monitors the extent and severity of food insecurity in U.S. households through an annual, nationally representative survey. Reliable monitoring of food security contributes to the effective operation of the federal food assistance programs, as well as that of private food assistance programs and other government initiatives aimed at reducing food insecurity. This report presents statistics from the survey covering households' food security, food expenditures, and the use of federal food and nutrition assistance programs in 2014. Key findings in the report include:

- The estimated percentage of U.S. households that were food insecure remained essentially unchanged from 2013 to 2014; however, food insecurity was down from a high of 14.9% in 2011. The percentage of

  In 2014, 86.0% of U.S. households were food secure throughout the year. The remaining 14.0% (17.4 million households) were food insecure.

  households with food insecurity in the severe range—described as very low food security—was unchanged.
- In 2014, 86.0% of U.S. households were food secure throughout the year. The remaining 14.0% (17.4 million households) were food insecure. Food-insecure households (those with low and very low food security) had difficulty at some time during the year providing enough food for all their members due to a lack of resources. The changes from 2013 (14.3%) and 2012 (14.5%) to 2014 were not statistically significant; however, the cumulative decline from 14.9% in 2011 was statistically significant.
- In 2014, 5.6% of U.S. households (6.9 million households) had very low food security, unchanged from 5.6% in 2013. In this more severe range of food insecurity, the food intake of some household members

  Children were food insecure at times during 2014 in 9.4% of U.S. households with children (3.7 million households).

  was reduced and normal eating patterns were disrupted at times during the year due to limited resources.
- Children were food insecure at times during the year in 9.4% of U.S. households with children (3.7 million households), essentially unchanged from 9.9% in 2013. These households were unable at times during the year to provide adequate, nutritious food for their children [55].

Food insecurity in the U.S. has healthcare consequences. A study by researchers at the Boston University School of Medicine used data from the USDA, Census Bureau, and research on food security journal publications between 2005 and 2015 to estimate these healthcare costs [56]. The investigators examined the costs of treating diseases and health conditions associated with household food insecurity. They included earnings lost when people took time off work because of these illnesses or to care for family members with illnesses related to food insecurity. The investigators estimated that the absence of food security in the U.S. carries enormous healthcare costs, more than $160 billion in 2014. For sake of comparison, this figure is about five times the whole year 2016 budget request of the U.S. National Institutes of Health, the country's foremost federal health research agency.

4.5   GLOBAL FOOD SECURITY

Food security is assured if food sustainability exists. In that regard, one source, the Barilla Center for Food & Nutrition, provides surveillance of national food sustainability. Their Food Sustainability Index (FSI) ranks 67 countries on food system sustainability. It is a quantitative and qualitative benchmarking model constructed from 38 indicators and 90 individual metrics that measure the sustainability of food systems across three categories: food loss and waste, sustainable agriculture, and nutritional challenges. The index has three key types of performance indicators—environmental, societal, and economic [57]. While each country is ranked on each of the three ranking categories, an overall ranking is also derived. The top five countries on their FSI scale were France, Netherlands, Canada, Finland, and Japan. The U.S. ranked 26th, a low ranking attributable primarily to a low score on nutritional challenges—in other words, poor nutritional diets. Food insecurity is a consequence of several factors, which are discussed in the following sections.

4.5.1   CHALLENGES OF GLOBAL FOOD INSECURITY

In 2018 the UN's Food and Agriculture Organization (FAO) released its annual summary of global food security [58]. The FAO reported "In 2017, the number of undernourished people is estimated to have reached 821 million—around one person out of every nine in the world. Undernourishment and severe food insecurity appear to be increasing in almost all subregions of Africa, as well as in South America, whereas the undernourishment situation is stable in most regions of Asia. The decline is more pronounced in developing regions, despite significant population growth. In recent years, progress has been hindered by slower and less inclusive economic growth as well as political instability in some developing regions, such as Central Africa and western Asia" [58].

Of special note in the FAO report is that in 2017, 7.5% of children less than five—50.5 million—were affected by wasting (low weight for height) consequently putting them at a higher risk of mortality.

> In 2018, the FAO reported around one person out of every nine in the world is undernourished.

The FAO cautions, "If we are to achieve a world without hunger and malnutrition in all its forms by 2030, it is imperative that we accelerate and scale up actions to strengthen the resilience and adaptive capacity of food systems and people's livelihoods in response to climate variability and extremes."

This food warning from the FAO was corroborated by an analysis of a 2018 United Nations Conference on Trade and Development report. "In general, countries in Latin America, East Africa and South Asia are net food exporters, while most of the rest of Asia and Africa remain net food importers" [59]. Asia is "unable to feed itself"—and needs to invest another $800 billion in the next 10 years to produce more food and meet the region's needs, according to a report of the Asia Food Challenge Report. The growing population of Asia is a major factor in food security investments.

Perspective: These FAO data are discouraging because the FAO estimates that 821 million people still lack adequate food nourishment. For sake of perspective, this number is approximately the 2017 combined populations of Indonesia, Brazil, Pakistan, and Russia [60]. And as will be described in this section, several factors are contributing to food insecurity in areas of the world, but climate change has become the dominant concern to the FAO.

4.5.2 THREATS TO GLOBAL FOOD SECURITY

While as noted in the prior section, according to FAO data, millions of people lack adequate food nourishment, some encouraging data exist about global food production, specifically, from a study conducted by researchers at the Potsdam Institute for Climate Impact Research in Germany [61]. The researchers were interested in the relationship between food waste and the waste's generation of greenhouse gases (GHGs). Their study provides a systematic approach to estimate consumer-level food waste on a country scale and globally, based on food availability and requirements. The study revealed that in the year 2010, food availability was 20% higher than was required on a global scale. Surplus between food availability and requirements of a given country was considered as food waste. The global food requirement changed from 2,300 to 2,400 kcal/cap/day during the last 50 years, while food surplus grew from 310 to 510 kcal/cap/day. Similarly, GHG emissions related to the food surplus increased from 130 Mt CO_2 eq/year to 530 Mt CO_2 eq/year, an increase of more than 300%. Moreover, the global food surplus might increase up to 850 kcal/cap/day, while the total food requirement will increase only by 2%–20% by 2050. Consequently, GHG emissions associated with the food waste may also increase tremendously to 1.9–2.5 Gt CO_2 eq/year.

Reflection on the FAO report and the Potsdam study leads to the conclusion that food security is greatly influenced by food distribution systems. Put into different words, surplus food isn't getting to those in need. Moreover, many factors contribute to the lack of food security domestically and globally. Six of these factors will be discussed herein. As will be evident, all six are factors derivative of anthropogenic causes.

4.5.2.1 COVID-19 Threat

As described in Volume 2, Chapter 1, the global pandemic of COVID-19, commencing in early 2020, caused great loss of life and associated illnesses, economic stress, and sociopolitical turmoil. Among the impacts are consequences to food supplies.

On a global level, regarding the impact of COVID-19 on food security, the Food and Agriculture Organization (FAO) observed, "Although disruptions in the food supply chain are minimal through mid-2020, challenges have been already experienced in terms of logistics. Food needs to move across borders with no restrictions and in compliance with existing food safety standards. To mitigate the pandemic's impacts on food and agriculture, FAO urges countries to meet the immediate food needs of their vulnerable populations, boost their social protection programmes, keep global food trade going, keep the domestic supply chain gears moving, and support smallholder farmers' ability to increase food production. Countries with existing humanitarian crises are particularly exposed to the effects of the COVID-19 pandemic. Even as their own domestic needs may be rising as a result of the pandemic, it is critical that donor countries ensure continued delivery of humanitarian assistance where food insecurity is already high. The disease does not recognize borders. If left unchecked in one place, the entire human community remains at risk" [62].

4.5.2.2 Unhealthful Diets

A form of food insecurity that may not be obvious is the problem of food consumed but inadequate for nutritional purposes, i.e., unhealthful diets. A Global Burden of Disease study evaluated the consumption between 1990 and 2017 of major foods and nutrients across 195 countries and quantified the impact of poor diets on morbidity and mortality from non-communicable diseases (specifically cancers, cardiovascular diseases, and diabetes) [63]. The study examined 15 dietary elements— diets low in fruits, vegetables, legumes, whole grains, nuts and seeds, milk, fiber, calcium, seafood omega-3 fatty acids, polyunsaturated fats, and diets high in red meat, processed meat, sugar-sweetened beverages, trans-fatty acids, and sodium.

> Diets high in sodium, low in whole grains, and low in fruit together accounted for more than half of all diet-related deaths globally in 2017.

Overall, an estimated 11 million deaths were attributable in 2017 to poor diet. Diets high in sodium, low in whole grains, and low in fruit together accounted for more than half of all diet-related deaths globally in 2017. The causes of these deaths included 10 million deaths from cardiovascular disease, 913,000 cancer deaths, and almost 339,000 deaths from type 2 diabetes. Deaths related to diet had increased from 8 million in 1990, largely due to increases in the population and population aging. The researchers observed that people in almost every region of the world could benefit from rebalancing their diets to eat optimal amounts of various foods and nutrients [63]. What then is a healthful food diet? The National Heart, Lung, And Blood Institute provides the following healthful eating plan:

- Emphasize vegetables, fruits, whole grains, and fat-free or low-fat dairy products
- Include lean meats, poultry, fish, beans, eggs, and nuts
- Limit saturated and *trans* fats, sodium, and added sugars
- Control portion sizes [64].

Concerning added sugars, a multinational team of European university researchers investigated associations between soft drink consumption and mortality in ten European countries [65]. In this population-based cohort study of 451,743 individuals from ten countries in Europe, greater consumption of total, sugar-sweetened, and artificially sweetened soft drinks was associated with a higher risk of all-cause mortality. Consumption of artificially sweetened soft drinks was positively associated with deaths from circulatory diseases, and sugar-sweetened soft drinks were associated with deaths from digestive diseases. Results of this study appear to support ongoing public health interventions to reduce the consumption of soft drinks and added sugars. One such intervention has occurred in Chile.

In 2016 the government of Chile, a country with some of the world's highest obesity rates, adopted regulations that included advertising restrictions on unhealthful foods, bold front-of-package warning labels, and a ban on junk food in schools [66]. Consumption of sugar-sweetened drinks dropped nearly 25% in the 18 months following the issuance of the regulations. Correspondingly a 5% increase in purchases of bottled water, diet soft drinks, and fruit juices without added sugar occurred in Chile. Since then, Peru, Uruguay, Israel have adopted Chilean-style front-of-package labels; Brazil and Mexico are expected to finalize similar labels in 2020, and a dozen other countries are considering them as well, according to an analyst [66]. The linkage of a major health problem, obesity, to a policy of consumer education represents a significant public health accomplishment.

4.5.2.3   Organic Food Supplies

There is emerging research that suggests a form of healthful diet includes the consumption of organic foods. Produce can be called organic according to the U.S. Department of Agriculture if it is certified to have grown on soil that had no prohibited substances applied for 3 years prior to harvest. Prohibited substances include most synthetic fertilizers and pesticides. A study of the consumption of organic food associated with health outcomes was the subject of interest to several French public health researchers [67]. Based on a cohort study of 68,946 French adults, the investigators found that people who reported eating more organic food were 25% less likely to develop cancer. It is noteworthy that those who ate mostly organic food were 73% less likely to develop non-Hodgkin lymphoma.

However, researchers cautioned that research investigating the underlying factors involved with this association is needed to implement adapted and targeted public health measures for cancer prevention. However, authors of a comprehensive review of published literature commented, "Depending on the outcome of interest, associations between organic vs conventional food consumption and health outcome therefore need to be carefully adjusted for differences in dietary quality and lifestyle factors, and the likely presence of residual confounding needs to be considered. In children, several studies have reported a lower prevalence of allergy and/or atopic disease in families with a lifestyle comprising the preference of organic food. However, organic food consumption is part of a broader lifestyle in most of these studies and associated with other lifestyle factors" [68].

4.5.2.4   The Healthy, Hunger-Free Kids Act, 2010

The Healthy, Hunger-Free Kids Act of 2010 was signed into law by President Barack Obama (D-IL) on December 13, 2010. The bill funded child nutrition programs and free lunch programs in U.S. schools through 2015 [69]. According to the U.S. Department of Agriculture

> For ten countries in Europe, greater consumption of total, sugar-sweetened, and artificially sweetened soft drinks was associated with a higher risk of all-cause mortality.

(USDA), for the 2012–2013 school year, 21.5 million U.S. children received a free lunch or reduced-price lunch at school. In addition, the bill set new nutrition standards for schools and allocated $4.5 billion for their implementation. The new nutrition standards were a point initiative of First Lady Michelle Obama in her campaign against childhood obesity. The Healthy, Hunger-Free Kids Act allowed the USDA, for the first time in 30 years, the opportunity to make reforms to the school lunch and breakfast programs by improving the critical nutrition and hunger safety net for millions of children. In particular, school food programs gave greater emphasis to diets containing more vegetables, fruit, skim milk, and grains and fewer saturated fats and sugar.

As a matter of public health, on January 17, 2020, Michelle Obama's birthday, the Trump administration's USDA announced proposed new rules for the Food and Nutrition Service that would allow schools to cut the amount of vegetables and fruits required at lunch and breakfasts while giving them license to sell more pizza, burgers, and fries to students. The agency is responsible for administering nutritional programs that feed nearly 30 million students at 99,000 schools [69]. Allowing schools more freedom to make food service choices was the stated rationale for the USDA's revised policy.

4.5.2.5 Human Population Growth

The human population continues to increase in both numbers and complexity of social structures (e.g., megacities). Although some disagreement exists on population forecasts, there is no disagree-

> The world's population is projected by the UN to reach 8.5 billion in 2030 and 11.2 billion by 2100.

ment that the 2016 world human population of ~7.3 billion will increase by billions during the 21st century. For the purposes of this book, population estimates developed by the UN will be utilized, "Currently, the world population continues to grow though more slowly than in the recent past. Ten years ago, world population was growing by 1.24% per year. Today, it is growing by 1.18% per year or approximately an additional 83 million people annually. The world population is projected to increase by more than one billion people within the next 15 years, reaching 8.5 billion in 2030, and to increase further to 9.7 billion in 2050 and 11.2 billion by 2100" [70].

Further, nine countries are expected to make up half of the world's population growth between now and 2050: India, Nigeria, Pakistan, Congo, Ethiopia, Tanzania, the U.S., Indonesia, and Ghana. Africa has the world's highest rate of population growth [70].

This projected increase in population presents numerous sociopolitical questions, not the least of which is, "Will there be enough food?" In consideration of this question, FAO concluded in 2009, "Political turmoil, social unrest, civil war and terrorism could all be on the table unless the world boosts its food production by 60% come mid-century. The world's population is expected to hit 9 billion people by 2050, which, coupled with the higher caloric intake of increasingly wealthy people, is likely to drastically increase food demand over the coming decades. [...] Exacerbating this problem is a convergence in diets worldwide, with reliance on an ever smaller group of crops leaving global food supplies increasingly vulnerable to inflationary pressure, insects and disease" [71].

While the FAO notes that progress has been made in the battle against global hunger, with vegetable production in Asia and the Pacific, where more than three-quarters of the world's vegetables are grown, increasing by 25% over the last decade. However, FAO estimates that 842 million people in the world remain under-nourished, with nearly two-thirds of them living in the Asia-Pacific region. One in four children less than the age of 5 years is stunted due to malnutrition.

To combat the problem, FAO has outlined two primary options: increasing arable land areas as well as productivity rates. A lack of available arable land and more sluggish growth rates in staple crops have complicated efforts to bolster these two pillars of food security. Over the past 2 years, productivity rates for rice and wheat have hovered around 0.6%–0.8%. Those rates would have to stabilize around 1% in order to offset serious shortages [71].

> In 2014, the FAO, the World Bank, and the World Resources Institute collectively estimated that the world is losing 25%–33% of the food it produces.

Environmentalists have also urged better food distribution methods. In February 2014, the FAO, the World Bank, and the World Resources Institute collectively esti-mated that the world is losing 25%–33% of the food it produces—nearly 4 billion metric tons annually. More efficient agricultural production, better means of stor-ing food and biologically diverse, and local food systems less susceptible to global changes have also been proposed as solutions to help tackle the growing threat of food insecurity [72].

Increased human population has contributed to greater interconnectivity between food-importing and food-exporting nations. This interconnectivity has been inves-tigated by researchers interested in the effects of disruptions (e.g., climate change) on food security. In one investigation, annual staple food production and trade data from 1992 to 2009 were used to analyze the changing properties of the global food system. "Over the 18-year study period, we show that the global food system is rela-tively homogeneous (85% of countries have low or marginal food self-sufficiency) and increases in complexity, with the number of global wheat and rice trade con-nections doubling and trade flows increasing by 42% and 90%, respectively. The increased connectivity and flows within these global trade networks suggest that the global food system is vulnerable to systemic disruptions [...]. To test this hypothesis, we superimpose continental-scale disruptions on the wheat and rice trade networks. We find greater absolute reductions in global wheat and rice exports along with larger losses in network connectivity as the networks evolve due to disruptions in European wheat and Asian rice production. Importantly, our findings indicate that least devel-oped countries suffer greater import losses in more connected networks through their increased dependence on imports for staple foods [73]."

A separate investigation was organized by Lloyd's of London, a global insurance company. The company was interested in the impacts of serious disruptions in food security in regard to the impact on insurance claims [74]. Research for the project was led by Anglia Ruskin University. The report explores the scenario of a near-term global food supply disruption, considered plausible on the basis of past events, espe-cially in relation to future climate trends. The global food system, the authors find, is "under chronic pressure to meet an ever-rising demand, and its vulnerability to acute disruptions is compounded by factors such as climate change, water stress, ongoing globalization and heightening political instability [75]."

Lloyd's scenario analysis shows that food production across the planet could be significantly undermined due to a combination of just three catastrophic weather events, leading to shortfalls in the production of staple crops, and ensu-

> A group of water scientists advocates switching to a vegetarian diet over the next 40 years to avoid catastrophic food shortages.

ing price spikes. In the scenario, which is "set in the near future," wheat, maize, and soybean prices "increase to quadruple the levels seen around 2000," while rice prices increase by 500%. This leads to rocketing stock prices for agricultural commodities, agricultural chemicals, and agriculture engineering supply chains, leading to [...] geopolitical mayhem as well as escalating terrorism and civil unrest [75]. While this report raises troubling issues of global import, it is important to understand that the model used in the research did not include any sociopolitical adjustments made over time that could mitigate the model's projected dire outcomes.

As described in this section, food security will need to adjust to increased numbers of humans. Food security specialists have begun to reflect on what adjustments will be needed and how to achieve them. An interesting reflection comes from a group of water scientists who stated that the world's population might have to switch almost completely to a vegetarian diet over the next 40 years to avoid catastrophic food shortages. Noting that humans now derive about 20% of their protein from animal-based products, a figure that may need to decrease to just 5% in order to feed the globe's extra 2 billion population by 2050. Animal protein-rich food consumes five to ten times more water than a vegetarian diet. The water scientists recommended adopting a vegetarian diet as one option to increase the amount of water available to grow more food in an increasingly climate-erratic world [76].

4.5.2.6 Food Waste

A major contributor to food insecurity is wasting of food. Food wastage is a major problem, especially in countries that are ill equipped to adequately grow, harvest, transport, distribute, and utilize food supplies. Shown in Figure 4.2 is an example of food wastage. The FAO has provided estimates of the globe's food wastage [77]:

FIGURE 4.2 Example of food waste. (FAO [71].)

- About one-third of the food produced in the world for human consumption every year—~1.3 billion tonnes—gets lost or wasted.
- Food losses and waste amount to about US\$ 680 billion in industrialized countries and US\$ 310 billion in developing countries.
- Industrialized and developing countries dissipate about the same quantities of food—respectively 670 and 630 million tonnes.
- Per capita waste by consumers is between 95 and 115 kg a year in Europe and North America, while consumers in sub-Saharan Africa, south and south-eastern Asia, each throw away only 6–11 kg a year.
- Every year, consumers in rich countries waste almost as much food (222 million tonnes) as the entire net food production of sub-Saharan Africa (230 million tonnes) [77].

An additional form of food waste not usually included in tallies by the FAO and other agricultural organizations occurs when on-farm food is lost due to factors such as problems in harvesting. There are several reasons why edible crops, e.g., melons, do not make it to market. In some cases, market prices for those crops might be too low to justify the cost of making additional passes through the field or orchard to harvest crops. In other cases, not being able to find labor to harvest the crops means that they are left to rot. National statistics concerning on-farm food loss are lacking, but one study of on-farm food loss focused on farms in northern California in 2019 [78]. Researchers conducted food loss measurements for 20 hand-harvested crops in 123 fields in 2016 and 2017. Results showed an average of 11,299 kg/ha of edible produce, or 31.3% of marketed yield, remained in fields after harvest. When walk-by (unharvested) field losses of 2.4% are included, total losses were 33.7% of marketed yields. It goes without saying that on-farm field loss of edible food is a major problem.

Adding to these FAO data are the findings from a study by researchers at the Potsdam Institute for Climate Impact Research in Germany [61]. As previously discussed, the researchers were interested in the relationship between food waste and the waste's generation of greenhouse gases (GHG). The study revealed that in the year 2010, food availability was 20% higher than was required on a global scale. Similarly, GHG emissions related to the food surplus increased from 130 Mt CO_2 eq/year to 530 Mt CO_2 eq/year, an increase of more than 300%. Moreover, the global food surplus might increase up to 850 kcal/cap/day, while the total food requirement will increase only by 2%–20% by 2050. Consequently, GHG emissions associated with the food waste may also increase tremendously to 1.9–2.5 Gt CO_2 eq/year [61]. Other aspects of food waste are described in Volume 2, Chapter 2 (Waste Generation and Management).

> Methane emissions from landfills represent one of the largest sources of greenhouse gas emissions from the waste sector.

Two countries, France and South Korea, have implemented national programs for purpose of reducing food waste. France is considered a global leader in curbing food waste. In February 2016, France became the first country in the world to prohibit supermarkets from throwing away unused food. Supermarkets of a certain size must donate unused food or face a monetary fine. Other policies require schools to teach students about food sustainability, companies to report food waste statistics in their environmental reports, and restaurants to make take-out bags available [79]. The people of France wasted 234 pounds of food per person annually, compared to about 430 pounds per capita thrown away a year in the U.S.

South Korea has also implemented assertive policies for purpose of food waste reduction. The South Korean government banned sending food to landfills in 2005 and in 2013 also prohibited the dumping of garbage juice (leftover water squeezed from food waste) into the sea. Today, a remarkable 95% of food waste is recycled a significant leap from <2% in 1995. Seoul has managed to cut the amount of food waste produced by 400 metric tons per day. Since 2013, South Koreans have been required by law to discard food waste in biodegradable bags, priced according to volume, and costing the average four-person family about $6 a month. By purchasing them from the local convenience store or supermarket, residents are effectively paying a tax on their food waste [80].

In the U.S., there is no national policy on food waste reduction, although one U.S. federal agency, the USDA, has a program that deals with food waste issues [81]. The 2018 Congressional Farm Bill created a fund for new composting programs. It was one part of a suite of actions in the Farm Bill designed to curb U.S. food waste. The actions encourage systemic change in the country's food system and support an ambitious government goal to slash domestic food waste by half in the next decade. The law set the stage for the creation of a high-level "food-waste liaison" under the Secretary, USDA, with specific duties for researching and cutting waste. The effectiveness of the 2018 Farm Bill's authorities for reducing food waste await evaluation.

One state, California, has promulgated policies that other states are considering. In 2016, California state legislators developed rigorous waste-disposal legislation that mandates a 50% reduction in organic waste disposal by 2020 and a 75% reduction by 2025. The state has some 25 composting yards that accept food waste and 14 anaerobic digesters, with as many as 100 new or expanded organics recycling facilities under consideration. Additionally, San Francisco is home to one leading composting program. In 2009, the city became the first in the U.S. to enact an ordinance making recycling and composting mandatory for businesses and residents alike. In 2012, city officials announced they had diverted 80% of all food waste from landfills. In 2018 that amounted to sending more than 650 tons of organic material to compost facilities every day [82]. Shown in Figure 4.3 is the preparation of food waste for composting. Composted food waste is a valuable commercial product useful for garden fertilizer.

Perspective: Food waste connotes two public health implications. First, food wasted is food unavailable for human consumption or some other positive outcome such as composting. Second, food waste that reaches burial in landfills causes emissions of methane, a powerful greenhouse gas, as it decays. For these and other reasons, national policies of food waste reduction are beneficial to human health and environmental quality. The examples from France and South Korea illustrate that strong-willed national policies on food waste are both possible and beneficial.

4.5.2.7 Climate Change
The effects of climate change on food security will be consequential, according to a modeling study conducted by University of Oxford investigators [83]. The

FIGURE 4.3 Composted food waste applied to farm field. (Buzby, J. 2021, USDA, Food and Nutrition, Health and Safety, https://www.usda.gov/media/blog/2021/05/07/composting-uneaten-food-interview-frank-franciosi-us-composting-council.)

investigators linked an agricultural modeling framework, the International Model for Policy Analysis of Agricultural Commodities and Trade (IMPACT), to a comparative risk assessment of changes in fruit and vegetable consumption, red meat consumption, and bodyweight for deaths from coronary heart disease, stroke, cancer, and an aggregate of other causes of mortality. The model was used to calculate the change in the number of deaths attributable to climate-related changes in bodyweight and diets. The model projected that by the year 2050, climate change will lead to per-person reductions of 3.2% in global food availability, 4.0% in fruit and vegetable consumption, and 0.7% in red meat consumption.

These changes will be associated with 529,000 climate-related deaths worldwide, representing a 28% reduction in the number of deaths that would be avoided because of changes in dietary and weight-related risk factors between the years 2010 and 2050. Twice as many climate-related deaths were associated with reductions in fruit and vegetable consumption than with climate-related increases in the prevalence of underweight individuals. The model predicted that most climate-related deaths would occur in south- and east-Asia. Adoption of climate-stabilization pathways would reduce the number of climate-related deaths by 29%–71%, depending on their stringency [83].

> A model projects that by year 2050, climate change will lead to per-person reductions of 3.2% in global food availability, 4.0% in fruit and vegetable consumption, and 0.7% in red meat consumption.
>
> These changes will be associated with 529,000 climate-related deaths worldwide.

In a different investigation, major "shocks" to global food production were investigated. Examples of major shocks would be protracted droughts, massive flooding, and prolonged high air temperatures. The study found these major shocks will be three times more likely within 25 years because of an increase in extreme weather brought about by global warming. The likelihood of such a shock, where production of the world's four major commodity crops (maize, soybean, wheat, and rice) falls by 5%–7%, is currently once-in-a-century. But such an event will occur every 30 years or more by 2040, according to the study by the UK-US Taskforce on Extreme Weather and Global Food System Resilience [84]. Such shocks could plausibly see the UN's food price index – which measures the international price of major commodities— rise by 50%, based on an analysis of how the market would likely respond. Increased food production volatility will mostly affect those developing countries experiencing high levels of poverty and political instability, such as countries in the Gulf or Sub-Saharan Africa. As climate change causes temperatures to rise even higher in the second half of the century, even more serious food shocks—where production drops by up to 10%—are also likely to occur much more often by the year 2070 [85].

Examples of loss of food security due to climate change are already present. For instance, UN bodies, international aid agencies, and governments have cautioned that droughts and floods triggered by one of the strongest El Niño weather events ever recorded have left nearly 100 million people in southern Africa, Asia, and Latin America facing food and water shortages and vulnerable to diseases including Zika [86]. For instance, in Mozambique, El Niño, the natural weather phenomenon that upturns normal weather patterns every few years in southern Africa, has caused the country to come to the end of another dry rainy season. For the second consecutive year, the town of Mbalavala's maize fields are empty, and the soil in vegetable gardens is like sand. The small amount of water from an emergency borehole must be shared between cattle and people. Mbalavala and 170,000 people in several hundred similar villages in Gaza and Inhambane, Mozambique's two most vulnerable provinces, will survive this year due to British food aid [87].

In January 2020, the UN's World Food Program (WFP) warned that an unprecedented number of people in 16 countries across southern Africa are gravely food insecure as climate change wreaks havoc on the region [88]. The agency estimated an impact of 45 million people—many of whom are women and children. The region has been hit hard by repeated droughts, widespread flooding, and economic hardship. Eswatini, Lesotho, Madagascar, Malawi, Mozambique, Namibia, Zambia, and Zimbabwe are among the hardest hit. Many families across the region are already skipping meals, taking children out of school, and falling into debt to stave off agricultural losses, the WFP said.

> A global food source, marine life, will be affected by climate change, which is postulated to seriously impair fish stocks.

Further, lack of food security poses major economic and social consequences beyond that of hunger and insufficient nutrition. For instance, the impact of a prolonged drought has forced Zimbabwe, because of widespread food shortages, to seek $1.6 billion in aid from global aid agencies to help pay for grain and other food [89]. Regarding social impacts of food insecurity, the UN's World Food Programme noted that women and children are bearing the brunt of a malnutrition and hunger crisis in Mauritania, while tens of thousands of Malian refugees face food shortages due to a lack of funding. In 2015 malnutrition reached emergency levels in six of Mauritania's

15 regions, affecting at least one in six people, and the proportion of malnourished children less than five across the country rose to 14% from 10% in 2014. Pregnant women were at special risk [90].

A review of the impacts of climate change on food quality and quantity concluded, "Greenhouse gas emissions are affecting the quantity and quality of our food in two ways. First, they are driving anthropogenic climate change, which decreases the yields of major cereal crops in some regions. Increased temperatures, changes in precipitation patterns, increased ozone concentrations, and more frequent and extreme heatwaves, floods, and droughts can reduce crop yields, particularly in the tropics, with risks increasing with additional warming depending on the region. Lower crop yields increase stunting and wasting, particularly in low-income and middle-income countries. Second, increased concentrations of carbon dioxide (CO_2)—by directly affecting plants—worsen the nutritional quality of food by decreasing protein and mineral concentrations by 5–15%, and B vitamins by up to 30%. Higher CO_2 concentrations increase photosynthesis in C3 plants (e.g., wheat, rice, potatoes, barley), which can increase crop yields. But those increases come at the cost of lower nutritional quality as plants accumulate more carbohydrates and less minerals (e.g., iron and zinc), which can negatively affect human nutrition" [91].

Climate change is projected by many scientists to also cause changes in the mix of food in the human diet. For example, the world is losing fish to eat as oceans warm [92]. A study by a team of marine biologists found that the amount of seafood that humans could sustainably harvest from a wide range of species shrank by 4.1% from 1930 to 2010, which they attributed to a casualty of human-caused climate change. As the oceans have warmed, some regions have been particularly hard hit. In the northeast Atlantic Ocean and the Sea of Japan, fish populations declined by as much as 35% over the period of the study. In a different study, researchers found a global shift to a "flexitarian" diet was needed to keep climate change even <2°C, let alone 1.5C [93]. This flexitarian diet means the average world citizen needs to eat 75% less beef, 90% less pork, and half the number of eggs while tripling consumption of beans and pulses (beans, lentils, and peas) and quadrupling nuts and seeds. This would halve emissions from livestock and better management of manure would enable further cuts.

> By the end of this century, less water and hotter air will combine to cut average yields of vegetables—which are crucial to a healthy diet—by nearly one-third.

While a global change to a flexitarian diet would make a contribution to mitigating some of the effects of climate change, adopting this form of diet would be complicated by another consequence of climate change: reduced crop yields and harvests of vegetables. A major study coordinated by the London School of Hygiene and Tropical Medicine cautions that global warming is expected to make vegetables significantly scarcer around the world, unless new growing practices and resilient crop varieties are adopted, researchers warned [94]. By the end of this century, less water and hotter air will combine to cut average yields of vegetables—which are crucial to a healthy diet—by nearly one-third, said the report in the Proceedings of the National Academy of Sciences. A 7.2°F (4°C) increase in temperature, which scientists expect by 2100 if global warming continues on its current trajectory, reduces average yields by 31.5%, said the report. The report concludes that for every degree Celsius that the Earth warms, corn yields will decrease an average of 7.4%, and wheat yields will decrease by 6%. Southern Europe, large parts of Africa, and South Asia might be particularly affected [95].

One consequence of climate change's impact on food production will attempt to adopt new agricultural methods to rely on less water. An example of one new technique is called "system of rice intensification" (SRI) [96]. Reports from India, Southeast Asia, and Africa suggest that average yield increases of 20%–50% are regularly being achieved by farmers adopting the SRI approach, which aims to stimulate the root system of plants rather than trying to increase yields in a conventional way by using improved seeds and synthetic fertilizers. Rather than growing rice in flooded patties, SRI, in contrast, involves the careful spacing of fewer but younger plants, keeping the topsoil around the plants well-aerated by weeding, using manure, and avoiding flooding. Increasing the production of rice is important because it is the major staple crop of nearly half the world.

Perspective: It surely is not hyperbole to state that how humankind deals with the food security challenges presented by unmitigated climate change will determine the fate of millions of people globally. And whether humans are malleable enough to make the dietary changes to convert to flexitarian diets is unknown, but will be a major challenge to policymakers.

4.5.2.8 Loss of Pollinators

Another factor that affects food security is the decline in numbers and diversity of pollinators. Farmers and gardeners know the vital value of the creatures that serve Nature as pollinators of flora, trees, and

> Loss of pollinators portends the loss of food sources and diminished food security.

others that require pollen transfer in order to reproduce. Loss of pollinators portends loss of food sources and diminished food security. The Intergovernmental Science-Policy Platform on Biodiversity and Ecosystem Services has reviewed the current situation in regard to the loss of pollinators. Excerpts from their findings follow:

1. More than three quarters of the leading types of global food crops rely to some extent on animal pollination for yield and/or quality. Pollinator-dependent crops contribute to 35% of global crop production volume.
2. Given that pollinator-dependent crops rely on animal pollination to varying degrees, it is estimated that 5%–8% of current global crop production, with an annual market value of $235–$577 billion worldwide, is directly attributable to animal pollination.
3. The vast majority of pollinator species are wild, including more than 20,000 species of bees, some species of flies, butterflies, moths, wasps, beetles, thrips,[1] birds, bats, and other vertebrates.

---

[1] Thrips: any of an order (Thysanoptera) of small to minute sucking insects many of which feed often destructively on plant juices.

4. Wild pollinators have declined in occurrence and diversity (and abundance for certain species) at local and regional scales in Northwest Europe and North America.

5. The International Union for Conservation of Nature (IUCN) Red List assessments indicates that 16.5% of vertebrate pollinators are threatened with global extinction (increasing to 30% for island species).

6. The abundance, diversity, and health of pollinators and the provision of pollination are threatened by direct drivers that generate risks to societies and ecosystems. Threats include land-use change, intensive agricultural management and pesticide use, environmental pollution, invasive alien species, pathogens, and climate change [97].

> In the U.S., an annual nationwide survey of U.S. beekeepers reported a loss of 33% of their honey bee colonies during the year spanning April 2016 to April 2017.

The global decline in bee populations has received special attention due to bees' vital role in food pollination. Two surveys illustrate the problem of loss of bee populations. In the U.S., an annual nationwide survey of U.S. beekeepers reported a loss of 33% of their honey bee colonies during the year spanning April 2016 to April 2017. This was a slight improvement over the previous year when the loss was reported as 40.5% [98]. The researchers noted that many factors are contributing to colony losses, with parasites and diseases at the top of the list. The number one culprit remains the varroa mite, a lethal parasite that can easily spread between colonies.

A second study of bee populations was conducted in 2019 by researchers at the University of Strathclyde [99]. An international survey found that the number of honey bee colonies declined by 16% in the winter of 2017–2018 across 38 countries. The study, based on the voluntarily submitted information, covered 33 countries in Europe—including the four nations of the UK—along with Algeria, Israel, and Mexico. The survey of 25,363 beekeepers found that, out of 544,879 colonies being managed at the start of winter, 89124 were lost, through a combination of circumstances including various effects of weather conditions, unsolvable problems with a colony's queen, and natural disaster.

Perspective: Declines in the prevalence of pollinators portend decreased food production at a time when global food diversity is a challenge. But as with other threats to food security, efforts in conservation, research on causes of declines, and policies on the protection of pollinating species will be necessary.

4.5.2.9   Soil Security and Arable Land

Humankind's harvesting of plants grown in soil has sustained our and other species for eons. Over time, humans learned how to cultivate soil, sow seeds or transplant seedlings, harvest crops, and consume the food grown for that purpose. In other

> Researchers assert that the world has lost a third of its arable land due to erosion or pollution in the past 40 years.

words, humans learned how to farm. Unfortunately, as with other threats to food security, there are problems with the condition and amount of arable soil globally. Of note, 40% of the planet's land is now devoted to human food production, up from 7% in 1700. In a study by the University of Sheffield's Grantham Centre for Sustainable Futures, researchers assert that the world has lost a third of its arable land due to erosion or pollution in the past 40 years. Their study calculated that nearly 33% of the world's adequate or high-quality food-producing land has been lost at a rate that far outstrips the pace of natural processes to replace diminished soil. Researchers attributed the continual ploughing of fields, combined with heavy use of fertilizers, as factors in the degradation of soils globally. They observed that erosion is occurring at a pace of up to 100 times greater than the rate of soil formation. This is a dire warning, since it takes around 500 years for just 2.5 cm of topsoil to be created amid unimpeded ecological changes [100].

In a 2018 UN-backed report, land degradation was cited as undermining the well-being of two-fifths of humanity, raising the risks of migration and conflict, according to a comprehensive global assessment [101]. The report reiterated that fertile soil was being lost at the rate of 24bn tonnes a year, largely due to unsustainable agricultural practices. The authors of the report estimated land degradation costs more than 10% of annual global GDP in lost ecosystem services such as carbon sequestration and agricultural productivity. The report encourages sustainable production and for the elimination of agricultural subsidies that promote land degradation.

> The problem of land degradation has prompted attempts to develop new farming approaches to soil conservation.

Farming practices and land use are under scrutiny for their impact on the environment and public health. In particular, what is called "intensive agriculture" has become the subject of debate among agriculturalist, environmentalists, and public health specialists. Intensive farming is an agricultural intensification and mechanization system that aims to maximize yields from available land through various means, such as heavy use of pesticides and chemical fertilizers. According to one environmental source, this intensification and mechanization have also been applied to the raising of livestock with billions of animals, such as cows, pigs, and chickens, being held indoors in what have become known as factory farms. Intensive farming practices produce more and less expensive food per acre and animal, which has helped feed an ever-increasing human population and may prevent surrounding land from being converted into agricultural land. However, intensive farming has become a major threat to the global environment through the loss of ecosystem services and contributor to global warming [102]. "Furthermore, intensive farming kills beneficial insects and plants, degrades and depletes the very soil it depends on, creates polluted runoff and clogged water systems, increases susceptibility to flooding, causes the genetic erosion of crops and livestock species around the world, decreases biodiversity, destroys natural habitats, and, significant contributors to the build-up of greenhouse gases in the atmosphere. However, certain aspects of intensive farming have helped ease climate change by helping boost yields in already cleared land that may be under-performing, which prevents the clearing of additional land. There are both pros and cons to intensive farming, but compared to the disadvantages, the advantages are fewer [...]" [102].

The problem of land degradation has prompted attempts to develop new farming approaches to soil conservation. As an example, the so-called "healthy soil movement" has led to regenerative agricultural approaches. This approach to farming is built around four basic rules: never till the soil; use cover crops so soil is never bare; grow a more diverse mix of plants; and graze livestock on fields after harvest or before planting. But regenerative farming is about more than cover crops and not tilling the soil. Farmers have found that increasing plants' diversity will lead to increased insect diversity, which in turn can reduce pesticide use because the "good" bugs eat the "bad" bugs, which helps keep plant pest populations down. The result is less reliance on chemical pesticides and less money spent on them. Additionally, soil fertility is enhanced via regenerative approaches [103].

An interesting alternative to farming of land is illustrated by an approach taken by the Netherlands [104]. The country is the second-largest vegetable exporter in the world, with exports totaling €6 billion ($6.8 billion) annually. Onions, potatoes, and some southern climate vegetables such as tomatoes, peppers, and chilies are among the country's top-selling products. Farmers use a greenhouse technology, termed "precision farming". The Netherlands is growing vegetables with far less water and pesticides than if production was happening in the soil or open air.

4.5.2.10 Regenerative Agriculture

Modern agriculture has evolved as a monoculture, or the focus on growing one type of crop. Farming this way increases yield but requires a great deal of tillage, a method of soil preparation that uses machinery to dig, rotate, and stir the earth. The more tillage that occurs, the weaker the soil becomes, and the greater the need for blasts of chemical pesticides. The inevitable end game of this cycle is the degradation, then erosion of farmland. If farms continue to operate as they currently are, the UN warns the planet's remaining topsoil will be fully depleted in 60 years, or—more chillingly—"60 harvests" [105].

> Regenerative agriculture is a system of principles designed to boost the farm ecosystem through the enhancement of soil health.

Farming methods that led to soil degradation led to **regenerative agriculture**, a system of principles designed to boost the farm ecosystem (emphasis added) through the enhancement of soil health. Regenerative agriculture incorporates permaculture and organic farming practices, including conservation tillage, cover crops, crop rotation, composting, mobile animal shelters, and pasture cropping to increase food production. This system is rooted in five pillars—better water management, low or no tillage (mechanical agitation of the soil), crop diversity, year-round cover crops, and livestock integration. The regenerative agriculture movement sprang out of a concern, in part, for preserving soil fertility, avoiding desertification, and combatting the effects of soil degradation on the environment. Regenerative agriculture is a way of growing food designed to enhance soil health. Proponents say the practice benefits farmers and food production while helping to mitigate climate change [105].

Perspective: As the world's human population grows, demands on food security will correspondingly increase. The production of food supplies will need to adapt to population increase, and to other realities of the 21st century, such as climate change.

Land use and farming practices, as well as reliance on fish stocks and ocean seafood, will continue to be stressed. Two policies, conservation and sustainability, will need to be globalized if food insecurity is to be avoided.

4.5.2.11 Genetically Modified Food

Food supplies are already being influenced by increased numbers of genetically modified foods. And as discussed elsewhere, controversy has accompanied this development, but suffice it to say here that GE food is already present in the food chain of the U.S. and many other countries. And it is also observed that GE food is considered by some as a threat to food supplies. In this section are four examples of this trend of adding GE food to food supplies.

> Genetically engineered food is already present in the food chain of the U.S. and many other countries.

In May 2015, FDA approved potatoes that won't bruise and apples that won't brown. The agency approved both GE foods, characterizing them as safe and nutritious as their conventional counterparts [106]. Also in 2015, FDA gave approval of genetically modified salmon for consumer use. This marked the first GE food animal endorsed for sale in the U.S. [107]. As the third example, in January 2015 USDA granted "non-regulated" status to genetically modified cotton and soybean plants. These are herbicide-tolerant crops to be used with a new herbicide intended to fight problematic weed resistance on farm fields [108].

In the fourth example, a simple genetic modification can triple the grain number of sorghum, a drought-tolerant plant that is an important source of food, animal feed, and biofuel in many parts of the world. Scientists at Cold Spring Harbor Laboratory (CSHL) figured out how that genetic change boosts the plant's yield: by lowering the level of a key hormone and by generating more flowers and more seeds. Their discovery points toward a strategy for significantly increasing the yield of other grain crops [109]. It remains to be seen whether any of these approved GE foods will become widespread in commercial food supplies. As a global issue, some countries (e.g., Scotland, Germany) have eschewed GE crops and produce based on concerns from farmers and food consumers [110].

In a different form of GE with implications for food production, researchers at the Oak Ridge National Laboratory discovered the specific gene that controls an important symbiotic relationship between plants and soil fungi, and successfully facilitated the symbiosis in a plant that typically resists it. The discovery could lead to the development of bioenergy and food crops that can withstand harsh growing conditions, resist pathogens and pests, require less chemical fertilizer, and produce larger and more plentiful plants per acre [111]. It remains to be seen whether GE food supplies will remain a threat to some governments and individuals or become a boon for dealing with issues of food insecurity.

4.6 GLOBAL FOOD SECURITY POLICIES

Food security has been presented as a global challenge to national governments and allied policymakers. The following sections will present some policies and resources that are intended as aids in combating food insecurity.

4.6.1   U.S. GLOBAL FOOD SECURITY ACT, 2016

In a political bipartisan action in 2016 between the U.S. Congress and the Obama administration, the Global Food Security Act of 2016 was enacted and signed into law by President Obama. This bill requires the President to develop and implement a Global Food Security Strategy to promote global food security, resilience, and nutrition. The key sections of the statute are as follows [55]:

Section 2: This section specifies that it is in the U.S. national security interest to promote global food security, resilience, and nutrition, consistent with national food security investment plans through programs and activities that:

- Accelerate inclusive, agricultural-led economic growth that reduces global poverty, hunger, and malnutrition;
- Increase the productivity, incomes, and livelihoods of small-scale producers;
- Build resilience to food shocks among vulnerable populations and households while reducing reliance upon emergency food assistance;
- Create an environment for agricultural growth and investment;
- Improve the nutritional status of women and children;
- Align with and leverage U.S strategies and investments in trade, economic growth, science and technology, agricultural research and extension, maternal and child health, nutrition, and water, sanitation, and hygiene;
- Strengthen partnerships between U.S. and foreign universities that build agricultural capacity; and
- Ensure the effective use of taxpayer dollars in achieving these objectives.

Section 3: This section sets forth definitions that apply to this bill.

Section 4: The President must coordinate a whole-of-government strategy to promote global food security, resilience, and nutrition, consistent with national food security investment plans. This section specifies required goals and criteria for the strategy. "The President must coordinate the efforts of federal departments and agencies to implement the strategy by establishing: (1) monitoring and evaluation systems, coherence, and coordination across federal departments and agencies; and (2) platforms for regular consultation and collaboration with key stakeholders and congressional committees" [55].

The USDA has the major responsibilities for the implementation of this law, with the law's impacts to be assessed in subsequent years.

4.7   SOCIAL JUSTICE ISSUES

Food insecurity in the U.S. varies by racial and ethnic factors. The last time the U.S. government formally measured food insecurity nationally was in 2018. At that time, about 25% of Black households with children were food insecure. In November 2020, the rate was about 39%, according to the latest analysis by Northwestern University economists. For Hispanic households with children, the rate was nearly 17% in 2018 and risen to 37% during the COVID-19 pandemic [112].

4.8   IMPACT OF COVID-19

As discussed in Volume 2, Chapter 1, the global COVID-19 pandemic, commencing in early 2020, caused unprecedented interruption of social systems such as education institutions and severe challenge to public health and healthcare policies and resources. The primary public health method for slowing the spread of the COVID-19 virus was "social distancing", which was implemented by U.S. state governors as a policy of "stay-in-place" restrictions. Locations where persons would normally congregate such as athletic events and restaurants were closed. Regarding food issues, the closure of food services impacted the demand for fresh food such as vegetables. Highlighted here are two impacts of the pandemic on food issues: food waste and diets associated with infected individuals' medical status.

Closure of restaurants, bars, and lesser patronage of grocery stores translated into a problem of food waste experienced by farmers and the food industry. Many of the nation's largest farms are being forced to destroy tens of millions of pounds of fresh food that they can no longer sell [113]. The amount of waste is staggering. The nation's largest dairy cooperative, Dairy Farmers of America, estimates that farmers are dumping as many as 3.7 million gallons of milk each day. A single chicken processor is smashing 750,000 unhatched eggs every week. Some effort was made to provide some of the food to food banks, but this effort had only a small effect on reducing the large volume of food waste.

As the second impact of the pandemic pertains to individuals' diet, a matter of both content and insecurity. As to food content, findings from the CDC indicate that about half of the people who have been hospitalized with COVID-19 are obese. CDC says those individuals who are obese are at "higher risk for severe illness" from COVID-19 [114]. Obesity, scientists know, is a diet-driven condition that contributes to many other serious health problems. Being overweight is strongly linked to the development of type 2 diabetes, for instance. As noted by CDC of the 48% of the COVID-19 victims who are obese, 28% of those hospitalized with the virus have diabetes.

As to food insecurity, a 2020 survey found a fifth of young children in the U.S. are not getting enough to eat [115]. The rate is three times higher than in 2008, at the worst of the Great Recession. A survey of households with children 12 and under by a Brookings researcher found that 17.4% reported the children themselves were not eating enough, compared with 5.7% during the Great Recession. Food insecurity has several causal factors, but surely the high rate of unemployment in the U.S. caused by the pandemic is a factor due to resources to purchase food. Inadequate nutrition can leave young children with permanent developmental damage.

4.9   HAZARD INTERVENTIONS

A number of hazard interventions are necessary if food safety and security are to be assured. As presented in this chapter, breakdowns in food safety can have serious consequences. As noted, in the U.S., annually one of every six Americans will experience a foodborne illness; 128,000 are hospitalized, and 3,000 die. This public health toll occurs in a country with a strong food safety history and legislation.

Globally, it is estimated that ~800 million people suffer from undernourishment. Some of the interventions that could reduce the hazard of unsafe food and food insecurity are as follows.

1. Food safety policies should be adopted and implemented at all levels of the food supply and consumption chain. In particular, food inspections and permits for food services are efficacious policies for the prevention of foodborne illnesses. The FDA Food Modernization Act of 2011 is an example of a comprehensive food safety statute, with prevention of illness as the operative policy.
2. Because a significant proportion of foodborne illness is due to inadequate preparation of food in home residences, education about food preparation is advocated. This should be a component of school curriculums.
3. Threats to global food security should be understood and appropriate policy actions are taken. In particular, the use of existing technology for enhancing food security should be encouraged as a matter of global food policy. For example, the adoption of the new generation of temperature- and humidity-controlled warehouses and silos in low-income countries will enhance food security [116].
4. The two primary threats to global food security are continued increases in global human population and the impacts of climate change. Resources such as the UN World Food Programme program will require both financial and policy support in order to overcome food shortages in geographic areas of famine and undernourishment. And international efforts to mitigate climate change will be necessary if global food insecurity is to be prevented.
5. Because water security and food security are intimately interrelated, policies to protect and conserve water supplies are necessary for food security.
6. Programs and policies to reduce food waste are encouraged in order to mitigate a source of greenhouse gases and to diminish the impact on land- and water-based food sources.
7. Programs and policies for purpose of conserving land and ocean quality from the impacts of unwise waste disposal methods are encouraged.
8. Individuals, who have the choice, should gravitate to ecofriendly food consumption. In particular, food consumption that contributes to climate change reduction is considered ecofriendly. Eating less food made of animal products and preferring organic food are both ecofriendly [117].
9. Individuals have a responsibility for ensuring that their food supply is healthful and is adequately prepared for consumption.

4.10 SUMMARY

Described in this chapter are three major federal environmental health statutes that are intended to enhance food safety in the U.S. The FDCAct, which dates from 1906, as public health policy prohibits the distribution of adulterated and mislabeled foods, drugs, cosmetics, and medical devices. Similarly, as public health policy, the FMIAct, which also dates from 1906, requires that meat and meat products are subject to federal inspection before entering the human food supply. The third major U.S. food safety statute is the FDA Modernization Act of 2011, which is oriented to the prevention of foodborne illness and other potential hazards, rather than responding to them. All three federal statutes therefore adopt a policy of federal government involvement in the inspection of food quality prior to the release of food products into commerce and for human consumption. By this process, adulterated or impure food is interdicted before entering the food chain. This, of course, is an example of the core principle of public health, the prevention of disease and disability.

In distinction to other environmental hazards, government involvement in food safety is rather limited and involves multiple partners in the public health effort to prevent foodborne illnesses. To be more specific, food safety requires the active participation of government, private sector entities, and individual food consumers to a degree not found in issues of air pollution, water contamination, toxics control, and waste management. Indeed, U.S. states have food quality responsibilities that exceed those of the federal government, as illustrated in this chapter by the State of Georgia's food quality statute. Moreover, private sector entities such as food producers, transporters, and food servers (e.g., restaurants) have quite significant roles and responsibilities for protecting against foodborne illness. However, in distinction to other environmental hazards, individuals play a critical role in protecting themselves against foodborne illness. For public health purposes, how individuals prepare food in the home is critical. After all, even the most wholesome food, if prepared under unsanitary conditions, has the portent to cause human illness.

Food security was described in this chapter as a complementary issue of food safety. Presented here were several factors, such as global population increase and climate change, as factors that will challenge global food security. And as discussed in this chapter, food safety and security are also the domain of concern by the EU, China, and India.

REFERENCES

1. FAO (UN Food and Agriculture Organization). 1996. World Food Summit. Rome: Office of Director-General, Media Centre.
2. WHO (World Health Organization). 2019. Food safety: Food facts. Geneva: Office of Director-General, Media Centre.
3. CDC (Centers for Disease Control and Prevention). 2014. Estimating foodborne illness: An overview. Atlanta: National Center for Emerging and Infectious Zoonotic Diseases.
4. Tack, D. M., E. P. Marder, P. M. Griffin, et al. 2019. Preliminary incidence and trends of infections with pathogens transmitted commonly through food: Foodborne diseases active surveillance network, 10 U.S. Sites, 2015–2018. *Weekly* 68(16):369–73.
5. AAP (American Academy of Pediatrics). 2019. Public policies to reduce sugary drink consumption in children and adolescents. Policy statement. Pediatrics, March.
6. Hawkes, C. 2018. Sugar tax: What you need to know. *The Conversation*, April 6.
7. Mullee, A., D. Romaguera, J. Pearson-Stuttard, et al. 2019. Association between soft drink consumption and mortality in 10 European countries. *JAMA Intern. Med.* doi: 10.1001/jamainternmed.2019.2478.
8. Staff. 2019. Eating healthy. Washington, D.C.: Science in the Public Interest.
9. Tyko, K. 2021. USDA issues public health alert for more than 211,000 pounds of ground turkey for possible salmonella risk. *USA Today*, April 11.

10. FAO (UN Food and Agriculture Organization). 2013. *Food Waste Footprint Impacts on Natural Resources*. Rome: Office of Director-General, Media Centre.
11. USDA (U.S. Department of Agriculture). 2016. Manure & nutrient management programs. Washington, D.C.: National Institute of Food and Agriculture.
12. EPA (Environmental Protection Agency). 2016. National Pollutant Discharge Elimination System (NPDES). Washington, D.C.: Office of Water.
13. CFSAN. 1981. The story of the laws behind the labels. Part I. 1906: The Federal Food, Drug, and Cosmetic Act. Silver Spring: Food and Drug Administration, Center for Food Safety and Nutrition.
14. CFSAN (Center for Food Safety and Nutrition, FDA). 1981. The story of the laws behind the labels. Part II. 1938–The Federal Food, Drug, and Cosmetic Act. Silver Spring: Food and Drug Administration, Center for Food Safety and Nutrition.
15. FDA (U.S. Food and Drug Administration). 2016. What does FDA regulate? Silver Spring: Office of Planning.
16. DoJ (U.S. Department of Justice). 2012. Abbott Labs to pay $1.5 billion to resolve criminal & civil investigations of off-label promotion of Depakote. Washington, D.C.: Office of Public Affairs.
17. FDA (Food and Drug Administration). 2015. Federal Food, Drug, and Cosmetic Act (FD&C Act). Silver Spring: Office of Food and Veterinary Medicine.
18. FDA (Food and Drug Administration). 2005. FDA food code. http://www.cfsan.fda.gov/~dms/foodcode.html.
19. FDA (Food and Drug Administration). 2005. Real progress in food code adoptions. Silver Spring: Office of Safety and Applied Nutrition.
20. Dye, J. 2012. FDA must act to remove antibiotics from animal feed: Judge. *Reuters*, March 23.
21. Wagner, L. 2015. Former Peanut Corp. exec gets 28 years for role in deadly salmonella outbreak. *National Public Radio: WABE 90.1*, September 21.
22. FDA (U.S. Food and Drug Administration). 2012. Document #213: Guidance for industry. Silver Spring: Division of Documents Management.
23. FDA (U.S. Food and Drug Administration). 2015. FDA's strategy on antimicrobial resistance: Questions and answers. Silver Spring: Office of Foods and Veterinary Medicine.
24. Polansek, T. 2015. U.S. sales of antibiotics for food animals rose over six years: FDA. *Reuters*, December 10.
25. Belluz, J. 2015. California enacts strictest animal antibiotic law in the U.S. *Vox*, October 11.
26. FDA (U.S. Food and Drug Administration). 2015. The FDA takes step to remove artificial trans fats in processed foods. *News Release*, June 16. Silver Spring: Office of Foods and Veterinary Medicine.
27. CCC (Calorie Control Council). 2003. Saccharin. http://www.saccharin.org/backgrounder.html.
28. Dulaney, C. 2015. Taco Bell, Pizza Hut to remove artificial flavors, coloring. *The Wall Street Journal*, March 26.
29. Ramakrishnan, S. 2015. Kellogg to stop using artificial products in cereals, snack bars. *Reuters*, August 4.
30. Choi, C. 2014. Kraft singles to lose artificial preservatives. *USA Today*, February 10.
31. House Agriculture Committee. 2002. Federal Meat Inspection Act of 1906. http://agriculture.House.gov/glossary/federal_meat_inspection_act_of 1906.htm.
32. WSDA (Washington State Department of Agriculture). 2002. Federal meat inspection act. www.wa.gov/agr/IBP/Federal%20meat%20inspection%20act.htm.
33. Chokshi, N. 2015. E. coli scare forces 167,427-pound ground beef recall. *The Washington Post*, November 2.

34. DoJ (U.S. Department of Justice). 2012. Abbott Labs to pay $1.5 billion to resolve criminal & civil investigations of off-label promotion of Depakote. Washington, D.C.: Office of Public Affairs.

35. Shapiro, R., M. L. Ackers, S. Lance, et al. 1999. Salmonella Thompson associated with improper handling of roast beef at a restaurant in Sioux Falls, South Dakota. *J. Food Prot.* 62(2):118–22.

36. USDA (U.S. Department of Agriculture). 2002. New measures to address *E. coli* 0157:H7 contamination. Backgrounder. Washington, D.C.: Congressional and Public Affairs.

37. USDA (U.S. Department of agriculture). 2014. Modernization of poultry slaughter inspection. Food Safety and Inspection Service. *Fed. Regist.* 79(162):49567.

38. FSIS. 2006. Acts & authorizing statutes. Washington, D.C.: U.S. Department of Agriculture, Food Safety and Inspection Service.

39. Curry, L. 2019. Years in the making, organic animal welfare rules killed by Trump's USDA. *Civil Eats*, May 6.

40. Fadula, L. Court strikes down Trump rollback of school nutrition rules. *The New York Times*, April 14.

41. Wartella, E. A., A. H. Lichtenstein, C. S. Boon, (eds.) 2010. Committee on examination of front-of-package nutrition rating systems and symbols. In: *Front-of-Package Nutrition Rating Systems and Symbols*, pp. 23–24. Washington, D.C.: Institute of Medicine.

42. Kenney, E. L., J. L. Barrett, S. N. Bleich, et al. 2020. Impact of the healthy, Hunger-Free Kids Act on obesity trends. *Health Affairs J.* 39(7). doi: 10.1377/hlthaff.2020.00133.

43. FDA (Food and Drug Administration). 2016. Background on the FDA Food Safety Modernization Act (FSMA). Silver Spring: Office of Media Affairs.

44. FDA (Food and Drug Administration). 2016. News release: FDA modernizes nutrition facts label for packaged foods, May 20. Silver Spring: Office of Media Affairs.

45. FDA (Food and Drug Administration). 2015. FDA news release: FDA issues final rule to add selenium to list of required nutrients for infant formula, June 22. Silver Spring: Office of Media Affairs.

46. USDA (Department of Agriculture). 2016. Food assistance programs. https://www.nutrition.gov/food-assistance-programs.

47. USDA (U.S. Department of Agriculture). 2015. Food security. http://www.usda.gov/wps/portal/usda/usdahome?navid=food-security.

48. State of Georgia. 2002. Rules and regulations: Food service. www.ph.dhr.state.ga/publications/foodservice.

49. GDA (Georgia Department of Agriculture). 2004. Homepage. http://www.agr.state.ga.us/html/food_safety_inquiries.html.

50. Anonymous. 2000. Should restaurant inspection reports be published? *J. Environ. Health* 62:27–32.

51. Simon, P. A., P. Leslie, G. Run, et al. 2005. Impact of restaurant hygiene grade cards on foodborne-disease hospitalizations in Los Angeles County. *J. Environ. Health* 67(7):32–6.

52. FSC (Food Secure Canada). 2019. What is food sovereignty? https://foodsecurecanada.org/who-we-are/what-food-sovereignty.

53. McLaughlin, K. 2017. Maine's new food sovereignty law gets a last-minute overhaul. *Civil Eats*, November 1.

54. Sibilla, N. 2019. Hundreds of homemade food businesses flourish under state food freedom laws. *Forbes*, January 22

55. House of Representatives. 2016. H.R.1567- Global Food Security Act of 2016. https://www.congress.gov/bill/114th-congress/house-bill/1567.

56. Grossman, E. 2015. American hunger-related healthcare costs exceeded $160 billion in 2014, According to New Study. *In These Times*, November 23.
57. BCFN (Barilla Center for Food & Nutrition). 2019. Food sustainability index. New York: The Economist, Intelligence Unit.
58. FAO (UN Food and Agriculture Organization). 2018. The state of food insecurity in the world 2018. Rome: Office of Director-General, Media Centre.
59. Soon, S. 2019. Asia is facing a food crisis and needs another $800 billion in the next 10 years to solve it. CNBC, November 27.
60. World Meters. 2016. Countries in the world by population. http://www.worldometers. info/world-population/population-by-country/.
61. Hiç, C., P. Pradhan, D. Rybski, and J. P. Kropp. 2016. Food surplus and its climate burdens. *Environ. Sci. Technol.* 50(8):4269–77.
62. FAO (Food and Agriculture Organization). 2020. Novel coronavirus (COVID-19). Rome: Headquaters.
63. GBD. 2017. Diet collaborators. 2017. Health effects of dietary risks in 195 countries, 1990–2017: A systematic analysis for the Global Burden of Disease study 2017. *Lancet*. doi: 10.1016/S0140-6736(19)30041-8.
64. NHLBI (National Heart, Lung, and Blood Institute). 2019. Healthy eating plan. https:// www.nhlbi.nih.gov/health/educational/lose_wt/eat/calories.htm.
65. Mullee, A., D. Romaguera, J. Pearson-Stuttard, et al. 2019. Association between soft drink consumption and mortality in 10 European countries. *JAMA Intern. Med.* 179(11):1479–90.
66. Jacobs, A. 2020. Sugary drink consumption plunges in Chile after new food law. *The New York Times*, February 11.
67. Baudry, J., K. E. Assmann, M. Touvier, et al. 2018. Frequency of organic food consumption with cancer risk. *JAMA Intern. Med.* 178(12):1597–606.
68. Mie, A., H. R. Andersen, S. Gunnarsson, et al. 2017. Human health implications of organic food and organic agriculture: A comprehensive review. *Environ. Health* 16:111. doi: 10.1186/s12940-017-0315-4.
69. Reiley, L. 2020. More pizza, fewer vegetables: Trump administration further undercuts Obama school-lunch rules. *The Washington Post*, January 17.
70. UN (United Nations). 2015. World population prospects: The 2015 revision. New York: Department of Economic and Social Affairs, Population Division.
71. FAO (UN Food and Agriculture Organization). 2016. World must produce 60% more food by 2050 to avoid hunger. Rome: Office of Director-General, Media Centre.
72. RT News. 2014. UN warns world must produce 60% more food by 2050 to avoid mass unrest. https://www.rt.com/news/world-food-security-2050-846/.
73. Puma, M., S. Bose, S. Y. Chon, and B. I. Cook. 2015. Assessing the evolving fragility of the global food system. *Environ. Res. Lett.* 10(2):024007. doi: 10.1088/1748-9326/ 10/2/024007.
74. Lloyd's of London. 2015. Emerging risk report: Food system shock. London: Office of Director.
75. Ahmeed, N. 2015. Society to collapse by 2040 due to catastrophic food shortages, environmental disaster. *Mint Press News,* June 22.
76. Vidal, J. 2012. Food shortages could force world into vegetarianism, warn scientists. *The Guardian*, August 25.
77. FAO (UN Food and Agriculture Organization). 2019. SAVE FOOD: Global initiative on food loss and waste reduction. Rome: Office of Media.
78. Baker, G. A., L. C. Gray, M. J. Harwood, et al. 2019. On-farm food loss in northern and central California: Results of field survey measurements. *Resour. Conserv. Recycl.* 149(October):541–9.

79. Hinckley, S. 2018. How France became a global leader in curbing food waste. *The Christian Science Monitor*, January 3.

80. Kim, M. S. 2019. The country winning the battle on food waste. *Huffpost News*, April 8.

81. Lydon, T. 2020. Can the U.S. slash food waste in half in the next ten years? *The Revelator*, February 14.

82. Richard, C. 2018. Already a climate change leader, California takes on food waste. *Civil Eats*, December 10.

83. Springmann, M., D. Mason-D'Croz, S. Robinson, et al. 2014. Global and regional health effects of future food production under climate change: A modelling study. *Lancet* 387(10031):1937–46.

84. Bailey, R., T. G. Benton, A. Challinor, et al. 2015. Extreme weather and resilience of the global food system. London: Foreign & Commonwealth Office.

85. Howard, E. E. 2015. Food production shocks' will happen more often because of extreme weather. *The Guardian*, August 13.

86. Vidal, J. 2016. El Niño is causing global food crisis, UN warns. *The Guardian*, February 16.

87. Vidal, J. 2016. As Mozambique's rivers dry up, the hopes of harvest evaporate too. *The Guardian*, February 17.

88. Staff. 2020. UN warns hunger crisis in southern Africa 'on scale we've not seen before.' *Deutsche Welle*, January 16.

89. Anonymous. 2016. Millions need aid as Zimbabwe battles drought. *Reuters*, July 26.

90. Guilbert, K. 2015. Women, children, refugees bear brunt of Mauritania food crisis. *Thomson Reuters Foundation News*, October 21.

91. Ebi, K. L. and I. Loladze. 2019. Elevated atmospheric CO_2 concentrations and climate change will affect our food's quality and quantity. *Lancet Planet. Health* 3(7):PE283–E284.

92. Free, C. M., J. T. Thorson, L. Malin, et al. Impacts of historical warming on marine fisheries production. *Science* 363(6430):979–83.

93. Springmann, M., M. Clark, D. Mason-D'Croz, et al. 2018. Options for keeping the food system within environmental limits. *Nature* 562:519–25.

94. Staff. 2018. Global warming will make veggies harder to find: Study. *The Straits Times*, June 12.

95. Gustin, G. 2017. How deeply will rising temperatures cut into crop yields? *Inside Climate News*, August 21.

96. Vital, J. 2018. A new farming technique using drastically less water is catching on. *Huffpost*, May 25.

97. IPBES (Intergovernmental Science-Policy Platform on Biodiversity and Ecosystem Services). 2016: Summary for policymakers of the assessment report of the Intergovernmental Science-Policy Platform on Biodiversity and Ecosystem Services on pollinators, pollination and food production. http://www.ipbes.net/sites/default/files/downloads/Pollination_Summary%20for%20policymakers_EN_pdf.

98. University of Maryland. 2017. American beekeepers lost 33 percent of bees in 2016–2017. *Science Daily News*, May 25.

99. University of Strathclyde. 2019. Honey bee colonies down by 16 percent. *Science Daily*, June 5.

100. Harvey, C. 2016. Our wasted food is a huge environmental problem – and it's only getting worse. *The Washington Post*, April 7.

101. Watts, W. 2018. Land degradation threatens human wellbeing, major report warns. *The Guardian*, March 26.

102. Everything Connects. 2016. Intensive farming. http://www.everythingconnects.org/intensive-farming.html.

103. Gunderson, D. 2018. Dirt rich: Healthy soil movement gains ground in farm country. *MPR News*, November 8.

104. Laurenson, J. 2019. Could hi-tech Netherlands-style farming feed the world? *Deutsche Welle*, January 23.

105. Feldman, L. 2020. Want a more sustainable food system? Focus on better dirt. *The Globe and Mail*, October 7.

106. Jalonick, M. C. and K. Ridler. 2016. FDA approves genetically engineered potatoes, apples as safe. http://phys.org/news/2015-03-gmo-potatoes-apples.html.

107. Duggan, T. 2015. Genetically modified salmon OKed, opponents object. *San Francisco Chronicle*, November 19.

108. Gillam, C. 2015. USDA approves Monsanto's new GMO soybeans, cotton. *St. Louis Post-Dispatch*, January 15.

109. Staff. 2018. Tripling the number of grains in sorghum and perhaps other staple crops. *Science Daily*, February 26.

110. Withnall, A. 2015. Germany follows Scotland's example with move to ban all GM crops and opt out of EU approvals. *Independent*, August 25.

111. Oak Ridge National Laboratory. 2019.Scientists make fundamental discovery to creating better crops. *Science Daily*, July 22.

112. Evich, H. B. 2020. Stark racial disparities emerge as families struggle to get enough food. *Politico*, July 6.

113. Corkery, M. and D. Yaffe-Bellany. 2020. 'We had to do something': Trying to prevent massive food waste. *The New York Times*, May 2.

114. Staff. 2020. What does junk food have to do with COVID-19 deaths? *Environment Health News*, April 28.

115. Staff. 2020. Coronavirus live updates: Nearly 1 in 5 children in U.S. not getting enough to eat. *The New York Times*, May 6.

116. Coclanis, P. A. 2017. The simplest way to improve the world's food systems requires no new science. *Quartz*, March 4.

117. O'Connor, L. 2015. 5 ways to be a climate-friendly eater. *Huffpost Impact*, December 30.

5 Tobacco Products, Vaping Devices, and Marijuana Smoking

5.1 TOBACCO PRODUCTS

In contrast to the other physical hazards described in this book, tobacco products, vaping devices, and marijuana smoking stand distinct as environmental health hazards. This is because these hazards are self-administered. In the instance

> The public health prevention paradigm is rather simple: don't use tobacco products and avoid contact with persons who do.

of tobacco products, they are accepted by their users even with knowledge of the adverse health consequences. In the situation of tobacco products, there is now global knowledge of the public health consequences of cigarette smoking, in particular, and companion knowledge about the health implications of other tobacco products as well. Further, in distinction to the other environmental hazards described in this book, the adverse effects of tobacco use are relatively amenable to traditional public health methods of prevention and policymaking. The public health prevention paradigm is rather simple: don't use tobacco products and avoid contact with persons who do. That dictum is, of course, easily expressed but difficult in practice.

Also to be discussed are so-called vaping devices. These are devices designed to vaporize a small container of some kind of liquid, thereby providing the user a vapor for inhaling. The liquids contain various kinds of flavoring substances and sometimes nicotine. The health consequences of regular use of vaping products and marijuana smoking are less clear than those for using tobacco products. But, as will be subsequently discussed, research has been published that suggests that these two forms of lung pollution are unwise for regular use, especially by young persons.

A discussion of the third subject of this chapter, marijuana smoking, concludes this chapter. This subject is topical, given the number of U.S. states that have approved the sale of recreational marijuana (also known as cannabis). The consequences of recreational smoking of marijuana are emerging, as will be discussed later in this chapter. But it can be asserted here as a general principle that some substances inhaled through frequent contact can be incompatible with good lung health.

This chapter commences with a description of the public health impacts and attendant policies pertaining to the uses of tobacco products. As illustrated in Figure 5.1, tobacco is an agricultural crop, most commonly used to make cigarettes and other consumer products. It is widely cultivated in warm regions, especially in the U.S. and China. In the U.S., tobacco is grown principally in the border states of Kentucky, Maryland, North Carolina, and Virginia. Tobacco plants comprise a vertical stalk,

DOI: 10.1201/9781003253358-5

FIGURE 5.1   Tobacco plants growing on a Kentucky farm. (With permission of University of Kentucky College of Agriculture, Department of Agronomy, Lexington, KY, http://www. uky.edu/Ag/Tobacco/Interns.htm, 2015.)

with leaves growing outward from the stalk, much like limbs on a tree trunk. At harvest, leaves are removed from the stalk, dried, and made into various tobacco products. The tobacco products are smoked, chewed, or snorted as a powder.

Covered in this chapter are several forms of tobacco products. Cigarettes are the most prevalent form of tobacco products, consisting of various tobacco blends encased in a paper cylinder. Cigarette smoking consists of placing one end of the cigarette into one's mouth and using a flame or other source of heat to ignite the tobacco at the other end of the cigarette. Another tobacco product is the cigar, which is a tightly rolled bundle of dried and fermented tobacco leaves, rolled in a series of types and sizes. As with cigarettes, one end of a cigar is placed into one's mouth, with the distal end set afire. Pipes are a third form of tobacco delivery system, where a mixture of ground tobacco leaves and additives are tamped into the bowl of the pipe and ignited. A third general product is smokeless tobacco, defined as tobacco that is chewed or snuffed rather than smoked by the user. Each of these tobacco delivery systems exposes the user to serious health consequences, which are elaborated subsequently in this chapter, accompanied by details of public health policies and actions to reduce the health toll of tobacco use.

5.1.1   Précis History of Tobacco Use

Tobacco has been with humankind for several millennia. According to one source, tobacco, a native plant of the Americas, was first discovered thousands of years ago. However, growing tobacco as a crop was pioneered by communities in the Andes at a much later time. Most estimates put this occurrence between 5000 and 3000 BCE [1–4], as indicated in Table 5.1, a summary of the history of tobacco.

The psychoactive ingredient of tobacco is nicotine. It is likely that over time nicotine evolved in tobacco as a defense mechanism, akin to how many other plants and animals evolved poisons for defensive measures. The tobacco plant should therefore

TABLE 5.1
Précis Summary of the History of Tobacco [2–4]

Date	Country	Event
5000–3000 BCE	Americas	First cultivation of the tobacco plant
Circa 1 BCE	Native Americans	Begin occasional smoking and using tobacco enemas
600–900 A.D.	Mexico	Mayans drew carved stone images depicting tobacco use
Early 1500s	Middle East	Tobacco first introduced when the Turks took it to Egypt
1492	Cuba	Columbus discovers tobacco smoking and takes the behavior to Europe
1531	Santo Domingo	European settlers begin tobacco cultivation
1530–1600	China	Tobacco introduced via Japan or the Philippines
1558	Europe	Tobacco plants brought to Europe. Attempts at cultivation fail
1600s	China	China philosopher Fang Yizhi points out years of smoking "scorches" one's lung
1612	Jamestown	First American settlers, Jamestown, Virginia, grew tobacco as a cash crop
1614	England	Seven thousand tobacco shops open with the first sale of Virginia tobacco
1633	Turkey	Death penalty imposed for smoking
1660	Africa	Portuguese and Spaniards ship tobacco to East Africa, spreading to Central and West Africa
1761	England	First study of effects of tobacco (John Hill); snuff users warned they risk nasal cancers
1769	New Zealand	Capt. James Cook arrives, smoking a pipe, and was doused in case he was a demon
1788	Australia	Tobacco arrives with the first fleet
1795	Germany	Samuel Thomas von Soemmering reports cancers of the lip in pipe smokers
1865	U.S.	First commercial cigarettes were made by Washington Duke, Raleigh, North Carolina
1881	U.S.	James Bonsack invents first cigarette-making machine, Raleigh, North Carolina
1950s	China	State monopoly took control of the tobacco business
1962	UK	First report of the British Royal College of Physicians on Smoking and Disease
1964	U.S.	Surgeon General Luther Terry reports that smoking causes lung cancer in men
1965	U.S.	Federal Cigarette Labeling and Advertising Act of 1965
1969	U.S.	Public Health Smoking Act of 1969
1984	U.S.	Comprehensive Smokeless Tobacco Health Education Act of 1984
1987	Australia-Victoria	First to use a tobacco tax to establish a health foundation to counter tobacco use
2001	Belgium	EU adopts its first Tobacco Products Directive
2003	China	Electronic cigarette invented by Chinese pharmacist Hon Lik as anti-smoking aid

(Continued)

TABLE 5.1 (*Continued*)
Précis Summary of the History of Tobacco [2–4]

Date	Country	Event
2003	Switzerland	WHO Framework Convention on Tobacco Control adopted as a global treaty
2009	U.S.	Family Smoking Prevention and Tobacco Control Act
2014	Belgium	EU amends and tightens its Tobacco Products Directive
2015	Switzerland	WHO estimates tobacco kills around 6 million people annually
2016	U.S.	FDA issues regulations on sales and use of e-cigarettes
2019	U.S.	FDA issues new graphic health warning cigarette labels

be considered poisonous. It is interesting that a recent new class of insecticides, neo-nicotinoids, is chemically related to nicotine. Nicotine is a stimulant. In small doses, nicotine can increase heart rate and blood pressure. When inhaled as a component of tobacco smoke, nicotine rapidly reaches the brain, crossing the blood-brain barrier within 8–20 seconds. Tobacco products that are chewed, placed inside the mouth, or snorted tend to release considerably larger amounts of nicotine into the body than smoking. Nicotine is a highly addictive stimulant drug, one that makes cessation of tobacco smoking and other uses of tobacco a challenging proposition.

5.1.2 PREVALENCE OF TOBACCO PRODUCTS AND USERS IN THE U.S.

As evident in the history of tobacco, the use of tobacco products in the Americas has a long history. The cultivation of tobacco for commercial purposes began in what later became the U.S., as was the invention of cigarette manufacturing equipment. In the U.S., as elsewhere, cigarette smoking is the most prevalent form of tobacco use, with other tobacco products rising and falling in use over the centuries. A considerable volume of tobacco use data is available from public health and business sources. This section will summarize the data on the prevalence in the U.S. of tobacco products and where data are available, the numbers of users of various tobacco products. The use of tobacco products by middle and high school students in the U.S. is an especially important public health concern, given the fact that many long-term tobacco users began as youth.

5.1.2.1 Cigarettes: Adults and Youth

CDC has assessed national estimates of smoking prevalence among adults aged ≥18 years using data from the 2015 National Health Interview Survey (NHIS). The percentage of U.S. adults who smoke cigarettes declined from 20.9% in 2005 to 15.0% in 2015. Higher tobacco taxes, tough anti-smoking messages, and smoke-free laws that ban smoking in indoor and outdoor areas appear to be factors in reductions in the prevalence of

> CDC: In 2019, an estimated 53.3% of high school students (8.0 million) and 24.3% of middle school students (2.9 million) reported having ever tried a tobacco product.

U.S. adults who smoke [5]. Similarly, CDC reported that from 2011 to 2015, current cigarette smoking declined among American middle and high school students. About 2 of every 100 middle school students (2.3%) reported in 2015 that they smoked cigarettes in the past 30 days—a decrease from 4.3% in 2011. About 9 of every 100 high school students (9.3%) reported in 2015 that they smoked cigarettes in the past 30 days—a decrease from 15.8% in 2011 [6].

In 2019 CDC's Office on Smoking and Health, in collaboration with FDA's Center for Tobacco Products, analyzed data from the 2019 National Youth Tobacco Survey (NYTS) to assess tobacco product use patterns and associated factors among U.S. middle and high school students [7]. Overall, 19,018 questionnaires were completed and weighted to represent ~27.0 million students. In 2019, an estimated 53.3% of high school students (8.0 million) and 24.3% of middle school students (2.9 million) reported having ever tried a tobacco product. Current (past 30-day) use of a tobacco product (i.e., electronic-cigarettes [e-cigarettes], cigarettes, cigars, smokeless tobacco, hookahs, pipe tobacco, and bidis [small brown cigarettes wrapped in a leaf]) was reported by 31.2% of high school students (4.7 million) and 12.5% of middle school students (1.5 million).

E-cigarettes were the most commonly cited tobacco product currently used by 27.5% of high school students (4.1 million) and 10.5% of middle school students (1.2 million), followed in order by cigars, cigarettes, smokeless tobacco, hookahs, and pipe tobacco. Tobacco product use also varied by sex and race/ethnicity. Among current users of each tobacco product, the prevalence of frequent tobacco product use (on ≥20 days of the preceding 30 days) ranged from 16.8% of cigar smokers to 34.1% of smokeless tobacco product users. Among current users of each individual tobacco product, e-cigarettes were the most commonly used flavored tobacco product (68.8% of current e-cigarette users). As a matter of public health, the FDA has been slow to act against the widespread prevalence of e-cigarettes used by young persons. As of November 2019, the agency has not vetted the vast majority of vaping devices or flavored liquids for safety [8].

5.1.2.2 Other Tobacco Products

There are several products in addition to cigarettes that contain tobacco and are used for smoking, chewing, or snuffing. These include cigars, pipes, chewing tobacco, and snuff. These products are less in commercial appeal than cigarettes and therefore will be described in lesser detail than cigarettes. Further, as will be subsequently related, policies on the use

In 2013, an estimated 12.4 million people in the U.S. aged 12 years or older (or 5.2%) were current cigar smokers. Percentages of U.S. adults who were current cigar smokers in 2013: 5.0% of all adults.

of these products vary according to federal, state, and local laws or ordinances and policies enforced by private sector groups, such as business operations.

Cigars: A cigar is defined as a roll of tobacco wrapped in leaf tobacco or in a substance that contains tobacco. Cigars differ from cigarettes in that cigarettes are a roll of tobacco wrapped in paper or in a substance that does not contain tobacco. The three major types of cigars sold in the U.S. are large cigars, cigarillos, and little cigars. The use of flavorings in some cigar brands has raised concerns that these products might be especially appealing to youth. Prevalence data from CDC provide cigar smoking rates and distribution data for the U.S. population [9]. In 2013, an estimated 12.4 million people in the U.S. aged 12 years or older (or 5.2%) were current cigar smokers. Percentages of U.S. adults who were current cigar smokers in 2013: 5.0% of all adults, with 8.2% of adult males and 2.0% of adult females, and for which White males (5%) and Hispanic adults (4%) are the population groups with the greatest cigar smoking rates.

The percentages of U.S. high school and middle school students who reported cigar smoking were as follows: U.S. high school students who were current smokers in 2014 were 8.2% of all students in grades 9–12. This aggregate figure represented 5.5% of female students and 10.8% of male students in grades 9–12. For middle school students, the percentage of U.S. students who were current cigar smokers in 2014 was 1.9% of all U.S. students in grades 6–8. This figure represented 1.4% of female students and 2.4% of male students in grades 6–8. The percentage of U.S. school students who smoke cigars is rather striking, indicating a need for further public health interventions targeted at school-age children and adolescents [9].

Pipes: Conventional tobacco pipes are a configuration consisting of a bowl attached to a tapered stem, which ends in a mouthpiece. Pipe bowls can be constructed of various woods, metal, clay, or corncobs. Mixtures of different varieties of tobacco leaves are shredded into flakes or crumbled by hand, often with some kind of flavoring added for sake of aroma. The tobacco mix is tamped into the pipe's bowl and ignited, with smoke being drawn through the stem and mouthpiece into the smoker's mouth. One special form of tobacco pipe is a water pipe, which typically consists of a head that is connected to a water jar, with an attached hose and mouthpiece. Tobacco and a moist fruit preparation are placed below burning charcoal in the head of the device, and the resulting smoke is inhaled through the hose into one's mouth [10]. A special form of water pipe is called a hookah, a water pipe that is used to smoke specially made tobacco that comes in different flavors [10].

Prevalence data on conventional pipe smokers in the U.S. is relatively sparse because their numbers have dwindled from the 18th and mid-19th centuries. While pipe smoking was fairly common in 1965 among men age 20 or older, the prevalence of pipe smoking over the following three decades has "declined drastically" across all races, regions, and education levels, as reported in the 1996 Preventive Medicine study. Current pipe smokers are typically men 45 years or older [11].

In a report from investigators with the American Cancer Society [12], "The prevalence of conventional pipe smoking among adult men in the U.S. has decreased from 14.1% in 1965 to 2.0% in 1991, and pipe smoking remains rare among U.S. women (<0.1% in 1991). The prevalence of pipe smoking is highest among men aged 45 or older and in the Midwest. Pipes are commonly used by some populations, including American Indians (male prevalence 6.9% in 1991), and by both men and women in parts of China (20% prevalence in 1996). The National Youth Tobacco Survey has measured the prevalence of pipe smoking among U.S. youth since 1999. The prevalence of current pipe smoking has increased from 2.4% to 3.5% of middle school students and from 2.8% to 3.2% of high school students from 1999 to 2002; prevalence was higher among boys than girls and varied by state and ethnicity" [12].

Water pipes are another area of health concern. These kinds of pipes are a form of tobacco pipe smoking that differs from conventional pipe smoking. This is a traditional smoking method going back centuries, known across various cultures as hookah, shisha, sheesha, and hubbledubble, among other names. "Hookah" is chosen for use in this chapter, since that is the name most often used by U.S. health agencies in their investigations. These pipes typically consist of a head that is connected to a water jar, with an attached hose and mouthpiece. Tobacco and a moist fruit preparation are placed below burning charcoal in the head of the device. When a smoker inhales through the mouthpiece, the air from the burning charcoal is pulled through the layer of tobacco and then through the water where it is cooled as bubbles, before being breathed in through the hose and mouthpiece [10]. Similar to cigarettes, hookah smoking delivers the addictive drug nicotine and hookah smoke is at least as toxic as cigarette smoke.

> From 2013 to 2014, hookah smoking about doubled for middle and high school students in the U.S.

The prevalence of hookah smoking is unknown but CDC reported "In recent years, there has been an increase in hookah use around the world, most notably among youth and college students. The Monitoring the Future survey found that in 2014, about 23% of 12th grade students in the U.S. had used hookahs in the past year, up from 17% in 2010. In 2014, this rate was slightly higher among boys (25%) than girls (21%). CDC's National Youth Tobacco Survey found that from 2013 to 2014, hookah smoking roughly doubled for middle and high school students in the U.S. Current hookah use among high school students rose from 5.2% (770,000) to 9.4% (1.3 million) and for middle school students from 1.1% (120,000) to 2.5% (280,000) over this period" [13]. A consideration of these prevalence data again reveals a concern that young persons are at particular risk of adverse health effects.

Smokeless tobacco: There are several kinds of tobacco products that are not smoked, but rather are placed into the user's mouth or nose. The latter method of using smokeless tobacco was popular in past centuries, a behavior called "snuffing." Currently, some of the more common products listed by the American Lung Association include [14]:

> Through the first half of the 20th century, containers called spittoons were a common fixture in many public places, including public buildings.

Chewing, oral, or spit tobacco: This tobacco comes as "wads" of loose leaves, plugs, or twists of dried tobacco leaves that may be flavored. A wad is chewed or placed between the cheek and gum or teeth. The nicotine in the tobacco is absorbed through the mouth tissues. The user spits out (or swallows) the brown saliva that has soaked through the tobacco. In centuries past, up through the first half of the 20th century, containers called spittoons were a common fixture in many public places, including public buildings. Cuspidor was a word used in lieu of spittoon in more formal settings such as legal documents. Tobacco chewers were expected to spit their wads of tobacco-laden saliva into the spittoons. In some rural areas, tobacco-spitting competitions were held as entertainment. This disgusting practice led to unhygienic areas in locales that permitted tobacco spitting.

Snuff or dipping tobacco: Snuff is finely ground tobacco packaged in cans or pouches. Snuff is sold as dry or moist and may have flavorings added. Snuff is used in "pinches," between the thumb and forefinger. Moist snuff is used by putting a pinch between the lower lip or cheek and gum. The nicotine in the snuff is absorbed through the tissues of the mouth. Moist snuff comes also in small, teabag-like pouches or sachets that can be placed between the cheek and gum. These are designed to be both "smoke-free" and "spit-free" and are marketed as a discreet way to use tobacco. Dry snuff is sold in a powdered form and is used by "snuffing the powder" up one or both nostrils of the user's nose.

Snus: Snus is a type of moist snuff first used in Sweden and Norway. The tobacco is often flavored with spices or fruit and is packaged like small tea bags. An American version of snus is similar in content to the Scandinavian variety, but with less moisture in the tobacco. Snus is held between the gum and mouth tissues and the nicotine-laden juice is swallowed. Prevalence of use data is scarce, with the principal use of snus occurring in Sweden and Norway. One report from Norway found that among young male adults, the prevalence of smoking (daily + occasional) was reduced from 50% in 1985 to 21% in 2013. Over the same period, the use of snus increased from 9% to 33%. The investigators suggested that the use of snus was a possible contributor to the reduced use of cigarettes [15]. In Sweden, the prevalence in 2011 of daily snus use among 16–84-year-old males and females was 19% and 4%, respectively. Occasional use of snus (less than daily use) was reported by 6% of Swedish males and 4% of Swedish females [16].

Smokeless tobacco sales in the U.S. are a $5–$6 billion annual business, with sales rising [17]. Convenience stores are the most commonplace to purchase these products within the U.S. In 2014, U.S. convenience stores generated ~$5.31 billion from chewing tobacco and snuff products, which amounted to about 1.26 billion units sold [18]. Smokeless tobacco products are highly addictive, owing to their higher content of nicotine [19]. Some users of smokeless tobacco mistakenly consider the products to be a safe or safer alternative to smoking tobacco products, particularly cigarettes. As will be described in the health effects section, there are no tobacco products without adverse health implications.

5.1.3 Global Prevalence of Tobacco Products and Users

As evident in the history of tobacco, the use of tobacco products is global. Cigarette smoking is the most prevalent form of tobacco use, with other tobacco products rising and falling in use over the

> WHO: Tobacco kills up to half of its users. Tobacco kills more than 8 million people each year.

centuries. The following excerpt from WHO [3] provides data on the global prevalence of tobacco users: "Tobacco kills up to half of its users. Tobacco kills more than 8 million people each year. More than 7 million of those deaths are the result of direct tobacco use while around 1.2 million are the result of non-smokers being exposed to second-hand smoke. Around 80% of the world's 1.1 billion smokers live in low- and middle-income countries" [19].

In December 2019, WHO released a report on global tobacco use trends over the preceding two decades [20]. The agency noted that overall global tobacco use had fallen, from 1.397 billion in 2000 to 1.337 billion in 2018, or by

> WHO: Overall global tobacco use has fallen, from 1.397 billion in 2000 to 1.337 billion in 2018, or by ~60≈million people.

~60 million people. This was largely driven by reductions in the number of females using these products (346 million in 2000 down to 244 million in 2018, or a decrease of about 100 million). Over the same period, male tobacco use had risen by around 40 million, from 1.050 billion in 2000 to 1.093 billion in 2018 (or 82% of the world's current 1.337 billion tobacco users in 2019). However, WHO noted that the number of male tobacco users has stopped growing and is projected to decline by more than 1 million fewer male users come 2020 (or 1.091 billion) compared to 2018 levels, and 5 million less by 2025 (1.087 billion). WHO attributed these decreases in the numbers of tobacco users to national government programs of tobacco use reduction.

Perspective: A consideration of the prevalence data on the use of tobacco products leads to substantial global health and economic concerns. Of the greatest gravity are data on the use of tobacco products by young people. Children in middle school and higher grades of school are often introduced to tobacco products via peer pressure. Education of youth on the morbidity and mortality consequences of tobacco products must be an integral part of youth education programs. Policies on controlling tobacco products will be introduced in subsequent sections, but, sadly, young smokers today will be many of tomorrow's adult victims of disease and premature death.

5.1.4 IMPACTS OF TOBACCO PRODUCTS ON HUMAN HEALTH

Each tobacco product must be considered a potential hazard to human health, including some persons who are not even using a product due to secondhand smoke or similar condition. This section will present the adverse effects of tobacco products on an individual's and public health.

5.1.4.1 Cigarettes

Prior to presenting a brief history of the public health investigations that led to the identification of cigarettes as a major global health epidemic, some statistics from the WHO will set the foundation for public health concerns. In 2019, WHO's

> WHO: Tobacco use and second-hand smoke kill 8 million people each year and leave many more in poor health.

director general stated, "Tobacco use and second-hand smoke kill 8 million people each year and leave many more in poor health; estimates suggest tobacco could kill up to 1 billion people this century," further commenting, "strong tobacco control policies, with a clear implementation strategy, are being adopted around the world. Twenty-three countries to date, representing 2.4 billion people, have adopted cessation support measures at best-practice level" [21].

As noted in the history of tobacco, concerns about the health consequences of tobacco products began in several countries where tobacco had come into common use. Included in the medical history of tobacco were studies by medical doctors in China, England, Germany, and the U.S. While other public and individual health studies followed, it remained for two leaders in public health, one British, one American, to lead the elucidation of the association between cigarette smoking and adverse health effects in smokers. Both men were medical doctors and well versed in medical epidemiology. Both held distinguished academic appointments in schools of medicine. Interestingly, both doctors were cigarette smokers until they independently began researching the adverse effects of cigarette smoking on human health and the consequent public health impacts. One can assert with confidence that both men's work has resulted in millions of lives spared of premature death due to cigarette smoking. Following is a précis history of each doctor's public health contributions toward cigarette smoking [22].

Sir Richard Doll: Richard Doll was the foremost medical epidemiologist of the 20th century, credited with turning epidemiology into a rigorous science. He was a leader in conducting research that associated cigarette smoking with lung cancer and heart disease. Richard Doll also did pioneering work on the relationship between radiation and leukemia as well as that between asbestos and lung cancer and alcohol consumption and breast cancer. His findings were further noteworthy because they have led to environmental health policies to prevent disease in at-risk populations [23].

William Richard Shaboe Doll studied medicine at St Thomas's Hospital Medical School, King's College, London, from where he graduated in 1937. He joined the Royal College of Physicians after the outbreak of World War II and served for much of the war as a member of the Royal Army Medical

> The work of Sir Richard Doll and U.S. Surgeon General Luther Terry has resulted in millions of lives spared of premature death due to cigarette smoking.

Corps as a medical specialist on a hospital ship. After his war service, Doll returned to St Thomas' to research asthma.

In 1950, Richard Doll and Austin Bradford Hill undertook a study of lung cancer patients in 20 London hospitals. The study had been prompted by hospital records of a 30-year period that showed a rapid, but unexplained, increase in lung cancer in men. The three suspects were inhalation of vehicle exhaust fumes, smuts (small flakes of soot or other dirt) from coal fires, or tarring of roads. The study involved patients with confirmed lung cancer and those without such diagnosis. Doll and Bradford Hill carefully recorded the lifestyle and personal habits of each person in their study. The investigators soon discovered that cigarette smoking was the only factor that the lung cancer patients had in common. Published in the *British Medical Journal* in 1950, the article stated: "The risk of developing the disease increases in proportion to the amount smoked. It may be 50 times as great among those who smoke 25 or more cigarettes a day as among non-smokers" [23].

Four years later, a longitudinal study of ~40,000 British medical doctors followed over 20 years confirmed Doll and Bradford Hill's report, from which the

British government issued advice that smoking and lung cancer rates were related. In 1969, Doll moved to Oxford University as the Regius Professor of Medicine. He continued work into carcinogens while at the Imperial Cancer Research Centre at the John Radcliffe Hospital, Oxford, working as part of the Clinical Trial Service Unit. This work notably included a study undertaken with Sir Richard Peto, in which it was estimated that tobacco, along with infections and diet, caused between them three-quarters of all cancers, which was the basis for much of WHO's conclusions on environmental pollution and cancer. Among numerous honors, Richard Doll was made a Fellow of the Royal Society in 1966 and knighted in 1971.

Surgeon General Luther Terry: As U.S. Surgeon General Luther L. Terry led the first public health campaign in the U.S. that warned of the dangers of cigarette smoking and his work persuaded millions to quit, thereby sparing them of morbidity and premature mortality

> On January 11, 1964, U.S. Surgeon General Luther Terry reported that cigarette smoking was an unmitigated health hazard, causing lung cancer and chronic bronchitis.

that comes with tobacco use. Luther Leonidas Terry received an M.D. degree at Tulane University in 1935. From 1940 to 1942 Dr. Terry served as instructor and assistant professor of preventive medicine and public health at the University of Texas at Galveston. In 1958 Terry became the Assistant Director of the National Heart Institute, Bethesda, Maryland. In 1961 President John F. Kennedy appointed Dr. Terry as U.S. Surgeon General, serving in that position until 1965 [24]:

As Surgeon General, Terry quit cigarette smoking in late 1963 and decided to make it his mission to urge millions of Americans who smoked cigarettes to do the same. Shortly after the release of a British report led by Richard Doll that was the first to warn the public of the health hazard of cigarette smoking, Terry established and chaired the Surgeon General's Advisory Committee on Smoking and Health tasked to produce a similar report for the U.S. On January 11, 1964, he delivered the first *Surgeon General's Report on Smoking and Health*, which reported that cigarette smoking was an unmitigated health hazard, causing lung cancer and chronic bronchitis. The report also noted that there was suggestive evidence, if not definite proof, for a causative role of smoking in other illnesses such as emphysema, cardiovascular disease, and various types of cancer. Based on more than 7,000 peer-reviewed articles, Terry's report concluded that cigarette smoking was a sufficient enough health problem to warrant "appropriate remedial action" [25].

The landmark Surgeon General's report of 1964 on smoking and health stimulated in the American public and government policymakers a greatly increased concern about tobacco and led to a broad-based anti-smoking campaign. It also motivated the tobacco industry to intensify its efforts to question the scientific evidence linking smoking and disease.

In June 1964, the U.S. Federal Trade Commission voted by a margin of 3-1 to require that cigarette manufacturers "clearly and prominently" place a

> The work of Richard Doll and Luther Terry in linking cigarette smoking with lung and other diseases has led to millions of lives saved from premature death.

warning on packages of cigarettes effective January 1, 1965, stating that smoking is dangerous to health, a warning in line with the warning issued by the Surgeon General's special committee. The same warning would be required in all cigarettes advertising effective July 1, 1965, as a consequence of the passage of the Cigarette Labeling and Advertising Act of 1965.

Surgeon General Terry changed this social value by clearly warning the public of the adverse health consequences of cigarette smoking. His association of smoking with the adverse outcomes of lung cancer and chronic bronchitis and other diseases changed how the American society viewed cigarette smoking. Terry set into motion anti-smoking campaigns that continue today. It is now rare to find public (and some private) buildings in the U.S. where smoking is permitted.

The work that Doll and Terry began in the 1960s quite literally lives on. The work of Doll and Terry stimulated programs of research on the health effects of tobacco, programs of research performed by investigators in academic and government programs of research. These programs of research have identified serious adverse consequences of tobacco use. The health toll is especially acute in cigarette smokers, given that this tobacco product is the most prevalent. A summary by CDC of the human health effects of cigarette smoking follows [25].

Cigarette Smoking and Death

- Cigarette smoking is the leading preventable cause of death in the U.S.
- Cigarette smoking causes more than 480,000 premature deaths annually in the U.S. This is nearly one in five deaths. Smoking causes more deaths each year than the following causes combined: human immunodeficiency virus (HIV), illegal drug use, alcohol use, motor vehicle injuries, and firearm-related incidents.
- More than ten times as many U.S. citizens have died prematurely from cigarette smoking than have died in all the wars fought by the U.S. during its history.

> Cigarette smoking causes more than 480,000 premature deaths annually in the U.S. This is nearly one in five deaths.

- Smoking causes about 90% of all lung cancer deaths in men and women. More women die from lung cancer annually than from breast cancer.
- About 80% of all deaths from chronic obstructive pulmonary disease (COPD) are caused by smoking.
- Cigarette smoking increases the risk for death from all causes in men and women. The risk of dying from cigarette smoking has increased over the last 50 years in men and women in the U.S.

Cigarette Smoking and Increased Health Risks

- Smokers are more likely than nonsmokers to develop heart disease, stroke, and lung cancer.

- Smoking is estimated to increase the risk: For coronary heart disease by two to four times; for stroke by two to four times;
- Of men developing lung cancer by 25 times; of women developing lung cancer by 25.7 times
- Smoking causes diminished overall health, increased absenteeism from work, and increased health care utilization and cost.

Cigarette Smoking and Cardiovascular Disease

- Smokers are at greater risk for diseases that affect the heart and blood vessels (cardiovascular disease).
- Smoking causes stroke and coronary heart disease, which are among the leading causes of death in the U.S.
- Even people who smoke fewer than five cigarettes a day can have early signs of cardiovascular disease.
- Smoking damages blood vessels and can make them thicken and grow narrower. This makes your heart beat faster and your blood pressure go up. Clots can also form.
- A stroke occurs when a clot blocks the blood flow to part of your brain or when a blood vessel in or around your brain bursts.
- Blockages caused by smoking can also reduce blood flow to your legs and skin.

Cigarette Smoking and Respiratory Disease

- Smoking can cause lung disease by damaging your airways and the small air sacs (alveoli) found in your lungs.
- Lung diseases caused by smoking include COPD, which includes emphysema and chronic bronchitis.
- Cigarette smoking causes most cases of lung cancer.
- If you have asthma, tobacco smoke can trigger an attack or make an attack worse. Further, long-term ambient ozone exposure has been associated with worse respiratory outcomes and increased emphysema and gas trapping, independent of smoking and workplace exposures, in smokers with or at risk for chronic obstructive pulmonary disease [26].
- Smokers are 12–13 times more likely to die from COPD than nonsmokers.

Cigarette Smoking and Cancer

- Smoking can cause cancer almost anywhere in one's body: bladder, blood (acute, myeloid leukemia), cervix, colon and rectum (colorectal), esophagus, kidney and ureter, larynx, liver, oropharynx, pancreas, stomach, trachea, bronchus, and lung.
- Smoking also increases the risk of dying from cancer and other diseases in cancer patients and survivors.

- If nobody smoked, one of every three cancer deaths in the U.S. would not occur.

Cigarette Smoking and Other Health Risks

- Smoking harms nearly every organ of the body and affects a person's overall health.
- Smoking can make it harder for a woman to become pregnant and can affect her baby's health before and after birth.
- Smoking increases risks for preterm (early) delivery, stillbirth (death of the baby before birth), low birth weight, sudden infant death syndrome, ectopic pregnancy, orofacial clefts in infants, and possibly, schizophrenia in children of mothers who smoked heavily during pregnancy [27].
- Smoking can also affect men's sperm, which can reduce fertility and also increase the risks for birth defects and miscarriage.
- Smoking can affect bone health. Women past childbearing years who smoke have weaker bones than women who never smoked and are at greater risk for broken bones.
- Smoking affects the health of your teeth and gums and can cause tooth loss.
- Smoking can increase your risk for cataracts (clouding of the eye's lens that makes it hard for you to see) and age-related macular degeneration (damage to a small spot near the center of the retina, the part of the eye needed for central vision).
- Smoking is a cause of type 2 diabetes mellitus and can make it harder to control. The risk of developing diabetes is 30%–40% higher for active smokers than nonsmokers.
- Smoking causes general adverse effects on the body, including inflammation and decreased immune function.
- Smoking is a cause of rheumatoid arthritis.

Of special note is the Surgeon General's estimate that 480,000 Americans die annually from cigarette smoking, with another 31,000 estimated to die annually from exposure to secondhand smoke.

The foregoing demonstrates the sweeping, adverse effects of cigarette smoking on the human body. Of special note is the Surgeon General's estimate that annually 480,000 Americans prematurely die from cigarette smoking, with another 31,000 estimated to die annually from exposure to secondhand smoke [28]. For comparison, the population of Atlanta, Georgia, is 498,044, a 2018 figure from the U.S. Census Bureau. Similarly, Miami's population in 2018 was estimated to be 470,914. Smoking-related deaths annually in the U.S. reflect the yearly loss of either of these two major cities.

5.1.4.1.1 Secondhand Tobacco Smoke

A particularly insidious public health hazard is called secondhand tobacco smoke. This form of environmental hazard is insidious because it affects persons who are not directly smoking tobacco. Secondhand smoke is a mixture of the smoke given off by the burning of tobacco products, such as cigarettes, cigars, or pipes, and the smoke exhaled by smokers. Secondhand smoke is also called environmental tobacco smoke (ETS), and exposure to secondhand smoke is sometimes called involuntary or passive smoking. Secondhand smoke contains more than 7,000 substances, several of which are known to cause cancer in humans or animals. EPA has concluded that exposure to secondhand smoke can cause lung cancer in adults who do not smoke. EPA estimates that exposure to secondhand smoke causes ~3,000 lung cancer deaths per year in nonsmokers.

During 2011–2012, about 58 million nonsmokers in the U.S. were exposed to secondhand smoke.

The 1992 EPA Risk Assessment, "Respiratory Health Effects of Passive Smoking," concluded that environmental tobacco smoke is causally associated with lung cancer in adults and designated ETS as a Group A (known human) carcinogen. Exposure to secondhand smoke has also been shown in a number of studies to increase the risk of heart disease and stroke [29].

The CDC has provided public health data on the consequences of exposure to secondhand tobacco smoke [30]. They note the following: Secondhand smoke is the combination of smoke from the burning end of a cigarette and the smoke breathed out by smokers. Secondhand smoke contains more than 7,000 chemicals. Hundreds are toxic and about 70 can cause cancer. Since the 1964 Surgeon General's Report, 2.5 million adults who were nonsmokers died because they breathed secondhand smoke.

- There is no risk-free level of exposure to secondhand smoke.
- Secondhand smoke causes numerous health problems in infants and children, including more frequent and severe asthma attacks, respiratory infections, ear infections, and sudden infant death syndrome (SIDS).
- Smoking during pregnancy results in more than 1,000 infant deaths annually.
- Some of the health conditions caused by secondhand smoke in adults include coronary heart disease, stroke, and lung cancer [30].

    Of special public health note is the CDC's statement about the adverse health effects of secondhand smoke on children [30]. In 2018, the agency stated the following:

- Studies show that older children whose parents smoke get sick more often. Their lungs grow less than children who do not breathe secondhand smoke, and they get more bronchitis and pneumonia.
- Wheezing and coughing are more common in children who breathe secondhand smoke.
- Secondhand smoke can trigger an asthma attack in a child. Children with asthma who are around secondhand smoke have more severe and frequent asthma attacks. A severe asthma attack can put a child's life in danger.
- Children whose parents smoke around them get more ear infections. They also have fluid in their ears more often and have more operations to put in ear tubes for drainage.

Perspective: Secondhand tobacco smoke is another deleterious route of exposure to the adverse consequences of smoking tobacco. This route of exposure can be characterized as a breach of ethics, owing to the risk that smokers present to nonsmokers.

5.1.4.1.2   Thirdhand Tobacco Smoke

In addition to the health hazard of secondhand smoke, a relatively new concept in environmental health is called thirdhand smoke (THS). According to the Mayo Clinic, "Thirdhand smoke is generally considered to be residual nicotine and other chemicals left on a variety of indoor surfaces by tobacco smoke. This residue is thought to react with common indoor pollutants to create a toxic mix. This toxic mix of thirdhand smoke contains cancer-causing substances, posing a potential health hazard to nonsmokers who are exposed to it, especially children. Studies show that thirdhand smoke clings to hair, skin, clothes, furniture, drapes, walls, bedding, carpets, dust, vehicles, and other surfaces, even long after smoking has stopped. Infants, children, and nonsmoking adults may be at risk of tobacco-related health problems when they inhale, ingest or touch substances containing thirdhand smoke" [31].

THS residue can accrue on surfaces, and its complex mixture resists normal cleaning. THS cannot be eliminated by airing out rooms, opening windows, using fans or air conditioners, or confining smoking to only certain areas of a home. Although THS is a relatively new public health concept, and researchers are still studying its possible dangers, knowledge of the health hazards presented by tobacco smoke suggests a precautionary approach to thirdhand smoke is warranted.

To investigate the presence of THS, an international team of scientists from the Max Planck Institute for Chemistry and Yale University conducted an experiment at a movie theater by measuring the real-time emissions of THS compounds from people into a non-smoking indoor environment [32]. Through four consecutive days of measurements with an online high-resolution mass spectrometer, 35 different volatile organic compounds (VOCs) previously associated with THS or tobacco smoke emissions were observed at significant concentrations in the theater. The gas emissions were equivalent to that of one to ten cigarettes of secondhand smoke in 1 hour. The investigators concluded that humans transport THS into indoor areas via their clothing and bodies.

5.1.4.1.3   Economic Costs of Smoking

Since 1964, smoking-related illnesses have claimed more than 20 million lives in the U.S., 2.5 million of which belonged to nonsmokers who developed diseases merely from secondhand-smoke exposure. However, the economic and societal costs of smoking are just as huge. Every year, smoking costs the U.S. more than $300 billion, which includes both medical care and lost productivity [33]. Unfortunately, some people pay more depending on the U.S. state in which they reside. To encourage the estimated 34.2 million tobacco users in the U.S. quit smoking, researchers looked into the true per-person cost of smoking in each of the 50 U.S. states and the District of Columbia. Calculated were the potential monetary losses—including both the lifetime and annual cost of a cigarette pack per day, health care expenditures, income losses, and other costs—brought on by smoking and exposure to secondhand smoke. Estimated total lifetime costs per smoker ranged from $3,171,757 in Massachusetts to $1,696,496 in Mississippi, with out-of-pocket expenses estimated as $174,850 and $95,834, respectively. These are significant sums of money if invested would have yielded valuable revenue.

5.1.4.2   Other Tobacco Products

Cigars: A description of the health risks and effects of smoking cigars must begin with the knowledge that cigars contain the same toxic and carcinogenic compounds found in cigarettes and are not a safe alternative to cigarettes. An early

> As the number of cigars smoked increases, the risk of premature death related to cigar smoking approaches that of cigarette smoking.

examination of the health effects of smoking cigars was performed in 2000 by CDC and American Cancer Society researchers [34]. Researchers examined the association between cigar smoking and death from tobacco-related cancers in a large, prospective cohort of U.S. men. Cox proportional hazards models were used to analyze the relationship between cigar smoking at baseline in 1982 and mortality from cancers of the lung, oral cavity/pharynx, larynx, esophagus, bladder, and pancreas over 12 years of follow-up of the American Cancer Society's Cancer Prevention Study II cohort. A total of 137,243 men were included in the final analysis. Women were not included because of no data on their cigar use. Current cigar smoking at baseline, as compared with never smoking, was associated with an increased risk of death from cancers of the lung, oral cavity/pharynx, larynx, and esophagus. There was an increased risk for current cigar smokers who reported that they inhaled the smoke for the pancreas and urinary bladder. The researchers concluded that their results from this large prospective study supported a strong association between cigar smoking and mortality from several types of cancer.

A more recent pronouncement from CDC in 2015 concludes that the health risks of regular cigar smoking include the following consequences [9]:

- Regular cigar smoking is associated with an increased risk for cancers of the lung, esophagus, larynx (voice box), and oral cavity (lip, tongue, mouth, and throat).
- Cigar smoking is linked to gum disease and tooth loss.
- Heavy cigar smokers and those who inhale deeply may be at increased risk for developing coronary heart disease.
- Heavy cigar smoking increases the risk for lung diseases, such as emphysema and chronic bronchitis.

Pipes: Although conventional (i.e., excluding water pipes) pipe smoking is much less prevalent than that for cigarettes and cigars, conventional pipe smoking brings serious health consequences to smokers and those exposed to secondhand pipe smoke. This assertion goes counter to belief by some pipe smokers that this form of tobacco smoking is less hazardous to health than that posed by cigarettes. Research on the health hazards of pipe smoking has been summarized by several sources. The seminal study of the association between the pipe smoking and adverse health effects was conducted by American Cancer Society (ACS) investigators [12]. They examined the association between the pipe smoking and mortality from tobacco-related cancers and other diseases in a cohort of U.S. men enrolled in the Cancer Prevention Study II, an American Cancer Society prospective study. The cohort of 138,307 men included those who reported, in their 1982 enrollment questionnaire, exclusive current or former use of pipes ($n = 15,263$ men) or never use of any tobacco product ($n = 123,044$ men). Analyses were based on 23,589 men who died during 18 years of follow-up.

Current pipe smoking, compared with never use of tobacco, was associated in the ACS study with an increased risk of death from cancers of the lung, oropharynx, esophagus, colorectum, pancreas, and larynx, and from coronary heart disease, cerebrovascular disease, and chronic obstructive pulmonary disease. These risks were generally smaller than those associated with cigarette smoking and similar to or larger than those associated with cigar smoking. Relative risks of lung cancer showed statistically significant increases with the number of pipes smoked per day, years of smoking, and depth of inhalation and decreases with years since quitting. This relative risk finding illustrates a dose-response outcome, a strong indicator of association between exposure and adverse health outcome.

Water pipes, while methodologically different from traditional tobacco pipes, present their own set of health concerns for smokers. In particular, while many hookah smokers might consider this practice less harmful than smoking cigarettes, hookah smoking carries many of the same health risks as cigarettes. The CDC has determined the cancer risks of hookah smoking as follows [13]:

- The charcoal used to heat tobacco in the hookah increases the health risks by producing smoke that contains high levels of carbon monoxide, metals, and cancer-causing chemicals.
- A typical 1-hour-long hookah smoking session involves 200 puffs, while an average cigarette is 20 puffs.
- The volume of smoke inhaled during a typical hookah session is about 90,000 mL, compared with 500–600 mL inhaled when smoking a cigarette.
- Using a hookah to smoke tobacco poses a serious potential health hazard to smokers and others exposed to the emitted smoke.

> The volume of smoke inhaled during a typical hookah session is about 90,000 mL, compared with 500–600 mL inhaled when smoking a cigarette.

A review of the health consequences of pipe smoking, be it traditional pipes or water pipes, reveals that this form of tobacco smoking is akin to other methods of smoking tobacco, that is, adverse health consequences accompany their use.

Smokeless tobacco products: Smokeless tobacco is one of the most addictive and potent ways of consuming tobacco. Half a can (17 g) of U.S.-style moist snuff contains 236 mg of nicotine; more than twice the daily nicotine consumption of the next potent tobacco product, Snus. Holding an average-size dip in the mouth for just 30 minutes can deliver as much nicotine as smoking three cigarettes. The CDC has summarized the adverse health of smokeless tobacco as follows [35]: "Smokeless tobacco is associated with many health problems. Using smokeless tobacco:

- Can lead to nicotine addiction,
- Causes cancer of the mouth, esophagus, and pancreas,
- Is associated with diseases of the mouth,
- Can increase risks for early delivery and stillbirth when used during pregnancy,
- Can cause nicotine poisoning in children,
- May increase the risk for death from heart disease and stroke."

These CDC findings indicate that regular use of smokeless tobacco can result in serious adverse health effects. Specifically, as noted above, there is an association between smokeless tobacco use and cancers of the mouth, esophagus, and pancreas and diseases of the mouth in general. Regrettably, such knowledge about these health effects has been ignored by some users of smokeless tobacco products. As an example, the use of smokeless tobacco has been a favorite habit of many professional baseball players. This practice was implemented using chewing tobacco, which was gradually replaced over time with snuff. Both products provided the player with a dose of nicotine, which was a stimulus when performing the various actions required of baseball play.

> Regular use of smokeless tobacco can result in serious adverse health effects: cancers of the mouth, esophagus, and pancreas, and diseases of the mouth.

In the late 20th century, the use of smokeless tobacco was discontinued in the minor league baseball system in the U.S. This policy was implemented by owners of the baseball teams, with the consent of players' representatives. However, the major league baseball system has eschewed any policy on controlling smokeless tobacco use by major league players, an outcome favored by players' representatives and their labor union. This was an arrangement that left the decision on smokeless tobacco use as a matter of individual preference by players. However, this laisse fare policy has begun to change due to no-tobacco use policies by some cities' banning tobacco in municipal parks and arenas. Examples are the cities of San Francisco, Los Angeles, and Boston [29]. Should this trend continue, municipal ordinances may ultimately be the remedy for preventing baseball players' health problems due to the use of smokeless tobacco.

5.1.5 IMPACTS OF TOBACCO PRODUCTS ON ECOSYSTEM HEALTH

The use of tobacco products has an impact on the health of ecosystems, albeit not as direct or evident as the effects on human health. Recall that healthy ecosystems contribute to clean air and water, fertile soil for crop production, pollination, and flood control, among many other benefits [35a]. Bearing this in mind, the impacts of tobacco products on ecosystems are several. Waste from tobacco products is a particular threat to ecosystems as pollutants in bodies of water. Fish and other marine life are exposed to nicotine and

> Cigarette butts are a special form of water pollutant, given that butts contain the residual of tobacco smoke drawn through the cigarette by the smoker.

other hazardous substances in tobacco products. Cigarette butts are a special form of water pollutant, given that butts contain the residual of tobacco smoke drawn through the cigarette by the smoker. This results in tars and other toxicants reaching water sources, exposing aquatic life to chemically contaminated water. Another impact of tobacco products is the cultivation of tobacco for commercial purposes. Because tobacco is not a food product, its cultivation removes the arable soil from food production. In a world of increasing human population, food security is already an issue of paramount importance due to climate change (Chapter 4).

One measured impact of tobacco waste impacting ecosystems is provided by volunteers who remove trash from coastal areas globally. In 2014, the total number of trash items picked up during the 29th year of Ocean Conservancy's International Coastal Cleanup, amounted to 15 million items, which weighed more than 16 million pounds. Cigarette butts were the most common item found, with more than 2 million collected. Ocean trash threatens ocean animals, fisheries, and tourism globally [36]. Another source observes that ~4.5 tn of the 6 tn cigarettes consumed annually are littered across the globe [37].

Cigarette butts also constitute a major source of urban street waste. Montreal is one city that has taken an active recycling approach to the removal of cigarette butts on its streets. In Montreal and other large cities, cigarette butts comprise about 30% of litter found on the ground. Given this amount of litter, since 2016 the city has installed 620 ashtrays spread out through the city. The collected butts are then recycled through a program that turns cigarette butts' paper and tobacco into compost and the plastic is used to make other commercial objects. Between April and November 2018, the City of Montreal recycled 459,719 cigarette butts [38]. This is a significant amount of urban litter that does not reach water and other ecosystems.

5.1.6   U.S. FEDERAL POLICIES ON TOBACCO USE AND CONTROL

There are several U.S. federal laws that pertain to the sales, marketing, and use of tobacco products. The primary laws are directed at cigarette smoking. As will be discussed, fewer legislative actions exist for control of other tobacco products, owing to their lesser prevalence of use in the U.S. A comprehensive description of the federal tobacco laws is beyond the purposes of this book and is available elsewhere [39]. Following is a distillation of the most significant federal actions targeted at public health implications of tobacco use [39].

There are several U.S. federal laws that pertain to the sales, marketing, and use of tobacco products, primarily directed at cigarette smoking.

Federal Trade Commission Act (FTCAct) of 1914 (amended in 1938)

- Empowers the Federal Trade Commission (FTC) to "prevent persons, partnerships, or corporations, from using unfair or deceptive acts or practices in commerce"
- On January 3, 1964, the FTC proposed a rule to strictly regulate the imagery and copy of cigarette ads to prohibit explicit or implicit health claims

The Jenkins Act of 1949

- The Jenkins Act requires any person who sells and ships cigarettes across a state line to a buyer, other than a licensed distributor, to report the sale to the buyer's state tobacco tax administrator. The act establishes misdemeanor penalties for violating the act. Compliance with this federal law by cigarette sellers enables states to collect cigarette excise taxes from consumers.

Federal Hazardous Substances Labeling Act (FHSAct) of 1960

- 1960—Authorized FDA to regulate substances that are hazardous (either toxic, corrosive, irritant, strong sensitizers, flammable, or pressure-generating). Such substances might cause substantial personal injury or illness during or as a result of customary use.
- 1963—FDA expressed its interpretation that tobacco did not fit the "hazardous" criteria stated previously and withheld recommendations pending the release of the report of the Surgeon General's Advisory Committee on Smoking and Health.

Federal Cigarette Labeling and Advertising Act of 1965

- Required package warning label—"Caution: Cigarette Smoking May Be Hazardous to Your Health" (other health warnings prohibited).
- Required no labels on cigarette advertisements (in fact, implemented a 3-year prohibition of any such labels).
- Required FTC to report to Congress annually on the effectiveness of cigarette labeling, current cigarette advertising, and promotion practices and to make recommendations for legislation.
- Required Department of Health, Education, and Welfare (DHEW) to report annually to Congress on the health consequences of smoking.

Public Health Cigarette Smoking Act of 1969

- Required package warning label— "Warning: The Surgeon General Has Determined that Cigarette Smoking Is Dangerous to Your Health" (other health warnings prohibited).
- Temporarily preempted the FTC requirement of health labels on advertisements.
- Prohibited cigarette advertising on television and radio (authority to Department of Justice [DOJ]).
- Prevents U.S. states or localities from regulating or prohibiting cigarette advertising or promotion for health-related reasons.

Consumer Product Safety Act of 1972

- Transferred authority from the FDA to regulate hazardous substances as designated by the Federal Hazardous Substances Labeling Act to the Consumer Product Safety Commission (CPSC).

- The term "consumer product" does not include tobacco and tobacco products.

Little Cigar Act of 1973

- Bans little cigar advertisements from television and radio (authority to DOJ).

1976 Amendment to the Federal Hazardous Substances Labeling Act of 1960

- The term "hazardous substance" shall not apply to tobacco and tobacco products.

Toxic Substances Control Act of 1976

- To "regulate chemical substances and mixtures which present an unreasonable risk of injury to health or the environment."
- The term "chemical substance" does not include tobacco or any tobacco products.

Comprehensive Smoking Education Act of 1984

- Requires four rotating health warning labels (all listed as Surgeon General's Warnings) on cigarette packages and advertisements (smoking causes lung cancer, heart disease and may complicate pregnancy; quitting smoking now greatly reduces serious risks to your health; smoking by pregnant women may result in fetal injury, premature birth, and low birth weight; cigarette smoke contains carbon monoxide) (preempted other package warnings).
- Requires Department of Health and Human Services (DHHS) to publish a biennial status report to Congress on smoking and health.
- Creates a Federal Interagency Committee on Smoking and Health.
- Requires the cigarette industry to provide a confidential list of ingredients added to cigarettes manufactured in or imported into the U.S. (brand-specific ingredients and quantities not required).

Comprehensive Smokeless Tobacco Health Education Act of 1986

- Institutes three rotating health warning labels on smokeless tobacco packages and advertisements (this product may cause mouth cancer; this product may cause gum disease and tooth loss; this product is not a safe alternative to cigarettes) (preempts other health warnings on packages or advertisements [except billboards]).
- Prohibits smokeless tobacco advertising on television and radio.
- Requires DHHS to publish a biennial status report to Congress on smokeless tobacco.

> Smokeless tobacco advertising on television and radio is prohibited.

- Requires FTC to report to Congress on smokeless tobacco sales, advertising, and marketing.
- Requires smokeless tobacco companies to provide a confidential list of additives and a specification of nicotine content in smokeless tobacco products.
- Requires DHHS to conduct public information campaign on the health hazards of smokeless tobacco.

Appropriations Act: Public Law 100–202 (1987)

- Banned smoking on domestic airline flights scheduled for 2 hours or less.

Public Law 101–164 (1989)

- Bans smoking on domestic airline flights scheduled for 6 hours or less.

Synar Amendment to the Alcohol, Drug Abuse, and Mental Health Administration (ADAMHA) Reorganization Act of 1992

- Requires all U.S. states to adopt and enforce restrictions on tobacco sales and distribution to minors.

Pro-Children Act of 1994

- Requires all federally funded children's services to become smoke-free. Expands upon 1993 law that banned smoking in Women, Infants, and Children (WIC) clinics.

Prevent All Cigarette Trafficking Act of 2009
The PACT Act amends the Jenkins Act of 1949. The act regulates the mailing of cigarettes and smokeless tobacco products to consumers through the U.S. Postal Service; adds new requirements for registration, reporting, delivery, and recordkeeping, including a List of Unregistered or Non-Compliant Delivery Sellers; increases penalties to a felony up to 3 years imprisonment; and gives ATF inspection authority to examine any records required to be maintained and any cigarettes or smokeless tobacco kept on the premises. In 2020, Congress amended the PACT Act to include the "Preventing Online Sales of E-Cigarettes to Children Act," which modifies the original definition of "cigarette" in the PACT Act to include Electronic Nicotine Delivery Systems (ENDS) [40].

Family Smoking Prevention and Tobacco Control Act of 2009

- Grants FDA the authority to regulate tobacco products.

The most recent U.S. tobacco law, the Family Smoking Prevention and Tobacco Control Act of 2009, merits additional description. The law, commonly called the Tobacco Control Act, was enacted by Congress in response to a prior decision from the U.S. Supreme Court. This decision is considered the most significant legal finding in U.S. tobacco litigation. The genesis of the litigation can be put simply, "FDA attempted to regulate tobacco products under its existing authorities to regulate food, drugs, and devices." This attempt prompted the lawsuit, which is described herein [41].

In March 2000, in a five to four opinion delivered by Justice Sandra Day O'Connor, the Supreme Court held that "Congress has not given the FDA the authority to regulate tobacco products as customarily marketed." The ruling was based on the Food, Drug, and Cosmetic Act (FDCAct) as a whole and in conjunction with Congress' subsequent tobacco-specific legislation. "By no means do we question the seriousness of the problem that the FDA has sought to address," Justice O'Connor wrote for the majority. Nonetheless, Justice O'Connor commented, "Congress, for better or for worse, has created a distinct regulatory scheme for tobacco products, squarely rejected proposals to give the FDA jurisdiction over tobacco, and repeatedly acted to preclude any agency from exercising significant policymaking authority in the area" [41]. Those words from the Court set the agenda for Congress to act, and the bipartisan result was the Tobacco Control Act of 2009.

The preface to the act states, "To protect the public health by providing the Food and Drug Administration with certain authority to regulate tobacco products, to amend title 5, United States Code, to make certain modifications in the Thrift Savings Plan, the Civil Service Retirement System, and the Federal Employees' Retirement System, and for other purposes." As summarized by the FDA (FDA Act 2016). Of particular note is the Act's definition of "tobacco product," a definition intended to respond to the Supreme Court's decision about FDA's lack of language to control tobacco products. As summarized by the FDA [41], the key provisions of the Tobacco Control Act are as follows:

The Tobacco Control Act

* Restricts tobacco marketing and sales to youth,
* Requires smokeless tobacco product warning labels,
* Ensures "modified risk" claims are supported by scientific evidence,
* Requires disclosure of ingredients in tobacco products,
* Preserves state, local, and tribal authority [42].

Restricts Tobacco Marketing and Sales to Youth

The Tobacco Control Act puts in place specific restrictions on marketing tobacco products to children and gives FDA authority to take further action in the future to protect public health. These provisions ban:

* Sales to minors
* Vending machine sales, except in adult-only facilities
* The sale of packages of fewer than 20 cigarettes
* Tobacco-brand sponsorships of sports and entertainment events or other social or cultural events

> The Tobacco Control Act puts in place specific restrictions on marketing tobacco products to children and gives FDA authority to take further action.

- Free giveaways of sample cigarettes and brand-name non-tobacco promotional items

Requires Smokeless Tobacco Product Warning Labels

The Tobacco Control Act requires that smokeless tobacco packages and advertisements have larger and more visible warnings. Smokeless tobacco includes tobacco products such as moist snuff, chewing tobacco, and snus. Every smokeless tobacco package and advertisement will include one of the following warning label statements:

- **WARNING**: This product can cause mouth cancer.
- **WARNING**: This product can cause gum disease and tooth loss.
- **WARNING**: This product is not a safe alternative to cigarettes.
- **WARNING**: Smokeless tobacco is addictive.

For smokeless tobacco packaging, the warning label statement must be located on the two principal sides of the package and cover at least 30% of each side. For advertisements, the warning label statements must cover at least 20% of the area of the ad. These changes aim to increase awareness of the health risks associated with smokeless tobacco use and improve public health.

Ensures "Modified Risk" Claims Are Supported by Scientific Evidence

The landmark law prohibits tobacco companies from making reduced harm claims like "light," "low," or "mild," without filing an application for a modified risk tobacco product and obtaining an order to market as such.

Requires Disclosure of Ingredients in Tobacco Products

Tobacco companies must provide FDA with detailed information about the ingredients in their products.

Preserves State, Local, and Tribal Authority

The Tobacco Control Act preserves the authority of state, local, and tribal governments to regulate tobacco products in specific respects.

Additional Authorities

The Tobacco Control Act further gives FDA authorities to:

> The Tobacco Control Act gives FDA the authority to regulate nicotine and ingredient levels in tobacco products.

- Require tobacco company owners and operators to register annually and open their manufacturing and processing facilities to be subject to inspection by FDA every 2 years.
- To implement standards for tobacco products to protect public health. For example, FDA has the authority to regulate nicotine and ingredient levels.
- Ban cigarettes with characterizing flavors, except menthol and tobacco.
- Fund FDA regulation of tobacco products through a user fee on the manufacturers of certain tobacco products sold in the U.S., based on their U.S. market share.

On May 5, 2016, FDA released its final rule on regulation of tobacco products under the agency's authorities in the Tobacco Control Act. In announcing its release, FDA stated, "The actions being taken today will help the FDA prevent misleading claims by tobacco product manufacturers, evaluate the ingredients of tobacco products and how they are made, as well as communicate their potential risks" [42]. FDA's rule also requires manufacturers of all newly regulated products, to show that the products meet the applicable public health standard set forth in the law and receive marketing authorization from the FDA, unless the product was on the market as of February 15, 2007.

The tobacco product review process gives the FDA for the first time the ability to evaluate important factors such as ingredients, product design, and health risks, as well as their appeal to youth and non-users. The FDA rule provides tobacco product manufacturers a staggered timetable for compliance with the new rule. Under the new rule, manufacturers can continue selling their products for up to 2 years while they submit—and an additional year while the FDA reviews—a new tobacco product application.

> Vape shops cannot give free samples to customers or sell to people younger than 18, under FDA regulations.

The same final rule extends its regulatory authority to cover all tobacco products, including vaporizers, vape pens, hookah pens, electronic-cigarettes (e-cigarettes), e-pipes, and all other ENDS. FDA now regulates the manufacture, import, packaging, labeling, advertising, promotion, sale, and distribution of ENDS. This includes components and parts of ENDS but excludes accessories. However, products marketed for therapeutic purposes (for example, marketed as a product to help people quit smoking) are regulated by the FDA through the Center for Drug Evaluation and Research (CDER).

These regulations went into effect on August 8, 2016. FDA will have to approve all e-cigarette products that have been available since February 2007. That means nearly every e-cigarette product on the market must go through an application process to deem whether it can continue to be sold. Manufacturers will be able to keep selling their products for up to 2 years while they submit a new production application, plus an additional year while the FDA reviews it.

Vape shops cannot give free samples to customers or sell to people younger than 18, under the new FDA regulations. Merchants will be required to ask for identification from customers who appear to be less than the age of 27. And vending machine sales of e-cigarettes are prohibited unless the machines are in adult-only facilities. Also covered are premium, hand-rolled cigars, as well as hookah and pipe tobacco. Before the new regulations, there was no federal law prohibiting retailers from selling e-cigarettes, hookah tobacco, or cigars to minors.

Defense Spending Bill of December 2019

On December 20, 2019, U.S. President Donald J. Trump (R-NY) signed into law the Department of Defense Spending Bill of 2019. The spending bill includes a measure that prohibits the sale of tobacco products in the U.S. to anyone under the age of 21 years old [43]. There already have been several states that have individually passed legislation to raise the tobacco-buying age to 21. As of December 2019, 19 states had raised the minimum age to buy tobacco products to 21. Raising the legal age to purchase tobacco products in the U.S. from 18 to 21 marks a major public health achievement for the Trump administration.

5.1.7 Trump Administration's Rules for New Tobacco Products

On January 21, 2021, the Trump administration announced the finalization of two key FDA rules for companies seeking to market new tobacco products [44]. As a reminder, new tobacco products require review and approval by the FDA. According to the agency, "Manufacturers or importers must demonstrate to the agency, among other things, that marketing of the new tobacco product(s) would be appropriate for the protection of the public health." That statutory standard requires the FDA to consider the risks and benefits to the population as a whole, including users and non-users of tobacco products. The agency's evaluation includes such things as reviewing a tobacco product's components, ingredients, additives, constituents, toxicological profile, and health impact, as well as how the product is manufactured, packaged, and labeled, findings from consumer perception research (if conducted), and the applicant's description of marketing plans for the product. The FDA asserts that the final rules help ensure new tobacco applications contain sufficient information for the agency's evaluation.

5.1.8 Labels on Tobacco Products

In a policy context, the placement of labels on tobacco products has engendered considerable challenges to government authorities, with concomitant litigation in the U.S. Labels on tobacco products have been considered for policy purposes as two kinds: health warnings and product advertisements. Health warning labels on cigarette containers have been required for many years in the U.S., with the 2009 Tobacco Control Act mandating that FDA strengthen such warning labels, including graphic images. FDA's attempts to require graphic images on cigarette packages were met with litigation filed by the tobacco industry, leading to a federal court decision that blocked large graphic health warnings on cigarette packages. A federal judge ruled that the requirement violated the U.S. Constitution's First Amendment free speech protections. An appeals court upheld that ruling, which was not further appealed by the U.S. government, leaving the FDA to develop graphic labels that would muster court review [45]. This development of revised graphic labels was completed in August 2019 [46]. On March 18, 2020, the FDA issued a ruling regarding health warnings on cigarette packs and advertisements. The 11 new warnings will fill 50% of the package with text and graphic imagery depicting the health consequences of smoking [47]. Shown in Figure 5.2 are illustrations of FDA's cigarette package graphic health warnings.

> In distinction to the U.S., some other countries have mandated and implement graphic labels on cigarette packages.

In distinction to the U.S., some other countries have mandated and implement graphic labels on cigarette packages. For example, shown in Figure 5.3 is a cigarette package from Canada, associating the oral disease with cigarette smoking. Interestingly, in November 2019, the Canadian government implemented regulations stipulating cigarette packaging to be a plain, drab brown color with standardized layouts and lettering. The government's goal is to reduce the 17% of Canadian adults who smoke cigarettes to 5% by 2035 [48]. Whether such graphic regulations will

FIGURE 5.2 Examples of FDA cigarette package health warnings. (FDA, FDA Newsroom, August 15, https://www.fda.gov/news-events/fda-newsroom/press-announcements [46].)

FIGURE 5.3 Health-warning cigarette package from Canada. (With permission of copyright holder, Prof. David Hammond, Department of Health Sciences, University of Waterloo, https://uwaterloo.ca/public-health-sciences/people-profiles/david-hammond.)

occur in the U.S. remains to be determined. In the UK, effective in 2017, cigarettes must be sold in standardized green packaging bearing graphic warnings of the dangers of smoking. Further, the UK directive extends to e-cigarettes, restricting tank sizes to no more than 2 mL and the nicotine strength of liquids to no more than 20 mg/mL, and there must be a 30% health warning on the front and back reading [49].

Regarding advertising labels, some governments consider product names placed on cigarette packages to be a form of product advertisement [50]. These governments consider product names to be a kind of inducement to purchase the product, and to reduce the appeal of cigarette smoking, no product labels are permitted. Shown in Figure 5.4 is a comparison of cigarette packages bearing product names, with the

FIGURE 5.4 Australian cigarette packages with no advertising (right side). (With permission of copyright holder, Prof. David Hammond, University of Waterloo.)

same product sold in Australia, where product names and advertising are replaced by grim health warnings.

In 2019, the WHO announced that more than half of the world's population—3.9 billion people living in 91 countries—benefit from large graphic tobacco pack warnings [51]. The agency further noted that tobacco use has declined proportionally in most countries, but population growth means the total number of people using tobacco has remained stubbornly high. As of 2019, there were an estimated 1.1 billion smokers, about 80% of whom reside in low- and middle-income countries (LMICs).

Perspective: Consideration of these federal laws provides an interesting perspective on policymaking directed toward a known health hazard, tobacco products. Of note, the legislative struggle over tobacco has endured for more than seven decades at the federal level, i.e., the U.S. Congress. Tobacco legislation by Congress has been difficult to achieve. This legislative inaction results from the strength of tobacco industry lobbying, as well as other factors such as the strong presence of members of Congress from tobacco-growing states in the southern U.S. It is noteworthy that tobacco has been expressly exempted from some federal laws pertaining to hazardous substance control. In lieu of outright banning tobacco products, federal laws have aimed at tobacco use reduction via education of the users.

> No federal law that outright bans tobacco products has been enacted by Congress.

A legislative philosophy of "education," not "regulation" has held sway. As a result, federal laws on tobacco have focused on controls of marketing, advertising, and distributing various tobacco products. For example, cigarette packages must by law contain health warning labels, with authority given to the FDA for implementation of content of such. Moreover, one observes that, even with the enactment of laws for the control of hazardous substances, no federal law that outright bans tobacco products has been enacted by Congress.

The most sweeping of U.S. tobacco control laws is the Tobacco Control Act of 2009. This act gave the FDA several new authorities to regulate tobacco products, including: restricting tobacco marketing and sales to youth; requiring smokeless tobacco product warning labels; ensuring "modified risk" claims are supported by scientific evidence; requiring disclosure of ingredients in tobacco products; and preserving state, territorial, local, and tribal authority over tobacco control and restrictions. Worthy of policy note is the act's language that preserves state, local, and tribal authorities. Sometimes an organization will attempt to add language that prevents states, local authorities, and indigenous tribes from enacting unwelcome policies. Federal law would override other levels of government, thereby giving lobbying groups only one target, Congress, rather than multiple targets.

5.1.9 STATE AND LOCAL TOBACCO POLICIES

The seminal work of Doll and Terry in the 1960s brought forth the association between cigarette smoking and adverse health effects. This led to education campaigns in the U.S. and Britain concerning the health hazard of smoking. But laws, regulations, and ordinances specific to limiting smoking did not follow; smoking was considered a personal choice and as such did not require legislation to limit a "personal" choice. In the 1980s, a new body of smoking research began to associate "secondhand smoke" with adverse health effects. Also sometimes called environmental tobacco smoke (ETS), this new information gave anti-smoking campaigners the ammunition needed to prod policymakers into action [52]. Secondhand smoke meant there were innocent victims of other persons' harmful behavior. This brought into policymaking play the notion of "victims' rights" and therefore required protective legislative action. In response, states and local governments began to promulgate laws and ordinances that limited tobacco smoking.

In the U.S., states, territories, tribes, and local governments assume the primary responsibility for policies on tobacco use, given that federal legislative policies (i.e., federal laws and federal agency regulations) relate only to marketing and advertising of tobacco products. Moreover, the U.S. Constitution delegates to states all authorities not specifically stated as the responsibility of federal government. As a consequence, all 50 states and territories have enacted laws that specify how tobacco products can be used within a state's borders.

Although there is considerable variability in tobacco laws across the states, the general intent is to specify under what conditions smoking tobacco can be permitted. Of the 50 states and territories/commonwealths, 24 states and two territories possess 100% smoke-free laws in all non-hospitality workplaces, restaurants, and bars [53]. An examination of the figure indicates that smoking bans are prevalent in New England and northern Mid-Western states. The remaining states and territories restrict smoking to designated areas.

A summary of the number of smoke-free and other tobacco-related laws was prepared by a nonsmokers foundation [54]. Excerpted from their list are the following highlight statistics:

- Across the U.S., 22,708 municipalities, representing 81.8% of the U.S. population, are covered by a 100% smoke-free provision in non-hospitality workplaces, and/or restaurants, and/or bars, by either a state, commonwealth, territorial, or local law. As an example, in December 2020 San Francisco's Board of Supervisors banned all tobacco smoking inside city apartments, citing concerns about secondhand smoke, but did not ban smoking marijuana [55].

- 43 states and the District of Columbia have local laws in effect that require non-hospitality workplaces and/or restaurants and/or bars to be 100% smoke-free.

43 states and the District of Columbia have local laws in effect that require non-hospitality workplaces and/or restaurants and/or bars to be 100% smoke-free.

- There are 3,788 states, commonwealths, territories, cities, and counties with a law that restricts smoking in one or more outdoor areas, including 1,819 that restrict smoking near entrances, windows, and ventilation systems of enclosed places.

- There are 2,356 colleges and universities with 100% smoke-free campuses. Of these, 1,986 are also 100% tobacco-free, and 1,965 prohibit the use of e-cigarettes anywhere on campus, 1,002 prohibit hookah use, and 448 prohibit smoking/vaping marijuana.

These data illustrate that considerable effort in all U.S. states and territories has occurred in regard to restrictions on tobacco use, although with considerable variance between states. It is instructive to illustrate how one state has chosen to enact tobacco products legislation. Excerpts from the Georgia state law are presented herein.

5.1.10 ILLUSTRATIVE STATE LAW ON TOBACCO CONTROL

The Georgia Smokefree Air Act was signed into law in May 2005. The act prohibits smoking inside most public places and sets guidelines for allowing smoking in and around public establishments. As stated by the Georgia Department of Health, the purpose of the act is to limit secondhand smoke exposure among children, adults, and employees and improve the health and comfort of the people of Georgia [56]. The law contains provisions that permit establishments to allow smoking if any person less than the age of 18 is prohibited from entry to or employment in the establishment and if smoking is allowed only in outdoor areas such as patios or in enclosed private rooms with independent air-handling systems [57].

Georgia Smokefree Air Act (Excerpts)
O.C.G.A. Sections 31-12A-1 through 31-12A-13
Title 31. Health
Chapter 12A. Smokefree Air
O.C.G.A. Section 31-12A-1. Short title
This chapter shall be known and may be cited as the "Georgia Smokefree Air Act of 2005."

O.C.G.A. Section 31-12A-2. Definitions

As used in this chapter, the term:

(1) "Bar" means an establishment that is devoted to the serving of alcoholic beverages for consumption by guests on the premises and in which the serving of food is only incidental to the consumption of those beverages, including, but not limited to, taverns, nightclubs, cocktail lounges, and cabarets.

[...] (8) "Local governing authority" means a county or municipal corporation of the state. [...]

(10) "Public place" means an enclosed area to which the public is invited or in which the public is permitted, including, but not limited to, banks, bars, [...]. A private residence is not a public place unless it is used as a licensed child care, adult day-care, or health care facility.

(11) "Restaurant" means an eating establishment, including, but not limited to, coffee shops, cafeterias, sandwich stands, and private and public school cafeterias, which gives or offers for sale food to the public [...]

(13) "Secondhand smoke" means smoke emitted from lighted, smoldering, or burning tobacco when the person smoking is not inhaling, smoke emitted at the mouthpiece during puff drawing, and smoke exhaled by the person smoking [...].

(16) "Smoking" means inhaling, exhaling, burning, or carrying any lighted tobacco product including cigarettes, cigars, and pipe tobacco.

(17) "Smoking area" means a separately designated enclosed room which need not be entered by an employee in order to conduct business that is designated as a smoking area [...]

O.C.G.A. Section 31-12A-3. Smoking prohibited in state buildings

Smoking shall be prohibited in all enclosed facilities of, including buildings owned, leased, or operated by, the State of Georgia, its agencies and authorities, and any political subdivision of the state, municipal corporation, or local board or authority created by general, local, or special Act of the General Assembly or by ordinance or resolution of the governing body of a county or municipal corporation individually or jointly with other political subdivisions or municipalities of the state.

Smoking shall be prohibited in all enclosed facilities of, including buildings owned, leased, or operated by, the State of Georgia, its agencies, and authorities.

O.C.G.A. Section 31-12A-4. Smoking prohibited in enclosed public places

Except as otherwise specifically authorized in Code Section 31-12A-6, smoking shall be prohibited in all enclosed public places in this state.

O.C.G.A. Section 31-12A-5. Smoking prohibited in an enclosed area within places of employment

(a) Except as otherwise specifically provided in Code Section 31-12A-6, smoking shall be prohibited in all enclosed areas within places of employment [...].

O.C.G.A. Section 31-12A-6. Areas exempt from smoking prohibitions

(a) Notwithstanding any other provision of this chapter, the following areas shall be exempt from the provisions of Code Sections 31-12A-4 and 31-12A-5: (1) Private residences, except when used as a licensed child care, adult day-care, or health care facility; (2) Hotel and motel rooms that are rented to guests and are designated as smoking rooms; [...]; (3) Retail tobacco stores, provided that secondhand smoke from such stores does not infiltrate into areas where smoking is prohibited under the provisions of this chapter; (4) Long-term care facilities as defined in paragraph (3) of Code Section 31-8-81; (5) Outdoor areas of places of employment; (6) Smoking areas in international airports, as designated by the airport operator; (7) All workplaces of any manufacturer, importer, or wholesaler of tobacco products, [...]; (8) Private and semiprivate rooms in health care facilities licensed under this title that are occupied by one or more persons, all of whom have written authorization by their treating physician to smoke; (9) Bars and restaurants, as follows: (A) All bars and restaurants to which access is denied to any person under the age of 18 [...]; or (B) Private rooms in restaurants and bars if such rooms are enclosed and have an air-handling system independent from the main air-handling system [...]; (10) Convention facility meeting rooms and public and private assembly rooms contained within a convention facility [...]; (11) Smoking areas designated by an employer which shall meet the following requirements: (A) The smoking area shall be located in a nonwork area [...]; (B) Air-handling systems from the smoking area shall be independent from the main air-handling system [...]; and (C) The smoking area shall be for the use of employees only [...]; (12) Common work areas [...] that are open to the general public by appointment only; [...], and (13) Private clubs, military officer clubs, and noncommissioned officer clubs [...].

Perspective: As matters of health policy, there are several features of this state's smoking law that merit comment. First, as stated separately by the state's Department of Human Services, "the purpose of the act is to limit secondhand smoke exposure among children, adults, and employees and improve the health and comfort of the people of Georgia." Second, the law commences with an extensive set of definitions. All such policy statements should commence similarly, since health policies, in particular, pertain to serious social subjects, e.g., smoking or operating motor vehicles, and are subject to enforcement. The gravity of such policies requires careful definition of terms and accompanying language, if for no other reason than for legal interpretation.

Third, Georgia's law contains exceptions to the no-smoking rule, e.g., tobacco shops and "open" bars are excluded from enforcement of no-smoking environments. Why are exceptions made to what is supposed to be a public health rule? The answer derives from the fact that policies, such as laws and ordinances, are the products of processes characteristic of democratic societies. And the involvement of interested parties and public participation are part of the democratic dialog that occurs in policymaking. In the case of Georgia's smoking law, the state has a large investment in tourism, leading to exempting bars, restaurants, hotels, and tobacco shops—all exempted under specific conditions—from the law's coverage of non-smoking conditions. In this specific instance, public health was balanced against the economic benefit of tourism. States that completely ban smoking without exemptions obviously had a different policymaking intent.

Subsequent to the enactment of Georgia's law, a study of the law's impact was conducted by one of the state's colleges of public health [58]. Investigators found that between the years 2006 and 2012, the percentage of Georgia establishments that permitted smoking without restriction in dining areas fell by more than half. However, during that same time, the percentage that permitted smoking in designated areas nearly doubled from about 9% in 2006 to almost 18% 6 years later. The investigators commented, "the most effective way to protect people from the dangers of second-hand smoke is to implement and enforce legislation that requires all indoor public places to be 100%" [58].

5.1.11 THE MASTER SETTLEMENT AGREEMENT

In addition to individual states' tobacco control legislation, such as the Georgia law, many of the states and U.S. territories banded together to litigate the tobacco industry for the recovery of states' costs associated with tobacco products. After several years of judicial actions and maneuvers by both states

> The Master Settlement Agreement is an accord reached in November 1998 between 46 states, five U.S. territories, the District of Columbia, and the five largest U.S. tobacco companies.

and the tobacco industry, a settlement occurred. The Master Settlement Agreement (MSA) is an accord reached in November 1998 between the attorneys general of 46 states, five U.S. territories, the District of Columbia, and the five largest U.S. tobacco companies [59]. The settlement addressed the companies' advertising, marketing, and promotion of tobacco products. In addition to requiring the tobacco industry to pay the settling states, ~$10 billion annually for the indefinite future, the MSA also set standards and restrictions on the sale and marketing of cigarettes by participating cigarette manufacturers. Among its many provisions, the MSA:

- Forbids participating cigarette manufacturers from directly or indirectly targeting youth;
- Imposes significant prohibitions or restrictions on advertising, marketing, and promotional programs or activities; and
- Bans or restricts cartoons, transit advertising, most forms of outdoor advertising, including billboards, product placement in media, branded merchandise, free product samples (except in adult-only facilities), and most sponsorships [59].

> Over the years, the litigating states and territories have collected record amounts of tobacco revenue from the MSA but are spending less of it on tobacco prevention programs.

Over the years, the litigating states and territories have collected record amounts of tobacco revenue from the MSA but are spending less of it on tobacco prevention programs. "According to the Campaign for Tobacco-Free Kids, which tracks state tobacco prevention spending vs. state tobacco revenues, only one state to date—North Dakota—currently funds a tobacco prevention program at even half the level recommended by the Centers for Disease Control and Prevention" [59]. Regrettably, in a public health sense, the states and territories have chosen to merge the tobacco settlement income into their general revenue stream and not for tobacco prevention programs.

5.1.12 SOCIAL JUSTICE ISSUES

Disparities exist across ethnic and racial groups in the U.S. in terms of tobacco use. According to CDC data for U.S. adults in 2019 who reported cigarette "use every day" according to racial/ethnic group and percent cigarette users: American Indian/Alaska Native, non-Hispanic (20.9); Other, non-Hispanic (19.7); White, non-Hispanic (15.5); Black, non-Hispanic (14.9); Hispanic (8.8); Asian, non-Hispanic (7.2) [60]. The reasons for these disparities between ethnic and racial groups are complex but would include cultural habits, targeted marketing by tobacco companies, socioeconomic factors, and health education.

There are also disparities reported between racial and ethnic groups exposed to secondhand smoke (SHSe) [61]. University of Texas researchers used data from four cycles of the National Health and Nutrition Examination Survey (NHANES) from 2011 to 2018 to estimate SHSe. For the 2017–2018 cycle, the SHSe prevalence continued to be twice as high among nonsmoker non-Hispanic Black individuals (48.02%) compared with non-Hispanic White individuals (22.03%) and among those living below the poverty level (44.68%) compared with those living above the poverty level (21.33%). Similarly, children aged 3–11 years continued to experience SHSe at high rates (38.23%). Given the adverse health effects of exposure to SHS, these racial disparities provide medical and public health specialists with additional data for conducting smoking cessation programs.

5.1.13 PRIVATE SECTOR TOBACCO POLICIES

The private sector in the U.S. and elsewhere in Europe and Asia has been a significant, positive policymaker in terms of control of tobacco use. These policies emerged from the recognition that employees who smoke cost businesses

> A policy of not hiring persons who smoked led to occasional complaints of denial of "smokers' rights" in the U.S.

more than non-smoking employees. These costs were incurred due to additional sick leave taken from work and additional health care costs on health insurance that were paid in part by employers. Simply put, cigarette smokers, in particular, added to the cost of business operations.

Companies gradually adopted human resource policies that encouraged tobacco smokers to participate in company-financed programs of smoking cessation. Companion policies consisted of not hiring persons who were smokers. This policy was dramatized by the findings from a longitudinal study by Stanford University researchers of persons seeking re-employment. In a study of 251 San Francisco area job seekers between 2013 and 2015, about half were smokers and half were not. After a year of follow-up, twice as many non-smokers had found employment. Further, the smokers' jobs paid about 25% less than jobs obtained by the non-smokers [62].

An example of a company's policy to not hire smokers is U-Haul, the moving and storage rental company, which announced on January 2, 2020, that, effective February 1 in U.S. 21 states, it will no longer hire nicotine users. Employees hired before that date will be unaffected by the new policy [63]. The nicotine-free policy will be enacted in those 21 U.S. states that lawfully allow the decline of nicotine users. U-Haul has a wellness program that includes nicotine cessation assistance for members, along with nutrition and fitness features. As a side note, according to the American Lung Association, 29 U.S. states and the District of Columbia have "smoker protection laws" that prevent employers from discriminating against employees for using tobacco products.

The policy of not hiring persons who smoked led to occasional complaints of denial of "smokers' rights" in the U.S. These kinds of complaints were generally unsuccessful when taken to litigation, depending on circumstances such as transparency of a company's smoking policies. Other policies restricted areas in business operations where smokers could congregate and smoke. In time, office building became "smoke free" due to policies set by employers or by government ordinance. In some instances, employee labor organizations petitioned, on behalf of their membership, for the establishment of smoke-free policies. One example of this kind of action was that taken by the airline flight attendants' union, which worked with airline management to establish no-smoking areas on commercial airlines. Later, all airlines became smoke-free, with no-smoking allowed while in flight.

5.1.14 GLOBAL CONTROL OF TOBACCO AND RELATED PRODUCTS

The human health toll of tobacco use is a global problem and therefore a worldwide challenge to policymakers and public health authorities. While many nations have implemented strong policies to reduce or prevent the use of tobacco products, much progress remains to be achieved. The global leader in reducing the use of tobacco products is WHO, which is an agency of the UN. As a UN agency, WHO has socio-political access to the nations of planet Earth, although not with equal impact due to geopolitical differences in national governments. The WHO has committed to a global program of action to prevent the health consequences of smoking. This program constitutes the agency's Framework Convention on Tobacco Control.

5.1.14.1 WHO Framework Convention on Tobacco Control, 2005

The WHO Framework Convention on Tobacco Control (WHO FCTC) is the first treaty negotiated under the auspices of WHO [64]. The FCTC went into effect on February 27, 2005. The WHO reports that 180 nations have signed the FCTC

> The WHO Framework Convention on Tobacco Control is the first treaty negotiated under the auspices of WHO.

treaty; the U.S. has not signed, owing to the treaty's expected opposition in the U.S. Senate, which has to ratify all international treaties under the U.S. Constitution.

The WHO FCTC was developed in response to the globalization of the tobacco epidemic. The spread of the tobacco epidemic is facilitated through a variety of complex factors with cross-border effects, including trade liberalization and direct foreign investment. Other factors such as global marketing, transnational tobacco advertising, promotion and sponsorship, and the international movement of contraband and counterfeit cigarettes have also contributed to the explosive increase in tobacco use. The core demand reduction provisions in WHO's FCTC are as follows:

- Price and tax measures to reduce the demand for tobacco, and
- Non-price measures to reduce the demand for tobacco, namely:
  - Protection from exposure to tobacco smoke;
  - Regulation of the contents of tobacco products;
  - Regulation of tobacco product disclosures;
  - Packaging and labeling of tobacco products;
  - Education, communication, training, and public awareness;
  - Tobacco advertising, promotion and sponsorship; and
  - Demand reduction measures concerning tobacco dependence and cessation.

The core supply reduction provisions in WHO's FCTC are contained in articles 15–17:

- Illicit trade in tobacco products;
- Sales to and by minors; and
- Provision of support for economically viable alternative activities.

The treaty has 180 signatories, including the European Community, which makes it one of the most widely embraced treaties in UN history [65]. However, the implementation of the FCTC by WHO's Member States has led to some criticism. In particular, one study reported "[...] we reviewed every first-cycle national implementation report and reconstructed the WHO database for the provisions most closely related to the six MPOWER priorities [...]. As of July 4, 2012, 361 (32.7%) of 1,104 countries' responses were misreported: 33 (3.0%) were clear errors, 270 (24.5%) were missing despite countries having submitted responses, and 58 (5.3%) were, in our opinion, misinterpreted by WHO staff" [66]. The authors of this paper imply that budget and staff reductions at WHO contributed to database problems in the FCTC program.

5.1.14.2   WHO's Key Facts on Global Tobacco Control

The WHO Framework Convention on Tobacco Control has broadened WHO's access to global databases on tobacco use, and these data provide a troubling image of the human and economic toll caused by tobacco products. The global consequences of tobacco use have been summarized by WHO in the following excerpts [67].

Key facts

- Tobacco kills up to half of its users;
- Tobacco kills around 6 million people annually. More than 5 million of those deaths are the result of direct tobacco use while more than 600,000 are the result of nonsmokers being exposed to secondhand smoke.

- Nearly 80% of the world's 1 billion smokers live in low- and middle-income countries. [...]

Secondhand smoke kills

- Secondhand smoke is the smoke that fills restaurants, offices, or other enclosed spaces when people burn tobacco products such as cigarettes, *bidis*, and water pipes. There are more than 4,000 chemicals in tobacco smoke, of which at least 250 are known to be harmful and more than 50 are known to cause cancer. There is no safe level of exposure to secondhand tobacco smoke.

> WHO: Tobacco kills around 6 million people annually. More than 5 million of those deaths are the result of direct tobacco use, while more than 600,000 are the result of nonsmokers being exposed to secondhand smoke.

- In adults, secondhand smoke causes serious cardiovascular and respiratory diseases, including coronary heart disease and lung cancer. In infants, it causes sudden death. In pregnant women, it causes low birth weight.
- Almost half of the children regularly breathe air polluted by tobacco smoke in public places.
- Secondhand smoke causes more than 600,000 premature deaths per year.
- In 2004, children accounted for 28% of the deaths attributable to secondhand smoke.
- Almost half of the children regularly breathe air polluted by tobacco smoke in public places.

Picture warnings work

Hard-hitting anti-tobacco advertisements and graphic pack warnings—especially those that include pictures—reduce the number of children who begin smoking and increase the number of smokers who quit. [...] Studies carried out after the implementation of pictorial package

> WHO: Anti-tobacco advertisements and graphic pack warnings reduce the number of children who begin smoking and increase the number of smokers who quit.

warnings in Brazil, Canada, Singapore, and Thailand consistently show that pictorial warnings significantly increase people's awareness of the harms of tobacco use.

Only 42 countries, representing 19% of the world's population, meet the best practice for pictorial warnings, which includes the warnings in the local language and cover an average of at least half of the front and back of cigarette packs. Most of these countries are low- or middle-income countries. The U.S. does not meet the best practices pictorial warning.

Ad bans lower consumption

- Bans on tobacco advertising, promotion, and sponsorship can reduce tobacco consumption.

- A comprehensive ban on all tobacco advertising, promotion, and sponsorship could decrease tobacco consumption by an average of about 7%, with some countries experiencing a decline in consumption of up to 16%.
- Only 29 countries, representing 12% of the world's population, have completely banned all forms of tobacco advertising, promotion, and sponsorship. The U.S. is not one of the countries.
- Around one country in three has minimal or no restrictions at all on tobacco advertising, promotion, and sponsorship.

Taxes discourage tobacco use

Tobacco taxes are the most cost-effective way to reduce tobacco use, especially among young and poor people. "A tax increase that increases tobacco prices by 10% decreases tobacco consumption by about 4% in high-income countries and about 5% in low- and middle-income countries. Even so, high tobacco taxes are a measure that is rarely implemented. Only 33 countries, with 10% of the world's population, have introduced taxes on tobacco products so that more than 75% of the retail price is tax. Tobacco tax revenues are on average 269 times higher than spending on tobacco control, based on available data" [67].

Tobacco taxes are the most cost-effective way to reduce tobacco use, especially among young and poor people.

Perspective: These findings from WHO characterize a global environmental health epidemic and corresponding challenge to policymakers. Worthy of repeating are a few salient facts gathered by WHO's Framework Convention on Tobacco Control: The use of tobacco kills around 6 million people annually. Nearly 80% of the world's 1 billion smokers live in low- and middle-income countries. Almost half of the children regularly breathe air polluted by tobacco smoke in public places. These sobering numbers present the calculus for needed national and international actions and policies for control of the tobacco epidemic.

5.2 VAPING DEVICES

The Center on Addiction defines "Vaping is the act of inhaling and exhaling the aerosol, often referred to as vapor, which is produced by an e-cigarette or similar device. The term is used because e-cigarettes do not produce tobacco smoke, but rather an aerosol, often mistaken for water vapor, that actually consists of fine particles. Many of these particles contain varying amounts of toxic chemicals. Vaping devices include not just e-cigarettes, but also vape pens and advanced personal vaporizers (also known as 'MODS')" [68].

The same sources state, "Generally a vaping device consists of a mouthpiece, a battery, a cartridge for containing the e-liquid or e-juice, and a heating component for the device that is powered by a battery. When the device is used, the battery heats up the heating component, which turns the contents of the e-liquid into an aerosol that is inhaled into the lungs and then exhaled. The e-liquid in vaporizer products usually contains a propylene glycol or vegetable glycerin-based liquid with nicotine, flavoring and other chemicals and metals, but not tobacco" [68]. As will be discussed in this section, there are various forms of vaping devices, with accompanying different implications for human health.

5.2.1 Précis History of Vaping Devices

A manufacturer of vaping devices and allied products has listed a generational history of their development [69]. According to this source, the following generation of vaping devices has occurred.

The First Generation: early generation products resembled cigarettes in appearance and were far simpler when compared to current devices. As simple products, many were disposable while others took refills.

The Second Generation: saw the introduction of vape pens. Vape pens were largely responsible for the growth of the industry and remain very prevalent. The reason for this is that pens are refillable and can make use of the multitude of e-liquids on the market.

The Third Generation: saw the introduction of MODs (short for modified or modification) to the marketplace. Like vape pens, MODs were more efficient and had a longer battery life. MODs were so named because they allowed users more control over the vaping experience.

Manufacturers of vaping devices have adjusted their products according to the targeted user audience. The early devices, e-cigarettes, were purposefully designed to appeal to smokers of tobacco cigarettes, with the commercial assertion that e-cigarettes were less harmful to human health. And it seems that some manufacturers have designed their products to appeal to younger audiences to make use of vaping devices.

5.2.2 Prevalence of Vaping Devices and Users in the U.S.

Researchers at Johns Hopkins University investigated vaping patterns in U.S. adults in 2018. A structured telephone survey revealed that about 10.8 million adults were currently using e-cigarettes, and more than half of them were <35 years old. Further, one in three e-cigarette users was vaping daily [70]. According to the investigators, "The most common pattern of use in the U.S. is dual use, i.e., current use of both traditional cigarettes and electronic-cigarettes." Twenty-somethings, smokers of traditional cigarettes, unemployed adults, and people who identify as lesbian, bisexual, gay, and transgender (LGBT) were reported to be more likely than other individuals to use e-cigarettes.

> A structured telephone survey revealed that about 10.8 million U.S. adults were using e-cigarettes, and more than half of them were <35 years old.

Regarding adolescents' vaping patterns in the U.S., a survey of adolescent drug trends in 2018 conducted by researchers at the University of Michigan found troubling increases in adolescents' vaping from 2017 to 2019 [71]. These results came from the annual Monitoring the Future survey, which has tracked national substance use among U.S. adolescents every year since 1975 for 12th grade students and since

1991 for 8th and 10th grade students. Monitoring the Future surveys nationally representative samples of 12th-, 10th-, and 8th-grade students annually. The project surveyed 43,703 respondents in 2017, 44,482 in 2018, and 42,531 in 2019. Overall response rates for these 3 years were 80% in 12th grade, 86% in 10th grade, and 88% in 8th grade.

Results showed significant increases in 30-day nicotine vaping in samples at each of the three grade levels from 2018 to 2019. As a result of these (and previously reported) annual increases, vaping prevalence more than doubled in each of the three grades from 2017 to 2019. In 2019, the prevalence of use during the previous 30 days was more than 1 in 4 students in the 12th grade, more than 1 in 5 in the 10th grade, and more than 1 in 11 in the 8th grade. Students who had vaped nicotine during the previous 12 months and those who had ever vaped nicotine also significantly increased in each grade from 2018 to 2019.

An additional study of vaping use and patterns was conducted by the CDC [72]. Data from the CDC's 2020 National Youth Tobacco Survey, researchers found: "In 2020, 19.6% of high school students (3.02 million) and 4.7% of middle school students (550,000) reported current e-cigarette use. Among current e-cigarette users, 38.9% of high school students and 20.0% of middle school students reported using e-cigarettes on 20 or more of the past 30 days; 22.5% of high school users and 9.4% of middle school users reported daily use. Among all current e-cigarette users, 82.9% used flavored e-cigarettes, including 84.7% of high school users (2.53 million) and 73.9% of middle school users (400,000)."

Perspective: Increases in vaping prevalence are both troubling and challenging to public health programs. In particular, the presence of a highly addictive drug, nicotine, as an outcome of adolescents drug behavior bodes ill for programs of cessation of vaping behaviors. Compounding the challenge of educating adolescents to forego vaping behaviors is the finding that teens are generally unaware that they are vaping nicotine [73]. Having given some sense of adults and adolescents vaping prevalence, some discussion of forms of vaping devices follows. Two forms are particularly common as vaping devices, as discussed in the ensuing sections.

5.2.2.1 Electronic-Cigarettes

An e-cigarette, also known as an electronic cigarette or e-cig, is a device used to inhale flavored liquid which can contain nicotine. Strictly speaking, electronic-cigarettes (e-cigarettes) are not a tobacco product, although some public

> An e-cigarette, also known as an electronic cigarette or e-cig, is a device used to inhale flavored liquid which can contain nicotine.

health specialists have advocated the classification. However, the product is closely associated with tobacco-based cigarettes and therefore included in this chapter. The association is due to the fact that both e-cigarettes and tobacco cigarettes contain nicotine. Vapor from both products can be inhaled into a smoker's lungs, thereby delivering a dose of nicotine and other substances [74]. While tobacco cigarettes have been a commercial product in the U.S. since 1865, by comparison, e-cigarettes are rather recent on the nicotine scene.

E-cigarettes are one of a class of electronic nicotine delivery systems (ENDS), of which electronic-cigarettes are the most common prototype [75]. E-cigarettes are devices that do not burn or use tobacco leaves but instead vaporize a solution the user then inhales. E-cigarettes are designed to simulate the act of tobacco smoking by producing an appealingly flavored aerosol that looks and feels like tobacco smoke and delivers nicotine but with less of the toxic chemicals produced by burning tobacco leaves [76]. E-cigarettes consist of three components: a cartridge that contains a liquid solution containing various levels of nicotine, flavoring, and other chemicals; a heating element (vaporizer); and a power source (usually a battery). Shown in Figure 5.5 are examples of typical forms of e-cigarettes. The main constituents of the solution, in addition to nicotine when nicotine is present, are propylene glycol, with or without glycerol and flavoring agents. ENDS solutions and emissions contain other chemicals, some of them considered to be toxicants [77].

According to a CDC study, e-cigarette use increased from 11.7% to 20.8% among high school students and from 3.3% to 4.9% among middle school students from 2017 to 2018 [78]. No change was found in the use of other tobacco products, including cigarettes, during this time.

Among youth:

- E-cigarettes are still the most commonly used tobacco product, ahead of cigarettes, cigars, smokeless tobacco, hookah, and pipes.
- E-cigarettes are the most commonly used product in combination with other tobacco products.
- E-cigarette use is highest for boys, whites, and high school students.

> U.S. Surgeon General: One JUUL pod or flavor cartridge contains about the same amount of nicotine as a whole package of cigarettes.

Of note, a federal judge on May 15, 2019, ordered FDA to speed up its reviews of thousands of electronic cigarettes currently on the market, siding with public health groups that sued the agency [79]. According to the judge, the FDA's decision to delay reviewing electronic cigarette products for several years amounted to a serious

FIGURE 5.5   Examples of e-cigarettes. (FDA, 2016.)

dereliction of duty. The agency gained the authority to regulate all tobacco products in 2016, but when the administration changed in 2017, FDA decided to delay enforcing the laws for vaping products until 2022, an action consistent with the Trump administration's stance against promulgating federal government regulations.

5.2.2.2 JUUL Vaping Device

The newest and most popular vaping product in the U.S. is the JUUL, which is a small, sleek device that resembles a computer flash drive [68]. Its subtle design makes it easy to hide, which helps explain why it has become so popular among middle and high school students. It now accounts for about 72% of the market share of vaping products in the U.S. It comes in several flavors like crème brûlée, mango, and fruit medley. Every JUUL product contains a high dose of nicotine, with one pod or flavor cartridge containing about the same amount of nicotine as a whole pack of cigarettes. The popularity of JUULs and the substantial burden of nicotine make this device a matter of concern to the public health of adolescents, in particular. The highly addictive nature of nicotine makes JUUL vaping a substantive concern and challenge to public health.

In 2018, U.S. Surgeon General Jerome Adams issued the following warming, "All JUUL e-cigarettes have a high level of nicotine. A typical JUUL cartridge, or 'pod,' contains about as much nicotine as a pack of 20 regular cigarettes. These products also use nicotine salts, which allow particularly high levels of nicotine to be inhaled more easily and with less irritation than the free-base nicotine that has traditionally been used in tobacco products, including e-cigarettes. This is of particular concern for young people because it could make it easier for them to initiate the use of nicotine through these products and also could make it easier to progress to regular e-cigarette use and nicotine dependence. However, despite these risks, approximately two-thirds of JUUL users aged 15–24 do not know that JUUL always contains nicotine" [80].

5.2.3 GLOBAL PREVALENCE OF VAPING DEVICES AND USERS

A market research group, Euromonitor, has observed that the global number of adult vapers has been increasing rapidly—from about 7 million in 2011 to 41 million in 2018. From that data, it can be shown that the annual global increase in the number of vapers is ~5 million per annum [81]. Further, the research group estimates that the number of adults who vape will reach almost 55 million by 2021. Furthermore, the global vapor products market was estimated to be worth $22.6 bn. The U.S., Japan, and the UK are the biggest markets. Vapers in the three countries spent a combined $16.3 bn on smokeless tobacco and vaping products in 2016. The vaping industry foresees a steady and profitable increase in sales of its products. How this increase becomes the subject of public health policies remains to be globally determined.

5.2.4 IMPACTS OF VAPING DEVICES ON HUMAN HEALTH

Smoking vaping devices has implications for causing adverse health effects. Given the fact that the key constituent of e-cigarette vapor is nicotine—indeed, nicotine delivery is the primary feature of e-cigarettes—CDC has published some cautionary advice about nicotine in general. As enunciated by that agency, "Nicotine from e-cigarettes is absorbed by users and bystanders. Nicotine is highly addictive; Nicotine is especially a health danger to youth who use e-cigarettes. It may have long-term, negative effects on brain growth; nicotine is a health danger for pregnant women and their developing babies. Using an e-cigarette and even being around someone else using an e-cigarette can expose pregnant women to nicotine and other chemicals that may be toxic" [82]. Depending on the manner of smoking the e-cigarette will determine the extent of lung and buccal exposure to nicotine and other contaminants.

In 2019, a study by Stanford University School of Medicine researchers found that vaping flavored nicotine liquids, including versions that don't contain nicotine, may cause "significant damage" to endothelial cells on the inside of blood vessels [83]. The endothelial cells lining the interior of blood vessels have a critical role in one's overall cardiovascular and heart health. The negative effects were found to vary in severity based on which flavorings were used with the liquids. The study involved exposing endothelial cells to both the blood of people who had vaped shortly before the samples were collected, as well as directly to e-liquids of varying nicotine strengths. The experiments involved human endothelial cells that were produced in a lab using pluripotent stem cells. This study raises significant issues with respect to potential public health outcomes. As a matter of personal and public health, avoidance of any form of lung pollution is advisable.

In a different kind of study of health effects of smoking e-cigarettes, University of California San Francisco researchers found that e-cigarette use significantly increases a person's risk of developing chronic lung diseases like asthma, bronchitis, emphysema, or chronic obstructive pulmonary disease [84]. This was the first longitudinal study linking e-cigarettes to respiratory illness in a sample representative of the entire U.S. adult population. Researchers found that for e-cigarette users, the odds of developing lung disease increased by about a third, even after controlling for their tobacco use and their clinical and demographic information. The study also found that people who used e-cigarettes and also smoked tobacco—by far the most common pattern among adult e-cigarette users—were at an even higher risk of developing chronic lung disease than those who used either product alone. The findings were based on an analysis of data from the Population Assessment of Tobacco and Health (PATH), which tracked e-cigarette and tobacco habits as well as new lung disease diagnoses in more than 32,000 American adults from 2013 to 2016.

A particularly noteworthy study of the adverse health effects of vaping was reported in 2020 [85]. Researchers from the Boston University School of Public Health and School of Medicine are one of the first to look at vaping in a large, healthy sample of the U.S. population over time, independently from other tobacco product use. The researchers used data on 21,618 healthy adult participants from the first four waves (2013–2018) of the nationally representative

> Researchers found that for e-cigarette users, the odds of developing lung disease increased by about a third, even after controlling for their tobacco use and their clinical and demographic information.

Population Assessment of Tobacco and Health (PATH), which is the most comprehensive national survey of tobacco and e-cigarette-related information available. Researchers found that participants who had used e-cigarettes in the past were 21% more likely to develop respiratory disease, and those who were current e-cigarette users had a 43% increased risk.

Regarding human health effects, in 2019 an outbreak in the U.S. of lung disease occurred in persons who were regular users of vaping devices [86]. In response, the CDC, the FDA, state and local health departments, and other clinical and public health partners commenced investigating a multistate outbreak of lung disease associated with e-cigarette product (devices, liquids, refill pods, and/or cartridges) use. By October 1, 2019, CDC announced there were 1,299 people across 49 states, Washington, D.C., and the U.S. Virgin Islands diagnosed with e-cigarette illness. Eighteen deaths have been reported from 15 states. All reported cases have a history of e-cigarette product use or vaping. Most patients have reported a history of using e-cigarette products containing THC. Most patients reported using THC-containing products. Some have reported the use of e-cigarette products containing only nicotine [87]. In October 2019, CDC coined the illness as EVALI, which is short for e-cigarette or vaping product use-associated lung injury [88]. Subsequent research on the Lung Injury Response Laboratory Working Group identified vitamin E acetate as being associated with EVALI in a convenience sample of 51 patients in 16 states across the U.S. [88].

Given adverse health effects in connection with some vaping devices, some governments have taken action. In September 2019, the government of India enacted a law that forbids the manufacture, import, sale, advertisement, and distribution of e-cigarettes, rather than their use. Those making or selling e-cigarettes could face fines of up to 100,000 rupees ($1,390) for a first offense, and either a larger fine or even a prison sentence for repeat offenses, the government said [89]. Another action stimulated by the report of vaping-associated lung illness was the decision in September 2019 by Walmart, the largest U.S. retailer to discontinue selling e-cigarettes [90].

In the U.S., in November 2019, Massachusetts enacted into law the U.S.'s most stringent ban on flavored vaping products, as well as menthol cigarettes. The new law immediately restricts the sale of all flavored nicotine vaping products and will ban menthol cigarettes starting June 1, 2020. Under the law, flavored vaping products will only be sold in licensed smoking bars, and they must be consumed onsite. It will also impose a steep 75% excise tax on e-cigarettes and require health insurers to cover tobacco cessation programs [91].

ENFORCEMENT EXAMPLE

In August 2019, the FDA demanded that four tobacco companies stop selling 44 different flavors of vaping liquid and hookah products that the agency said are being sold illegally. The FDA said the companies have been selling products that were introduced to the market after the effective date of a rule that gave the agency the authority to regulate all tobacco products [92].

Regarding flavored e-cigarettes, known for their flavors, such as bubblegum, banana, and strawberry, a multinational team of academic researchers reported in 2020 they had found that the flavoring chemicals of e-cigarette vapor alone can measurably damage the lungs, regardless of the presence of nicotine [93]. The researchers found changes in certain inflammatory proteins known to cause disease. In each individual who used e-cigarettes, they discovered irregular protein levels within their saliva and airways compared to individuals who did not vape. The researchers said the data suggest people who use flavored e-cigarettes are damaging their lungs every time they vape. Among the most toxic: chemical profiles for some chocolate and banana flavors.

Some proponents of electronic-cigarettes have argued their use can aid in reducing the smoking of tobacco products. A cohort study by French researchers found that, among daily smokers in France, regular (daily) e-cigarette use was associated with a statistically significantly **higher decrease** (emphasis added) in the number of cigarettes smoked per day as well as an increase in smoking cessation attempts. However, among former smokers, electronic cigarette use was associated with an increase in the rate of smoking relapse [94]. Overall, this study lends putative credence to the assertion that e-cigarettes are beneficial for tobacco smoking cessation, but that assertion must be cautioned by the type and frequency of use of an e-cigarette.

5.2.5 IMPACTS OF VAPING DEVICES ON ECOSYSTEM HEALTH

The principal implication of vaping devices for any impact on ecosystem health derives from the devices' potential as hazardous waste [95]. As a reminder, the U.S. federal management of hazardous waste is discussed in Volume 2, Chapter 2. In particular, the primary federal statute concerning hazardous waste is the Resource Conservation and Recovery Act. Recall that e-cigarettes have three main components: a cartridge that holds the flavored nicotine "vaping" solution, a heating element, and a lithium-ion battery. Two "vape" components—the cartridge and the lithium-ion battery—require hazardous waste removal and management.

Once expended, "vape" cartridges that held the flavored nicotine solution are typically discarded. However, without proper hazardous waste disposal, cartridges with residual solutions pose a threat to domestic pets and wildlife, should the residual nicotine find its way into the environment, and even to children, should they find the expended cartridges and play with them. To address this concern, some e-cigarette manufacturers and vendors offer recycling programs, with the incentive of discounted or free products for so many "vapes" returned for reclamation. For commercial and industrial enterprises, many jurisdictions stipulate that any discarded waste containing any concentration of nicotine requires hazardous waste management, and the rules can be severe. For example, in Minnesota, there is no minimal amount or concentration of nicotine-containing material that is exempt from regulation [95].

Disposing of "vape" lithium-ion batteries presents a problem. For retail consumers, the EPA doesn't regulate the disposal of batteries in small quantities, although local jurisdictions might have specific rules for electronic waste disposal. This is because it is dangerous to place lithium-ion batteries into a trash or recycling bin. This is because once in the back of a refuse or recycling truck, surrounded by dry paper and cardboard, pressure and heat can cause lithium-ion batteries to spark, causing a rolling inferno. Lithium-ion batteries are one of the leading causes of recycling truck fires. Again, the easier path is to take advantage of recycling programs offered by e-cigarette manufacturers and vendors. Alternatively, local refuse services of local municipality can provide guidelines for electronic waste disposal. Also, some electronic retail stores offer battery recycling as part of their brand.

For commercial and industrial users, the recycling of large quantities of batteries is regulated under the Universal Rules of Hazardous Waste regulations (40 CFR PART 273), which are imprecise enough about batteries to require careful interpretation and expert advice.

5.2.6   U.S. States Policies on Vaping Devices

With the absence of federal regulations, many states and cities adopted their own e-cigarette regulations, most commonly to prohibit sales to minors, including Maryland, Kentucky, Minnesota, New Jersey, New Hampshire, Tennessee,

> Several U.S. cities and states have enacted laws that increased the legal age to purchase e-cigarettes to age 21.

Utah, and Wisconsin. In July 2019, Colorado made vaping indoors illegal, with some exceptions. People can no longer vape inside any public building or any business, with the exception of cigar and tobacco bars and cannabis clubs. Vaping is also forbidden within 25 feet of a building's main doorway. Other states are considering similar legislation. Several U.S. cities and states have enacted laws that increased the legal age to purchase e-cigarettes to age 21. As of 2014, some states in the U.S. permit e-cigarettes to be taxed as tobacco products, and some state and regional governments in the U.S. had extended their indoor smoking bans to include e-cigarettes [96].

A review of regulations in 40 U.S. states found that how a law defines e-cigarettes is critical, with some definitions allowing e-cigarettes to avoid smoke-free laws, taxation, and restrictions on sales and marketing. Fewer policies have been created to restrict vaping indoors than cigarette smoking. Many local and state jurisdictions have begun enacting laws that prohibit e-cigarette usage everywhere that smoking is banned, although some state laws with comprehensive smoke-free laws will still allow for vaping to be permitted in bars and restaurants while prohibiting e-cigarettes in other indoor places. A 2017 report stated "As of 2 October 2015, five U.S. states and more than 400 counties have implemented some form of restriction of ECIG use indoors." In June 2019, San Francisco became the first U.S. city to effectively ban e-cigarette sales. The ordinance says "no person shall sell or distribute an electronic cigarette to a person in San Francisco unless that product has undergone premarket review by the U.S. Food and Drug Administration" [96]. As of June 2019, none have.

A different form of litigation regarding e-cigarettes was commended in October 2019. Three public school systems filed suit against Juul, the e-cigarette manufacturer, accusing it of endangering students and forcing educators to divert time and money to fight an epidemic of nicotine addiction [97]. The school systems in St. Charles, Missouri, Olathe, Kansas, and a Long Island school were believed to be the first in the U.S. to sue Juul, which dominates the e-cigarette market with devices that look like thumb drives and that have become wildly popular with American teenagers. The districts say Juul explicitly marketed its products to youths, leaving schools to shoulder the costs of stopping students from vaping, disciplining them when they break school rules, and providing support services when they become addicted. The litigation is likely to be lengthy, given the possibility of litigation appeals.

5.2.7 FEDERAL POLICIES ON VAPING DEVICES

The U.S. federal government articulated in December 2019 a partial initiative to reduce the use of e-cigarettes by teenagers. The Trump administration announced that U.S. sales of most flavored cartridges popular with youth will be forbidden, but flavored liquid nicotine for open tank devices will be exempt [98]. On January 2, 2020, the FDA announced enforcement regulations against illegally marketed electronic nicotine delivery systems (ENDS) products by focusing on the following groups of products that do not have premarket authorization: Any flavored, cartridge-based ENDS product (other than a tobacco- or menthol-flavored ENDS product); all other ENDS products for which the manufacturer has failed to take (or is failing to take) adequate measures to prevent minors' access; and any ENDS product that is targeted to minors or likely to promote the use of ENDS by minors.

A further policy action regarding vaping devices was taken by Congress via language inserted as a rider into pandemic economic stimulus legislation. Buried in the enormous spending/COVID-19 relief package that Congress approved in late December 2020 is a bill that imposes new restrictions on the distribution of all vaping equipment, parts, and supplies, including a ban on mailing them [99]. Title VI of the 2021 Consolidated Appropriations Act, which appears on page 5,136 of the 5,593-page bill, is called the Preventing Online Sales of E-Cigarettes to Children Act. U.S. Senator Dianne Feinstein (D–CA.), joined by seven cosponsors: six Democrats and Senator John Cornyn (R–TX) sponsored the act. The act will require every product related to vaping, whether of nicotine, THC, CBD, lavender, or anything else, to be subject to the Jenkins Act's burdensome requirements. The Jenkins Act of 1949 requires that vendors who sell cigarettes to customers in other states register with the tax administrators in those states and notify them of all such sales so they can collect the taxes that the buyers are officially obligated to pay. Feinstein's bill also requires the U.S. Postal Service to "clarify" that the ban on mailing cigarettes covers all vaping products.

5.2.8 INTERNATIONAL POLICIES ON VAPING DEVICES

A comprehensive study of international policies on the use of vaping devices was conducted by Johns Hopkins University public health researchers [100]. National policies regulating e-cigarettes were identified by (1) conducting web searches on Ministry of Health websites, and (2) broad web searches. The mechanisms used to regulate e-cigarettes were classified as new/amended laws or existing laws. The policy domains identified include restrictions or prohibitions on product: sale, manufacturing, importation, distribution, use, product design including e-liquid ingredients, advertising/promotion/sponsorship, trademarks, and regulation requiring:

> Researchers identified 68 countries that regulate e-cigarettes.

taxation, health warning labels, and child-safety standards. The classification of the policy was reviewed by a country expert.

Researchers identified 68 countries that regulate e-cigarettes: 22 countries regulate e-cigarettes using existing regulations; 25 countries enacted new policies to regulate e-cigarettes; 7 countries made amendments to existing legislation; 14 countries use a combination of new/amended and existing regulation. Common policies include a minimum age of purchase, indoor-use (vape-free public places) bans, and marketing restrictions. Few countries are applying a tax to e-cigarettes. A range of regulatory approaches are being applied to e-cigarettes globally; many countries regulate e-cigarettes using legislation not written for e-cigarettes. Findings from this study suggest the regulation of e-cigarettes is increasing globally, seemingly in concert with regulations pertaining to the control of tobacco products.

5.3 MARIJUANA SMOKING

The marijuana plant, much as the tobacco plant, is a weed. In fact, in street vernacular, weed is a word often spoken in reference to marijuana. Illustrated in Figure 5.6 are marijuana plants being grown illegally on a plot of public land in Oregon. And also similar to tobacco, leaves of the marijuana plant are smoked, with the outcome being the inhalation of products of combustion of marijuana's elements and byproducts of combustion. Because of policy changes in several U.S. states that allow recreational use of marijuana, the number of persons potentially exposed to marijuana smoke is likely to increase. This increase and its potential adverse effects on public health give rise to a discussion of the prevalence and consequences of marijuana smoking and

> The marijuana plant, much as the tobacco plant, is a weed. In fact, in street vernacular, weed is a word often spoken in reference to marijuana.

FIGURE 5.6 Marijuana plants growing illegally on public land in Oregon. (Drug Enforcement Agency, Seattle Division, https://www.dea.gov/seattle/seattle-contacts, 2018.)

policies related to its prevalence. Before proceeding, the beneficial medical uses of marijuana and its main chemical constituent are not discussed herein, only the recreational smoking of marijuana.

5.3.1 PRÉCIS HISTORY OF MARIJUANA SMOKING

Marijuana, also known as cannabis, weed, or pot, has a long history of human use [101]. While marijuana and cannabis are synonyms, this chapter will use "marijuana" most often as the preferred nomenclature. A published history of marijuana cites that most ancient cultures didn't grow the plant to get high, but as herbal medicine, likely starting in Asia around 500 BCE [101]. The history of cannabis cultivation in the U.S. dates back to the early colonists, who grew hemp for textiles and rope.

The cannabis or hemp plant originally evolved in Central Asia before people introduced the plant into Africa, Europe, and eventually the Americas, although an analysis of fossil pollen from ancient plants convinced researchers that cannabis might have originated high on the Tibetan Plateau [102]. Hemp fiber was used to make clothing, paper, sails and rope, and its seeds were used as food. Because it's a fast-growing plant that's easy to cultivate and has many uses, hemp was widely grown throughout colonial America and at Spanish missions in the Southwest. In the early 1600s, the Virginia, Massachusetts, and Connecticut colonies required farmers to grow hemp. These early hemp plants had very low levels of tetrahydrocannabinol (THC), the chemical responsible for marijuana's mind-altering effects [101].

> In the 1830s, an Irish doctor studying in India found that cannabis extracts could help lessen stomach pain and vomiting in people suffering from cholera.

In the 1830s, an Irish doctor studying in India found that cannabis extracts could help lessen stomach pain and vomiting in people suffering from cholera. By the late 1800s, cannabis extracts were sold in pharmacies and doctors' offices throughout Europe and the U.S. to treat stomach problems and other ailments. Scientists later discovered that THC was the source of marijuana's medicinal properties. As the psychoactive compound responsible for marijuana's mind-altering effects, THC also interacts with areas of the brain that are able to lessen nausea and promote hunger. In fact, the U.S. Food and Drug Administration has approved two drugs with THC that are prescribed in pill form (Marinol and Syndros) to treat nausea caused by cancer chemotherapy and loss of appetite in AIDS patients [101].

In the U.S., marijuana wasn't widely used for recreational purposes until the early 1900s. Mexicans that immigrated to the U.S. during the tumultuous years of the Mexican Revolution introduced the recreational practice of smoking marijuana to American culture. Massive unemployment and social unrest during the Great Depression stoked resentment of Mexican immigrants and public fear of the "evil weed." As a result—and consistent with the Prohibition era's view of all intoxicants—29 states had outlawed cannabis by 1931. No federal government policy was in place at that time.

5.3.2 Prevalence of Marijuana Smoking and Users in the U.S.

A reliable estimate of the number of marijuana smokers in the U.S. is unavailable. However, opinion surveys have been conducted that included questions about marijuana use. In one 2017 survey, conducted by *Yahoo News* and Marist University, more than half of American adults had tried marijuana at least once in their lives, with nearly 55 million of them, or 22%, currently used it, where the survey defined "current use" as having used marijuana at least once or twice in the past year. About 35 million were what the survey called "regular users," or people who used marijuana at least once or twice a month [103].

> A federal survey on drug use found about 33 million adults in the U.S. used marijuana in the past year.

In a second survey, a Gallup poll released in 2016 found that more than 33 million adults identified as "current" marijuana users, although it didn't specify a time frame [104]. A federal survey on drug use found about 33 million adults in the U.S. used marijuana in the past year, considerably lower than the Marist poll's 55 million figure [105]. A digest of these data suggests ~33 million U.S. adults use marijuana to some extent. This number will surely increase as more U.S. states begin permitting sales and recreational use of marijuana.

Adding to the public health gravity of the cited numbers of marijuana users are the results of a study of U.S. teens. A study of nearly 50,000 8th, 10th, and 12th graders in Arizona found that one-third of participants said they had used marijuana, and nearly a quarter said they had used marijuana concentrates at least once in their lives [106]. Marijuana concentrates are substances with very high levels of THC (tetrahydrocannabinol), the active ingredient in marijuana responsible for the drug's intoxicating effects.

5.3.3 Global Prevalence of Marijuana Smoking and Users

The WHO is the international health authority on matters of global health. In addition to the organization's responsibilities in preventing the spread of pandemics such as AIDS and in responding to regional epidemics of infectious diseases

> WHO estimates that about 147 million people, 2.5% of the world's population, consume marijuana (annual prevalence).

such as Ebola in central Africa, WHO is also concerned with and about the global presence of substance-abusing drugs such as heroin and marijuana. In 2019, WHO opined that marijuana is by far the most widely cultivated, trafficked, and abused illicit drug globally, further noting that half of all drug seizures worldwide are cannabis seizures. The geographical spread of those seizures is also global, covering practically every country of the world.

The WHO estimates that about 147 million people, 2.5% of the world's population, consume marijuana (annual prevalence) compared with 0.2% consuming cocaine and 0.2% consuming opiates [107]. It is noteworthy that WHO considers marijuana as a global illicit drug when contemporaneously recreational marijuana policies in the U.S. are permitting increased use. As of November 2020, 37 U.S. states allow for at least the medical use of cannabis.

5.3.4 Impacts of Marijuana Smoking on Human Health

Smoking marijuana is not without adverse health effects, according to the WHO. Quoting from that agency's review of the health and scientific literature concerning marijuana smoking [107]:

Acute health effects of cannabis use

The acute effects of cannabis use have been recognized for many years, and recent studies have confirmed and extended earlier findings. These may be summarized as follows:

- Cannabis impairs cognitive development (capabilities of learning), including associative processes; free recall of previously learned items is often impaired when cannabis is used both during learning and recall periods.
- Cannabis impairs psychomotor performance in a wide variety of tasks, such as motor coordination, divided attention, and operative tasks of many types; human performance on complex machinery can be impaired for as long as 24 hours after smoking as little as 20 mg of THC in cannabis; there is an increased risk of motor vehicle accidents among persons who drive when intoxicated by cannabis.
- A single dose of the main psychoactive ingredient (THC) in cannabis—equal to one joint—in otherwise healthy people, can temporarily induce psychiatric symptoms, including those associated with schizophrenia, according to a 2020 study by researchers at King's College London [108].

Chronic health effects of cannabis use

- "Selective impairment of cognitive functioning that includes the organization and integration of complex information involving various mechanisms of attention and memory processes.
- Prolonged use may lead to greater impairment, which may not recover with cessation of use, and which could affect daily life functions.
- Development of a cannabis dependence syndrome characterized by a loss of control over cannabis use is likely in chronic users.
- Cannabis use can exacerbate schizophrenia in affected individuals.
- Epithelial injury of the trachea and major bronchi is caused by long-term cannabis smoking.
- Airway injury, lung inflammation, and impaired pulmonary defense against infection from persistent cannabis consumption over prolonged periods.

Cannabis used during pregnancy is associated with impairment in fetal development leading to a reduction in birth weight.

- Heavy cannabis consumption is associated with a higher prevalence of symptoms of chronic bronchitis and a higher incidence of acute bronchitis than in the non-smoking cohort.
- Cannabis used during pregnancy is associated with impairment in fetal development leading to a reduction in birth weight.
- Cannabis use during pregnancy may lead to the postnatal risk of rare forms of cancer although more research is needed in this area" [107].
- Whether regular smoking of marijuana is associated with an increased incidence of lung cancer is unknown, with medical authorities advocating additional research on the subject [109].
- Heavy cannabis users—who are prone to destructive lung disease sometimes known as "bong lung"—can suffer irreversible lung damage, a New Zealand 2020 study suggests [110]. The researchers expressed concern that policy debates about social acceptance of marijuana were not considering the lung health effects.

> Heavy cannabis users—who are prone to destructive lung disease sometimes known as "bong lung"— can suffer irreversible lung damage.

In 2019, the U.S. Surgeon General Jerome Adams issued a public health advisory concerning marijuana and the developing brain [111]. He noted that the marijuana available today is much stronger than previous versions. The THC concentration in commonly cultivated marijuana plants has increased 3-fold between 1995 and 2014 (4% and 12%, respectively). Marijuana available in dispensaries in some states has average concentrations of THC between 17.7% and 23.2%.

In addition to smoking marijuana, another form of exposure to marijuana smoke is via vaping of marijuana, perhaps in the belief that vaping is a less hazardous route of exposure to marijuana. However, a recent study by the University of Southern California researchers found participants who had vaped cannabis any number of times from within the last month to their overall lifetime had a stronger link to symptoms of bronchitis (daily cough, congestion, and phlegm) in comparison to people who had never vaped cannabis [112]. The more times those participants had vaped cannabis in the last month, the stronger the risk.

The Surgeon General's advisory was intended to raise awareness of the known and potential harms to developing brains, posed by the increasing availability of highly potent marijuana in multiple, concentrated forms. The advisory observed that the human brain continues to develop from before birth into the mid-20s and is vulnerable to the effects of addictive substances. Frequent marijuana use during adolescence is associated with changes in the areas of the brain involved in attention, memory, decision-making, and motivation. Further, marijuana use during pregnancy can affect the developing fetus. THC can enter the fetal brain from the mother's bloodstream and may disrupt fetal brain development. The advisory concludes with this

> An increase of 2.1 traffic fatalities per billion vehicle miles traveled occurred in recreational marijuana states relative to control states.

statement, "No amount of marijuana use during pregnancy or adolescence is known to be safe."

A different question regarding an association between marijuana use and potential adverse public health outcomes relates to fatal vehicle crashes. Put more simply, does marijuana smoking increase fatal vehicle crashes? This question has been researched by several investigators. The most comprehensive of these studies assessed the change in traffic fatality rates in the first four U.S. states (Colorado, Washington, Oregon, and Alaska) to legalize recreational marijuana [113]. Fatality rates in the 20 U.S. states that did not legalize recreational or medical marijuana served as comparison data. Results showed an increase of 2.1 traffic fatalities per billion vehicle miles traveled in experimental (i.e., recreational marijuana) states relative to control states in the postcommerialization study period. The investigators extrapolated this increased fatality rate to 6,800 excess roadway deaths annually in the instance of national legalization of recreational marijuana.

Perspective: A review of this set of adverse effects associated with marijuana indicates the absence of cancer, with the exception of the use of marijuana during pregnancy perhaps being associated with postnatal risk of some rare cancers. This absence of cancer in the WHO list does not necessarily translate into a lack of causation of cancer through regular smoking of marijuana. But given the presence in marijuana smoke of many of the same products of combustion found in tobacco smoke, it would be a propitious behavior to reduce or cease smoking marijuana in order to prevent lung or other cancers.

5.3.5 IMPACTS OF MARIJUANA SMOKING ON ECOSYSTEM HEALTH

Marijuana plants are grown both indoors and outdoors, depending on the motives of the grower. Indoor marijuana horticulture often occurs in secretive greenhouses under conditions of ambient light sources and water supplies. The advantages of indoor horticulture include control of climate conditions and avoidance of pests. In the U.S., indoor sites are illegal under federal government drug enforcement law.

Marijuana is grown outdoors in the U.S. and elsewhere where the plant is considered an agricultural crop. The ecological consequences of marijuana farming are similar, but probably more extreme, to agricultural practices for food crops. In one review of the ecological impact of illegal marijuana growing, investigators of illegal marijuana sites in northern California found that growers' abundant use of pesticides and rodenticides had caused adverse health impacts on local fish and mammals. Rodenticides, used to kill rats, were particularly damaging to local ecosystems, causing fatalities to burrowing animals like fishers and loss of fish that were damaged by pesticides in water supplies contaminated by runoff from the marijuana fields [114].

5.3.6 U.S. POLICIES ON USES OF MARIJUANA

The Marijuana Tax Act of 1937 was the first federal U.S. law to criminalize marijuana nationwide. The Act imposed an excise tax on the sale, possession, or transfer of all hemp products, effectively criminalizing all but industrial uses of the plant [101].

As part of the "War on Drugs," the Controlled Substances Act of 1970, signed into law by President Richard Nixon (R-CA), repealed the Marijuana Tax Act and listed marijuana as a Schedule I drug—along with heroin,

> The Controlled Substances Act of 1970 listed marijuana as a Schedule I drug—along with heroin, LSD, and ecstasy.

LSD, and ecstasy—with no medical uses and a high potential for abuse. However, in 1972, the National Commission on Marijuana and Drug Abuse released a report titled "Marijuana: A Signal of Misunderstanding." The report recommended "partial prohibition" and lower penalties for possession of small amounts of marijuana. However, the Nixon administration chose to ignore the report's findings.

California, in the Compassionate Use Act of 1996, became the first state to legalize marijuana for medicinal use by people with severe or chronic illnesses. As of 2019, 29 U.S. states, Washington, D.C., and the U.S. territories of Guam and Puerto Rico allow the use of cannabis for limited medical purposes. As of January 2018, nine states and Washington, D.C., have legalized marijuana for **recreational** (emphasis added) use. Colorado and Washington became the first states to do so in 2012. At the federal level, cannabis is still illegal under U.S. federal law, however, and the evolving legal status of marijuana is a subject of ongoing controversy in the U.S. and globally [101]. Of public health relevance, in 2019 the Washington state legislature passed a bill that will build a standard for cannabis lab accreditation. Because cannabis is federally illegal, labs in Washington and other states have no established federal guidelines for testing cannabis [115]. The information that the labs provide has an impact on the quality and price of the cannabis product, which also provides a public health service by guarding against harmful varieties of cannabis.

5.3.7 INTERNATIONAL POLICIES ON MARIJUANA SMOKING

The European Monitoring Centre for Drugs and Drug Addiction (EMCDDA) reports that under international laws, cultivation, supply, and possession of cannabis should be allowed only for "medical and scientific purposes" [116]. "In general, possession of the drug for personal use should be a crime, to deter use, and most countries make this punishable by imprisonment. In recent years, however, several jurisdictions have reduced their penalties for cannabis users, and some have permitted supply of the drug, allowing us to observe different control models and their consequences. Policy discussions are complicated by conflicting claims—decriminalization or legalization, medical or recreational use, policy success or failure…"

> While all EU Member States treat possession of cannabis for personal use as an offence, more than one-third do not allow prison as a penalty for minor offenses.

The EMCDDA further observes that while all EU Member States treat possession of cannabis for personal use as an offence, more than one-third do not allow prison as a penalty for minor offenses. In many of the countries where the law allows imprisonment for such cannabis possession, national guidelines advise against it.

Uruguay and Canada are the only sovereign states that have fully legalized the consumption and sale of recreational cannabis nationwide.

5.4   EFFECTS OF COVID-19 INFECTION ON SMOKERS' HEALTH

According to the WHO, "there are no peer-reviewed studies that have evaluated the risk of SARS-CoV-2 infection associated with smoking. However, tobacco smokers (cigarettes, waterpipes, bidis, cigars, heated tobacco products) may be more vulnerable to contracting COVID-19, as the act of smoking involves contact of fingers (and possibly contaminated cigarettes) with the lips, which increases the possibility of transmission of viruses from hand to mouth. Smoking waterpipes, also known as shisha or hookah, often involves the sharing of mouthpieces and hoses, which could facilitate the transmission of the COVID-19 virus in communal and social settings" [117].

> Stanford University: Teens and young adults who use e-cigarettes were five to seven times more likely to test positive for COVID-19.

However, Stanford University researchers found that teens and young adults who use e-cigarettes were five to seven times more likely to test positive for the virus [118]. Researchers found that teenagers and young adults ages 13–24 who use e-cigarettes are five times more likely to be diagnosed with COVID-19 than their non-vaping peers. Those who are dual users—people who smoke both traditional and electronic-cigarettes—are seven times more likely to test positive for the virus, the researchers found. The researchers gathered their data through an online survey posted on social media and gaming sites. More than 4,000 teens and young adults from all 50 U.S. states responded, completing the roughly 15-minute survey. More research would be needed to verify if vaping enhances contracting COVID-19.

A different kind of study examined the associations of the COVID-19 pandemic with patterns of youth e-cigarette use, drawing on a national sample of 2,167 underage youth and young adults sampled between May 6 and 14, 2020 [119]. Among the key findings of the study were that more than half of participants (1,198) reported changes in their e-cigarette use since the beginning of the COVID-19 pandemic. Among 1,197 participants reporting on the type of change, one-third of youth quit vaping, and another one-third reduced their use of e-cigarettes, with the remaining youth either increasing their use or switching to other nicotine or cannabis products. It might be speculated that youths who practiced social distancing during the pandemic became less influenced by peer pressure to smoke e-cigarettes.

5.5   HAZARD INTERVENTIONS

As detailed in this chapter, the use of tobacco products, vaping devices, and marijuana smoking are hazards to human and ecosystem health. Tobacco is particularly a global public health hazard. Bearing in mind that one-half of tobacco users will die from use of the product, there are interventions known to reduce the hazard of using tobacco products. These interventions include the following:

- The most effective intervention is to forego using any tobacco products—cigarettes, smokeless tobacco, and tobacco-imitation products. Education of potential tobacco users can be an aid to the prevention of tobacco use—especially among young persons.
- For persons who have chosen to use tobacco products, there are programs of assistance in quitting tobacco's grip. Contacting one's medical doctor or public health agency for advice is a good first step. Additionally, a tobacco user can obtain assistance from government sources (e.g., CDC.gov) and an internet search will identify local Quit Smoking organizations.
- Avoid secondhand smoke. This can be facilitated by not patronizing businesses that allow smoking on their premises. Non-tobacco users should not reside with those who use tobacco.
- Support government agencies and NGOs that endeavor to restrict tobacco use.
- Do not financially support organizations that invest in tobacco companies.
- Make known to policymakers your concerns about tobacco products and their use.

These interventions for reduction or elimination of tobacco products as a hazard to human and ecological health are similarly relevant for application to vaping devices and marijuana smoking. For all three forms of lung pollution discussed in this chapter, the single most effective intervention is to not smoke and avoid inhaling the smoke from other persons.

5.6 SUMMARY

Of the three forms of lung pollution discussed in this chapter, tobacco products stand out as a hazard to millions of people globally. WHO estimates that tobacco kills ~6 million people annually. This figure does not include the premature deaths attributable to smokeless tobacco. As a perspective, this annual toll is equal to the genocide of the Holocaust, or the regional metropolitan populations of Dallas-Ft. Worth, USA, Madrid, Spain, or Nanjing, China [120]. Of equal gravity, as WHO observes, tobacco kills up to half of its users. This is a health risk level without equal. For the person who decides to smoke tobacco, the risk is the same as playing Russian roulette with half of the weapon's chambers containing live ammunition.

On a more promising note, health research has clearly identified the adverse health effects of tobacco use, and focused policymaking has been generally successful in lowering smoking rates in many countries, although much remains to be achieved. Such intervention strategies, particularly: controls on advertising of tobacco products, limits on smoking in public places, taxes on tobacco products, prosecution of illicit distribution of tobacco products, and youth education programs have been beneficial as policy elements in tobacco control policies. Future generations will ask with incredulity why a poisonous plant, tobacco, was allowed to kill so many of their ancestors.

Concerning vaping devices and marijuana smoking, much remains to be learned about the adverse consequences to human and ecological health. However, enough is already known about the addictive nature of nicotine and the problem of products of combustion of vapors and other materials in both vaping devices and marijuana to warrant counseling potential users about the potential adverse effects on their health.

REFERENCES

1. JTI Global. 2016. The history of tobacco. http://www.jti.com/about-tobacco/history-of-tobacco/.
2. Jacobs, M. 1997. From the first to the last ash: The history, economics, & hazards of tobacco. Cambridge: Cambridge Department of Human Service Programs, Cambridge Tobacco Education Program.
3. WHO (World Health Organization). 2000. The history of tobacco. Atlas 2. http://www.who.int/tobacco/en/atlas2.pdf.
4. Randall, V. R. 1999. History of tobacco. Dayton: The University of Dayton School of Law.
5. Thompson, D. 2015. U.S. smoking rate falls to 15 percent: CDC. *Health Day News*, September 1.
6. CDC (Centers for Disease Control and Prevention). 2016. Youth and tobacco use. http://www.cdc.gov/tobacco/data_statistics/fact_sheets/youth_data/tobacco_use/.
7. Wang T. W., A. S. Gentzke, M. R. Creamer, et al. 2019. Tobacco product use and associated factors among middle and high school students: United States, 2019. *MMWR Surveill. Summ.* 68(SS-12):1–22.
8. Thomas, K. and S. Kaplan. 2019. E-cigarettes went unchecked in 10 years of federal inaction. *The New York Times*, November 1.
9. CDC (Centers for Disease Control and Prevention). 2015. Cigars. http://www.cdc.gov/tobacco/data_statistics/fact_sheets/tobacco_industry/cigars/.
10. McNamee, D. 2014. Water pipe smoking is 'less safe' than commonly thought. *Health Day News*. http://www.medicalnewstoday.com/articles/276881.php.
11. Stein, L. 2016. Pipe smoking. *Health Day News*. https://consumer.healthday.com/encyclopedia/cancer-8/mis-cancer-news-102/pipe-smoking-645343.html.
12. Henley, S. J., M. J. Thun, A. Chao, and E. E. Calle. 2004. Association between exclusive pipe smoking and mortality from cancer and other diseases. *J. Natl. Cancer Inst.* 96(11):853–61.
13. CDC (Centers for Disease Control and Prevention). 2015. Dangers of hookah smoking. http://www.cdc.gov/features/hookahsmoking/.
14. ALA (American Lung Association). 2015. The emergence of new smokeless tobacco products. http://www.lung.org/assets/documents/tobacco/new-smokeless-tobacco-products.pdf.
15. Lund, I. and K. E. Lund. 2014. How has the availability of snus influenced cigarette smoking in Norway? *Int. J. Environ. Res. Public Health*. 11(11):11705–17.
16. Swedish Match. 2013. Tobacco use behaviors for Swedish snus and U.S. smokeless tobacco. http://www.accessdata.fda.gov/Static/widgets/tobacco/MRTP/19%20appendix-6b-environ-tub-report-2013.pdf.
17. Gorenstein, D. 2014. The market for smokeless tobacco keeps on growing. *Market Place*. http://www.marketplace.org/2014/06/17/business/market-smokeless-tobacco-keeps-growing.
18. The Statistica Portal. 2015. Statistics and facts on smokeless tobacco in the U.S. http://www.statista.com/topics/2500/smokeless-tobacco-in-the-united-states/.
19. WHO (World Health Organization). 2019. Tobacco: Key facts. https://www.who.int/news-room/fact-sheets/detail/tobacco.
20. WHO (World Health Organization). 2019. WHO launches new report on lobal tobacco use trends. *News Release*, December 19. Geneva: Media Center.

21. Ghebreyesus, T. A. 2019. Progress in beating the tobacco epidemic. *Lancet* 394(10198):548–9.
22. Johnson, B. L. 2019. Legacies of hope: Profiles of people whose deeds saved lives and prevented disabilities. Amazon Kindle Publishing.
23. Tucker, A. 2005. Sir Richard Doll. *The Guardian*, July 24.
24. Flannery, M. A. 2007. Luther Terry. *Encyclopedia Alabama*. Auburn: Auburn University, University Outreach.
25. CDC (Centers for Disease Control and Prevention). 2015. Health effects of cigarette smoking. http://www.cdc.gov/tobacco/data_statistics/fact_sheets/health_effects/effects_cig_smoking/.
26. Paulin, L. M., A. J. Gassett, N. E. Alexis, et al. 2020. Association of long-term ambient ozone exposure with respiratory morbidity in smokers. *JAMA Intern. Med.* 180(1):106–15.
27. Niemelä, S., A. Sourander, H.-M. Surcel, et al. 2016. Prenatal nicotine exposure and risk of schizophrenia among offspring in a national birth cohort. *Am. J. Psychiatry* 173:799–806. doi: 10.1176/appi.ajp.2016.15060800.
28. U.S. Surgeon General. 2014. The health consequences of smoking: 50 years of progress: A report of the Surgeon General. Rockville: Office of Surgeon General.
29. EPA (Environmental Protection Agency). 2016. Secondhand tobacco smoke and smoke-free homes. https://www.epa.gov/indoor-air-quality-iaq/secondhand-tobacco-smoke-and-smoke-free-homes.
30. CDC (Centers for Disease Control and Prevention). 2018. Health effects of secondhand smoke. Atlanta: Office of Smoking and Health.
31. Dale, L. and Mayo Clinic. 2016. Healthy lifestyle, what is thirdhand smoke, and why is it a concern? https://www.mayoclinic.org/healthy-lifestyle/adult-health/expert-answers/third-hand-smoke/faq-20057791.
32. Sheu, R., C. Stönner, J. C. Ditto, et al. 2020. Human transport of thirdhand tobacco smoke: A prominent source of hazardous air pollutants into indoor nonsmoking environments. *Sci. Adv.* 6(10):eaay4109. doi: 10.1126/sciadv.aay4109.
33. McCann, A. 2021. The real cost of smoking by state. *WalletHub*, January 13.
34. Shapiro, J. A., E. J. Jacobs, and M. Thun. 2000. Cigar smoking in men and risk of death from tobacco-related cancers. *J. Nat. Cancer Inst. (JNCI)* 92(4):333–7.
35. CDC (Centers for Disease Control and Prevention). 2014. Smokeless tobacco: Health effects. http://www.cdc.gov/tobacco/data_statistics/fact_sheets/smokeless/health_effects/#addiction.
35a. EPA (Environmental Protection Agency). 2019. Ecosystem Services. Office of Research and Development. https://www.epa.gov/eco-research/ecosystem-services.
36. Ocean Conservancy. 2015. More than 16 million pounds of trash collected during international coastal cleanup. http://www.oceanconservancy.org/who-we-are/newsroom/2015/16-million-pounds-icc.html.
37. Gould, H. 2015. Why cigarette butts threaten to stub out marine life. *The Guardian*, June 9.
38. Olivier, A. 2019. Montreal redefines the term 'butthead' in bid to cut down on pollution. *Global News: Environment*, 19 May.
39. CDC (Centers for Disease Control and Prevention). 2016. Selected actions of the U.S. government regarding the regulation of tobacco sales, marketing, and use. Atlanta: Office of Smoking Prevention.
40. Chowdhury, A. and G. D. Rende. 2021. Congress amends the PACT Act to apply to all vaping products, placing huge burden on small manufacturers as third-party common carriers refuse to ship products. *National Law Rev. XI (66)*, March 7, 2021.
41. Oyez (Chicago-Kent College of Law at Illinois Tech). 1999. Food and Drug Administration v. Brown & Williamson Tobacco Corporation. https://www.oyez.org/cases/1999/98-1152.

42. FDA (Food and Drug Administration). 2015. Tobacco Control Act. http://www.fda.gov/TobaccoProducts/Labeling/RulesRegulationsGuidance/ucm246129.htm.

43. Howard, J. 2019. US raises legal age to buy cigarettes, vapes to 21. *CNN News*, December 20.

44. FDA (Food and Drug Administration). 2021. FDA marks historic public health milestone with finalization of two key rules for companies seeking to market new tobacco products. *FDA News Release*, January 19.

45. Reinberg, S. 2013. U.S. Supreme Court rejects challenge to new cigarette labeling. *Health Day News*, April 22.

46. FDA (Food and Drug Administration). 2019. FDA proposes new required health warnings with color images for cigarette packages and advertisements to promote greater public understanding of negative health consequences of smoking. Washington, D.C.: FDA Newsroom, August 15, https://www.fda.gov/news-events/fda-newsroom/press-announcements.

47. Haridy, R. 2020. Graphic tobacco warnings coming to US after courts forced FDA action. *New Atlas*, March 18.

48. Anonymous. 2019. Cigarette packages will be the 'ugliest colour in the world' starting in November. *The Canadian Press*, May 1.

49. Press Association. 2017. Stricter cigarette packaging rules come into force in UK. *The Guardian*, May 19.

50. Belluz, J. 2016. Cigarette packs are being stripped of advertising around the world. But not in the U.S. *Vox*, June 2.

51. WHO (World Health Organization). 2019. WHO launches new report on the global tobacco epidemic. *News Release*, July 26. Geneva: Communications Office.

52. American Cancer Society. 2019. Health risks of secondhand smoke: What is secondhand smoke? https://www.cancer.org/cancer/cancer-causes/tobacco-and-cancer/secondhand-smoke.html.

53. ANRF (American Nonsmokers Rights Foundation). 2016. Smokefree lists, maps, and data. http://www.no-smoke.org/goingsmokefree.php?id=519#ords.

54. ANRF (American Nonsmokers Rights Foundation). 2019. Overview list-Number of smokefree and other tobacco-related laws. https://no-smoke.org/wp-content/uploads/pdf/mediaordlist.pdf.

55. Maldonado, S. 2020. San Francisco bans smoking inside apartments; pot smoking OK. *SFGate*, December 3.

56. DHS (Georgia Department of Human Services). 2005. The Georgia Smokeless Air Act of 2005 becomes law. Atlanta: State of Georgia, Division of Public Health.

57. Georgia Public Library Service. 2016. Georgia Smokefree Air Act O.C.G.A. §§ 31-12A-1 through 31-12A-13. Athens: University of Georgia, Department of Public Library Services.

58. Chandora R. D., C. F. Whitney, S. R. Weaver, and M. P. Eriksen. 2015. Changes in Georgia restaurant and bar smoking policies from 2006 to 2012. *Prev. Chronic Dis.* 12:140520.

59. Public Health Law Center. 2016. Master settlement agreement. St. Paul: Mitchell Hamline School of Law.

60. CDC (Centers for Disease Control and Prevention). 2020. Current cigarette smoking among U.S. adults aged 18 years and older. https://www.cdc.gov/tobacco/data_statistics/fact_sheets/adult_data/cig_smoking/index.htm.

61. Shastri, S. S., R. Talluri, S. Shete. 2020. Disparities in secondhand smoke exposure in the United States. *JAMA Intern. Med.* 181(1):134–7.

62. Prochaska, J. J., A. K. Michalek, C. Brown-Johnson, et al. 2016. Likelihood of unemployed smokers vs nonsmokers attaining reemployment in a one-year observational study. *AMA Internal Med.* 176(5):662–70.

63. Goodwin, J. 2020. U-Haul will no longer hire smokers in 21 states. *USA Today*, January 2.

64. WHO (World Health Organization). 2005. WHO framework convention on tobacco control. Publication details. https://www.who.int/fctc/text_download/en/.

65. EP (European Parliament). 2001. Manufacture, promotion, and sale of tobacco products. Summary. Directive 2001/37/EC. EUR-Lex. http://eur-lex.europa.eu/legal-content/EN/TXT/?uri=URISERV%3Ac11567.

66. Hoffmanemail, S. J. and Z. Rizvi. 2012. WHO's undermining tobacco control. *Lancet* 380(9843):727–8.

67. WHO (World Health Organization). 2015. Fact sheet 339, July 6. Geneva: Office of Director-General, Media Centre.

68. Center on Addiction. 2019. What is vaping? https://www.centeronaddiction.org/e-cigarettes/recreational-vaping/what-vaping.

69. Vaporfi. 2019. The evolution of the vape industry. http://www.vaporfi.com/.

70. Rapaport, L. 2018. Almost one in 20 U.S. adults now use e-cigarettes. *Health News*, August 27.

71. Miech, R., L. Johnston, P. M. O'Malley, et al. 2019. Trends in adolescent vaping: 2017–2019. *NEJM*, September 18. doi: 10.1056/NEJMc1910739.

72. Wang, T. W., L. J. Neff, E. Park-Lee, et al. 2020. E-cigarette use among middle and high school students: United States, 2020. *MMWR* 69:1310–2.

73. FDA (Food and Drug Administration). 2020. 2018 NYTS data: A startling rise in youth E-cigarette use. May 4. https://www.fda.gov/tobacco-products/youth-and-tobacco/2018-nyts-data-startling-rise-youth-e-cigarette-use.

74. Mundell, E. J. 2019. Survey finds teens unaware they are vaping nicotine. *Health News*, April 22.

75. NIDA (National Institute on Drug Abuse). 2016. Electronic cigarettes (e-cigarettes). https://www.drugabuse.gov/publications/drugfacts/electronic-cigarettes-e-cigarettes.

76. NIDA (National Institute on Drug Abuse). 2016. Drug facts: Electronic cigarettes (e-cigarettes). https://www.drugabuse.gov/publications/drugfacts/electronic-cigarettes-e-cigarettes.

77. WHO (World Health Organization). 2015. Electric cigarettes (e-cigarettes) or nicotine delivery systems. *Tobacco Free Initiative Statement*, March 30. Geneva: Office of Director-General, Media Centre.

78. CDC (Centers for Disease Control and Prevention). 2018. Tobacco use by youth is rising: E-cigarettes are the main reason. https://www.cdc.gov/vitalsigns/youth-tobacco-use/index.html.

79. Weixel, N. 2019. Federal judge orders FDA to start regulating e-cigarettes. *The Hill*, May 15.

80. Adams J. 2018. Surgeon general's advisory on e-cigarette use among youth. Rockville, MD: Office of Surgeon General.

81. Jones L. 2018. Vaping: The rise in five charts. Business report, *BBC News*, May 31.

82. CDC (Centers for Disease Control and Prevention). 2018. About electronic cigarettes (e-cigarettes). Atlanta: Office on Smoking and Health.

83. Lee, W. H., S.-G. Ong, Y. Zhou, et al. 2019. Modeling cardiovascular risks of e-cigarettes with human-induced pluripotent stem cell-derived endothelial cells. *J. Am. Coll. Cardio.* 73(21). doi: 10.1016/j.jacc.2019.03.476.

84. Bhatta, D. N. and S. A. Glantz. 2019. Association of e-cigarette use with respiratory disease among adults: A longitudinal analysis. *Am. J. Prev. Med.* 58(2):182–90.

85. Boston University School of Medicine. 2020. Vaping may increase respiratory disease risk by more than 40%, study finds. *Medical Press*, November 12.

86. CDC (Centers for Disease Control and Prevention). 2019. Outbreak of lung disease associated with e-cigarette use, or vaping. *News Update*, October 3. Atlanta: Office of Media Relations.

87. Thielking, M. 2019. Vaping-related illness has a new name: EVALI. *Stat*, October 11.
88. Blount, B. C., M. P. Karwowski, P. G. Shields, et al. 2019. Vitamin E acetate in bronchoalveolar-lavage fluid associated with EVALI. *NEJM*, December 20.
89. Staff. 2019. India announces ban on e-cigarettes. *DW News*, September 19.
90. Selyukh, A. 2019. Walmart to stop selling e-cigarettes. *NPR*, September 20.
91. Weixel, N. 2019. Mass. governor signs groundbreaking vaping flavor ban into law. *The Hill*, November 27.
92. Weixel, N. 2019. FDA orders four companies to stop selling flavored e-cigarette, hookah products. *The Hill*, August 8.
93. University of California San Diego. 2020. Liquid e-cigarette flavorings measurably injure lungs. *Medical Press*, December 14.
94. Gomajee, R. 2019. Association between electronic cigarette use and smoking reduction in France. *JAMA Intern. Med.* 179(9):1193–1200.
95. PEGEX. 2018. Do e-cigarettes require hazardous waste management? https://www.hazardouswasteexperts.com/hazardous-waste-management-e-cigarettes/.
96. Nedelman, M. 2019. San Francisco passes ban on e-cigarette sales, a US first. *CNN News*, June 25.
97. Hassan, A. 2019. Juul is sued by school districts that say vaping is a dangerous drain on their resources. *The New York Times*, October 7.
98. FDA (Food and Drug Administration). 2020. FDA finalizes enforcement policy on unauthorized flavored cartridge-based e-cigarettes that appeal to children, including fruit and mint. *FDA News Release*, January 2.
99. Staff. 2020. To protect 'children' from e-cigarettes, Congress imposes new restrictions on everything related to vaping of any kind. *News Break*, December 23.
100. Kennedy R. D., A. Awopegba, E. De Leon, and J. E. Cohen. 2017. Global approaches to regulating electronic cigarettes. *Tobacco Control* 26(4):440–5.
101. Anonymous. 2019. History of marijuana. *History*. https://www.history.com/topics/crime/history-of-marijuana.
102. Weisberger, M. 2019. We may finally know where the cannabis plant originated. *Live Science*, May 20.
103. Romano, A. 2017. Weed hits home: In a new Yahoo News/Marist Poll, parents and children are surprisingly open about pot use. *Yahoo News*, April 17.
104. McCarthy, J. 2016. One in eight U.S. adults say they smoke marijuana. *Gallup News*, August 8.
105. SAMHSA (Substance Abuse and Mental Health Services Administration). 2016. Results from the 2015 national survey on drug use and health: Detailed tables. Rockville: Office of Communications.
106. Rettner, R. 2019. Many teens are using ultra-potent 'marijuana concentrates'. *Live Science*, August 26.
107. WHO (World Health Organization). 2019. Management of substance abuse: Cannabis. Geneva: Office of Director-General, Media Centre.
108. Hunt, K. 2020. Single joint linked with temporary psychiatric symptoms, review finds. *CNN*, March 17.
109. Charles, S. and L. Carroll. 2019. Does marijuana cause lung cancer? Doctors call for more research. *Health News*, April 13.
110. Thomas, R. 2020. Heavy cannabis smokers can have badly damaged lungs for life: Study. *Radio New Zealand*, May 18.
111. Adams, J. 2019. U.S. Surgeon General's advisory: Marijuana use and the developing brain. Rockville: Office of the Surgeon General.
112. Braymiller, J. L., J. L. Barrington-Trimis, A. M. Leventhal, et al. 2021. Assessment of nicotine and cannabis vaping and respiratory symptoms in young adults. *JAMA Netw Open.* 3(12):e2030189.

113. Kamer, R. S., S. Warshafsky, and G. C. Kamer. 2020. Change in traffic fatality rates in the first 4 states to legalize recreational marijuana. *JAMA Research Letter*, June 22.
114. Wheelright, J. 2019. Marijuana: An environmental buzzkill. *Discover Magazine*, January 3.
115. Secaira, M. 2019. Without federal guidelines, Washington is creating its own standard for testing weed. *Crosscut*, September 2.
116. EMCDDA (European Monitoring Centre for Drugs and Drug Addiction). 2018. Cannabis legislation in Europe: An overview. http://www.emcdda.europa.eu/system/files/publications/4135/TD0217210ENN.pdf.
117. WHO (World Health Organization). 2020. Q&A: Tobacco and COVID-19. May 27.
118. Harrison, S. 2020. A new survey links vaping to higher Covid-19 risk. *Wired*, August 11.
119. Stokes, A. C. 2020. Declines in electronic cigarette use among US youth in the era of Covid-19: A critical opportunity to stop youth vaping in its tracks. *JAMA Network*, December 3.
120. World Atlas. 2016. Populations of 150 largest cities in the world. http://www.worldatlas.com/citypops.htm.

6 Toxic Substances in the Environment

6.1 INTRODUCTION

This chapter describes the five major U.S. federal government policies on the control of hazardous chemical substances in the general environment, referred to herein as hazardous environmental toxicants. The word *toxicant* means any toxic substance. The term covers substances that may be anthropogenic or naturally

> The word *toxicant* means any toxic substance. The term covers substances that may be anthropogenic or naturally occurring. A *toxin* is a poison produced naturally by an organism.

occurring. In distinction, a *toxin* is a poison produced naturally by an organism (e.g., plant, animal, insect). While other chapters have discussed chemical pollutants in air, water, food, and waste, this chapter deals with policies that are specific to hazardous substances found in general commerce. The five U.S. federal policies specific to control of toxic substances will be discussed, along with those of the EU, China, India, Brazil, and the WHO. Impacts of hazardous substances on human and ecosystem health are presented herein. It needs to be noted that the terms *hazardous* and *toxic* are distinct terms with somewhat different meanings, but are often used as synonyms by policymakers.

As background, humankind has known since antiquity that some substances possessed harmful properties. For instance, ancient peoples gradually learned which noxious plants to avoid toughing or eating; in effect, practicing the core principle of public health, prevention of disease and disability via hazard avoidance. Similarly, humankind learned to avoid venomous creatures whose bites could cause harmful health effects. The common factor between noxious plants and venomous creatures would over time become revealed to be chemical substances that possess toxic properties. In time, the study of chemical substances' harmful properties would be called *toxicology*.

The Industrial Revolution led to the production of machines, products, and services that involved especially the use of metals. In the process, metals had to be mined, smelted, forged, and fabricated into machinery for uses in agri-

> The Industrial Revolution led to the production of machines, products, and services that involved especially the use of metals.

culture, industrialization, transportation, and consumer goods. In the 19th century, through the mid-20th century, industrial processes often exposed workers to metal fumes and other harmful substances, and if exposure levels were sufficiently great, adverse health consequences occurred. While acute exposures to high levels of toxic

DOI: 10.1201/9781003253358-6

substances certainly occurred and could cause immediate adverse effects, there was also a gradual growth in knowledge about how exposure to substances over long periods of time could manifest their toxicity. For example, lead poisoning and metal fume fever were occupational health outcomes for many workers. As workplace conditions gradually improved in the industrialized countries, workers' exposure to metals lessened but did not disappear. The toxicity of metals had not changed, but exposure levels had decreased owing to improved work conditions, lessening the adverse health effects in workers.

In the mid-20th century, the manufacture of synthetic chemicals became a significant economic force and commercial reality, in part, due to the resource demands of World War II. The chemical industry had arrived, generating products

> In the mid-20th century, the manufacture of synthetic chemicals became a significant economic force and commercial reality.

such as therapeutic drugs, pesticides, herbicides, plastics, synthetic rubber, and cosmetics. In a sense, the Chemical Age, a term coined here, had arrived. The production and use of these products brought exposure to new, synthesized substances for which toxicology information was lacking. Moreover, the exposures were experienced by persons in the general environment, not solely confined to workplace environments. Exposure occurred at lower levels through contamination of environmental media such as outdoor ambient air and community drinking water supplies. The toxicological implications had grown from those of dealing with the consequences of short-term, high to medium levels of chemical substances, to the condition of long-term exposure to low concentrations of substances found in essential environmental media, i.e., air, water, and food.

> The cost of harmful chemical and heavy metal exposure globally has been estimated to ~10% of the global Gross Domestic Product (GDP).

The cost of harmful chemical and heavy metal exposure globally has been estimated to ~10% of the global Gross Domestic Product (GDP), a consequential amount [1]. To arrive at this estimate, researchers looked at exposures to toxicants such as lead, mercury, pesticides, endocrine-disrupting chemicals, flame retardants, and those found in polluted air and water sources. From the exposure data, they calculated how much they end up costing societies due to illnesses, reduced brain function, health care bills, and lost wages and productivity for employers. Gross domestic product is a measure of a country's total economy and includes the value of all goods and services produced. In 2016, the global GDP was $75 trillion. While there is some uncertainty about the accuracy of global exposure data for specific toxicants, the study does portray a considerable cost associated with the release of toxic substances.

6.2   U.S. POLICIES ON PESTICIDES/HERBICIDES

Before proceeding, some definitions are in order. Pesticides are chemicals that may be used to kill fungus, bacteria, insects, plant diseases, snails, slugs, or weeds among others. Types of pesticides and their targets include:

Insecticides – insects
Herbicides – plants
Rodenticides – rodents (rats and mice)
Bactericides – bacteria
Fungicides – fungi
Larvicides – larvae

In recognition of the need to control environmental releases of hazardous substances and to inform potential at-risk populations, the U.S. Congress has enacted five major statutes: The Federal Hazardous Substances Act; the Federal Insecticide, Fungicide, and Rodenticide Act (FIFRAct); the Toxic Substances Control Act (TSCAct); the Food Quality Protection Act (FQPAct); and the Lautenberg Chemical Safety Act. The last-named act is a major revision of the TSCAct and is therefore considered a separate act for this chapter. Each of these statutes is discussed in the following sections, commencing with a discussion of pesticides, the first class of toxic substances brought under policymaking authority by the U.S. Congress.

6.2.1 Impacts of Pesticides/Herbicides on Human Health

Pesticides are chemical substances evolved by nature or synthetically produced to be biologically active. As such, pesticides are intentionally harmful to living organisms, often with biological specificity. Given the mortal purpose of pesticides, their public health implications might seem obvious. However, the implications are a complicated public health proposition. For example, it can be argued that pesticides have benefited the public's health by reducing mosquito infestation, thereby reducing the number of persons at risk of contracting malaria or West Nile disease. However, some pesticides used to control mosquitoes are environmentally persistent and can cause serious harm to ecological systems. An example is the use of the pesticide DDT (dichlorodiphenyltrichloroethane) used in the tropics for malaria control, even though it causes ecological degradation.

> Pesticides are intentionally harmful to living organisms, often with biological specificity.

DDT and other chemicals are called Persistent Organic Pollutants, and their use and management are the subject of an international treaty. As an example of DDT's persistence, a 2019 Canadian study examined sediment samples from five major lakes in New Brunswick, Canada, an area that is estimated to have been sprayed with nearly 5.7 million kilos of DDT between 1952 and 1968. Researchers discovered alarming levels of DDT within these samples and determined that the dangers of DDT were still observable in these environments. The study found that small but vital populations of invertebrates, such as water fleas and zooplankton, both of which are food for frogs and some marine life, have experienced steep declines in populations due to the presence of DDT [2].

The FIFRAct provides some human and ecological health protection by requiring EPA to register pesticides, control their uses, and remove those found harmful from the U.S. market. In this regard, the FIFRAct serves as a gatekeeper over which pesticides get into the general environment. But this gatekeeping does not provide a complete prohibition of pesticides and similar chemicals from migrating into the U.S. environment. This is because many pesticides are approved for use in the U.S. because of their desirable properties of pest eradication, which can increase crop yields and improve food quality. Are the pesticides in the environment potentially harmful to human and ecological health? And if harmful, does this necessitate further effort to reduce pesticide levels and public health action?

6.2.1.1 General Toxicity of Pesticides

Research on the chronic exposure of adults to pesticides has produced features of Parkinson's disease, with ongoing research using animal models being used to conduct basic science on the etiology of the disease [3,4]. Concerning the general toxicity of pesticides, a coalition for alternatives to pesticides examined the scientific literature for evidence of pesticides' carcinogenicity and reproductive toxicity [5]. The investigators used EPA data on the carcinogenicity of chemicals in one of EPA's carcinogenesis categories.[1] Chronic exposure at lower levels has been associated with adverse neurological and behavioral conditions in young children [6].

Of the 250 pesticides evaluated by EPA, 12 of the 26 of greatest annual use in the U.S. had been classified as carcinogens.

A study conducted by Columbia University investigators in 2004 found that insecticide exposures were widespread among minority women in New York City during pregnancy [7]. The study consisted of 314 mother-newborn pairs and insecticide measurements in maternal ambient air during pregnancy as well as in umbilical cord plasma at delivery. For each log unit increase in cord plasma chlorpyrifos levels, birth weight decreased by 42.6 g and birth length decreased by 0.24 cm. Combined measures of cord plasma chlorpyrifos and diazinon (adjusted for relative potency) were also inversely associated with birth weight and length. Birth weight averaged 186.3 g less among newborns possessing the highest compared with lowest 26% of exposure levels. Further, the associations between birth weight and length and cord plasma chlorpyrifos and diazinon were highly statistically significant among newborns born before the years 2000–2001 when EPA phased out residential use of these insecticides. Among newborns born after January 2001, exposure levels were substantially lower, and no association with fetal growth was apparent. This investigation affirms the toxicological adage, "The dose makes the poison."

Regarding a commonly used insecticide, University of Iowa researchers reported the findings from a study examining the association of pyrethroid insecticide exposure with all-cause and cause-specific mortality among adults in the U.S. [8]. The nationally representative cohort included 2,116 adults aged 20 years and older who participated in the U.S. National Health and Nutrition Examination Survey conducted from 1999 to 2002 and who had provided urine samples for pyrethroid metabolite measurements. Researchers reported that environmental exposure to pyrethroid

---

[1] *Atrazine, metolachlor,* 2-4-D, *metan sodium,* methyl bromide, glyphosate, *dichloropropene,* chlorpyrifos, *cyanazine, pendimethalin, trifluralin, acetochlor,* alachlor, dicamba, EPTC, *chlorothaloni,* copper hydroxide, propanil, terbfos, *mancozeb, fluometuron,* MSMA, bentazone, diazinon, *parathion,* sodium chlorate. The 12 pesticides underlined have been classified by EPA as carcinogenic in one of EPA's carcinogenesis categories.

insecticides was associated with an increased risk of all-cause and cardiovascular disease mortality, with further studies being needed to replicate the findings and determine the underlying mechanisms.

In a study with dose-dependent results, investigators from the National Cancer Institute examined cancer rates in a large cohort of pesticide applicators [6]. The study involved a total of 54,383 pesticide applicators in Iowa and North Carolina. Exposure to the widely used pesticide

> The incidence of lung cancer was found statistically significantly associated with chlorpyrifos lifetime exposure days, suggesting a dose-dependent effect.

chlorpyrifos was found to be associated with an increased rate of lung cancer. The incidence of lung cancer was found statistically significantly associated with chlor-pyrifos lifetime exposure days, suggesting a dose-dependent effect. This study and the one from Columbia University imply that environmental health policies about pesticide use and application should be further strengthened to mitigate or decrease exposure to pesticides. Illustrated in Figure 6.1 is a pesticide applicator. Notable is the presence of protective clothing and face mask, factors that suggest the applicator has received training in the safe application of pesticides to field crops.

Regarding chlorpyrifos, in the late stages of the Barack Obama administration, the EPA was in the process of banning chlorpyrifos. However, the Donald Trump administration's EPA in 2017 reversed the proposed ban and announced it would allow farmers to keep

> In December 2020 the Oregon Department of Agriculture submitted rules to significantly limit the use of chlorpyrifos and phase out all its use by December 31, 2023.

using chlorpyrifos. Later, in April 2017 in a 2-1 decision, the U.S. 9th Circuit Court ruled unanimously that the EPA must decide by mid-July 2019 whether to reverse the Trump administration's overturn of a scheduled ban on chlorpyrifos. The ban had been strongly supported by EPA scientists [9]. In July 2019, EPA's Administrator

FIGURE 6.1   Field applicator of pesticides. (Seminole County, Office of Public Affairs, Florida, 2019.)

Andrew Wheeler announced his decision to allow chlorpyrifos to continue to be used on conventionally grown food crops, like peaches, cherries, apples, oranges, and corn. In contrast to the EPA approval action, in December 2020 the Oregon Department of Agriculture submitted rules to significantly limit the use of chlorpyrifos and phase out all its use by December 31, 2023 [10]. Some of the most notable changes under the new rules will be a prohibition on the pesticide on golf course turfgrass and certain types of enclosed structures. The rules also establish a 4-day restricted entry interval after use for all crops including nursery and Christmas trees and a ban on aerial applications for all crops except Christmas trees. It is interesting in a public health sense to ask why two government agencies review the same scientific findings on chlorpyrifos, yet arrive at quite opposite public health mandates.

6.2.1.2 Toxicity of the Herbicide Glyphosate

An herbicide, glyphosate, has become the subject of research in regard to adverse human health effects. Glyphosate is the active ingredient in the herbicide Roundup® and is the most widely used herbicide in the U.S. and globally [11]. One source notes that since 1974, when Roundup® was first commercially sold, more than 1.6 billion kilograms (or 3.5 billion pounds) of glyphosate has been used in the U.S., making up 19% of the 18.9 billion pounds of glyphosate used globally [12]. It is used in agriculture and forestry, on lawns and gardens, and for weeds in industrial areas. Some products containing glyphosate control aquatic plants.

There is a difference of opinion as to its toxicity and extent of hazard to human health. According to a university agriculture extension service, "Animal and human studies were evaluated by regulatory agencies in the USA, Canada, Japan, Australia, and the European Union, as well as the Joint Meeting on Pesticide Residues of the United Nations and World Health Organization (WHO). These agencies looked at cancer rates in humans and studies where laboratory animals were fed high doses of glyphosate. Based on these studies, they determined that glyphosate was not likely to be carcinogenic. However, a committee of scientists working for the International Agency for Research on Cancer (IARC) of the WHO evaluated fewer studies and reported that glyphosate is probably carcinogenic" [13].

> IARC: There is convincing evidence that glyphosate also can cause cancer in laboratory animals.

However, a different opinion about the hazard of glyphosate was registered by the IARC. Quoting from the IARC monograph on glyphosate, "A Working Group of 17 experts from 11 countries met at the International Agency for Research on Cancer (IARC) on March 3–10, 2015, to review the available published scientific evidence and evaluate the carcinogenicity of five organophosphate insecticides and herbicides: diazinon, glyphosate, malathion, parathion, and tetrachlorvinphos. The herbicide glyphosate and the insecticides malathion and diazinon were classified as probably carcinogenic to humans (Group 2A). The insecticides tetrachlorvinphos and parathion were classified as possibly carcinogenic to humans (Group 2B). For the herbicide glyphosate, there was limited evidence of carcinogenicity in humans for non-Hodgkin's lymphoma.[2] The evidence in humans is from studies of

---

[2] Non-Hodgkin's lymphoma: cancer that originates in the lymphatic system. In non-Hodgkin's lymphoma, tumors develop from lymphocytes, a type of white blood cell.

exposures, mostly agricultural, in the USA, Canada, and Sweden published since 2001. In addition, there is convincing evidence that glyphosate also can cause cancer in laboratory animals" [13]. The chemical industry has disputed the IARC's classification and in 2016 undertook actions to reverse the classification, but without success with IARC.

The IARC classification of glyphosate has resulted in many countries issuing an outright ban on its use, with other countries imposing restrictions or have issued statements of intention to ban or restrict due to health concerns. The IARC classification has also been the foundation of numerous lawsuits in the U.S. that allege litigants' exposure to glyphosate caused their non-Hodgkin's lymphoma. A partial list of countries that have issued outright bans include the following [14]: Austria, Belgium, Columbia, France, Greece, India, Netherlands, and Portugal. This list is likely to increase as other countries consider the health implications of continued approval of glyphosate.

Despite the IARC report's 2015 conclusion that glyphosate is a probable human carcinogen, linked to the contracting of non-Hodgkin's lymphoma, the EPA maintained in 2019 that glyphosate is not likely to be carcinogenic to humans. As such, glyphosate is not banned by the U.S. government. Of public health note, a study found the chemical in all of the 21 oat-based products it tested, with a breakfast cereal exhibiting the highest level of glyphosate at 833 ppb. In drinking water, EPA's maximum contaminant level for the chemical is 700 ppb [15]. This finding was forwarded in 2019 to the FDA for review under its food safety regulations and authorities.

On January 30, 2020, the EPA announced its completion of a regulatory review that found glyphosate is not a carcinogen. "EPA has concluded that there are no risks of concern to human health when glyphosate is used according to the label and that it is not a carcinogen," the agency said in a statement [16]. Notwithstanding EPA's regulatory determination, in June 2020 the German Bayer Company, which purchased the Monsanto Company, agreed to pay more than $10 billion to settle tens of thousands of cancer claims while continuing to sell the product without adding warning labels about its safety [17]. Whether the proposed settlement is accepted by U.S. courts and potential litigant remains to be determined.

6.2.1.3 Toxicity of the Herbicide Atrazine

Atrazine is one of the most commonly used herbicides in the U.S. Farmers spray it on crops such as corn, sorghum, and sugarcane to control grasses and broadleaf weeds. Consumers apply it to residential lawns to kill weeds. Atrazine persists in the environment and is a widespread drinking water contaminant. The herbicide and its breakdown products are linked to developmental and reproductive toxicity in people and aquatic organisms. In 2016, EPA found reproductive risks to wildlife and concluded in 2018 that combined exposure to atrazine from food, drinking water, and residential lawns poses developmental risks to children. However, on September 21, 2020, following 7 years of study by the agency, the EPA Administrator announced the agency's reapproval of atrazine [18]. The announcement was given during an event in Missouri attended by farm-group leaders and local lawmakers. EPA announced that to address the risks to children, the agency was lowering the amount of atrazine that can be applied to residential lawns. The EPA is also requiring workers who apply the herbicide to wear respirators to minimize their exposure. Whether these precautions are sufficient to prevent adverse health effects in children and field applications remain to be determined.

6.2.1.4 Toxicity of the Herbicide Dicamba

The herbicide dicamba has been commonly used agriculturally and residentially. Approval of genetically engineered dicamba-resistant crops is expected to lead to increased dicamba use, and there has been growing interest in potential adverse human health effects. Researchers at the National Cancer Institute conducted in 2020 a prospective cohort study of pesticide applicators in Iowa and North Carolina [19]. At enrollment (1993–1997) and follow-up (1999–2005), participants reported dicamba use. Exposure was characterized by cumulative intensity-weighted lifetime days, including exposure lags of up to 20 years. Among 49,922 applicators, 26,412 had used dicamba. Compared with applicators reporting no dicamba use, those in the highest quartile of exposure had an elevated risk of liver and intrahepatic bile duct cancer and chronic lymphocytic leukemia and decreased risk of myeloid leukemia. The associations for liver cancer and myeloid leukemia remained after lagging exposure of up to 20 years.

In 2020 the EPA asserted that dicamba was not likely to be carcinogenic, saying that it identified "several deficiencies" in the NCI study. The EPA previously reapproved dicamba in 2018 to control weeds on cotton and soybeans that have been genetically engineered to tolerate it, but at the time, opponents argued that other crops that are not resistant to dicamba may be impacted by its usage. In response to challenges, a federal court ruled in early 2020 that in the 2018 reapproval, the EPA "substantially understated" certain risks posed by the chemical's use. In October 2020, EPA reapproved the use of dicamba, stating the new approval circumvents issues with the prior approval by increasing the size of buffer zones between dicamba-sprayed crops and other crops and also increasing the buffer size for endangered species [20]. It remains whether the new approval by EPA will be challenged in court.

Perspective: The implications of pesticides and similar chemicals in community environments are of continuing concern to environmental and public health authorities, given the purpose of the chemicals. The FIFRAct provides the main federal framework for managing the hazard of pesticides. For EPA-approved pesticides, more than 1 billion pounds are used annually in various agricultural and other commercial applications in the U.S. Given the commercial value of pesticides, there will be continued releases of them into environmental media. This reality emphasizes the importance of policies that are committed to monitoring of pesticide levels in water, food, and human tissues and for conducting research on potential human and ecological impacts.

6.2.2 Impacts of Pesticides/Herbicides on Ecosystem Health

Pesticide toxicants can adversely affect ecosystems, as described in this section. The principal problem is that pesticides and herbicides migrate from their points of application, e.g., a garden. For instance, a spring rain will carry a portion of applied herbicides such as glyphosate from a field crop into a drain, ditch, stream, or an adjacent body of water, thus becoming a potential toxicant to fish and other marine life.

The classic example of a pesticide toxicant impacting an ecosystem is the experience of DDT causing ecological damage, as described in the book *Silent Spring* by Rachel Carson (1907–1964). A trained marine biologist, Carson rose within the Fish and Wildlife Service, supervising a small writing staff and in 1949 becoming chief editor of publications, later leaving government service to become a full-time writer. A gifted writer, Carson became interested in ecological consequences of the application of pesticides, focusing on DDT, a focus that led to her most famous book. *Silent Spring*, a work that many consider the most influential book of the 20th century. Shown in Figure 6.2 is an image of Carson taken circa publication of *Silent Spring*.

The presence of pesticides, herbicides, and rodenticides in U.S. environmental media raises questions about the potential impact on ecosystems. The presence of pesticides in U.S. waterbodies is of special relevance for ecological concerns. The U.S. Geological Survey [21] observes that about 1 billion pounds of conventional pesticides are used each year in the U.S. In 2006, the USGS reported the findings from a 10-year program of surveillance of pesticide levels in U.S. rivers, fish, and private

FIGURE 6.2 Rachael Carson, author of *Silent Spring*, circa 1962. (blog.response.restoration.noaa.gov, 2020.)

wells. The report is based on data from 51 major river systems from Florida to the Pacific Northwest, Hawaii, and Alaska, and a regional study conducted in the High Plains aquifer system. The USGS study, which covers the years 1992–2001, found that pesticides seldom occurred alone but almost always as complex mixtures. Most stream samples and about half the well samples contained two or more pesticides, and frequently more [21].

Findings showed pesticides were present throughout the year in most streams in urban and agricultural areas of the U.S. When the USGS measurements were compared with EPA drinking water standards and guidelines, pesticides were

> More than 1 billion pounds of EPA-approved pesticides are used annually in various agricultural and other commercial applications in the U.S.

seldom found at concentrations likely to affect humans. Concentrations of individual pesticides were almost always lower than the standards and guidelines, representing fewer than 10% of the sampled stream sites and about 1% of domestic and public supply wells.

As concerns fish tissues, organochlorine pesticides and their degradants were found in >90% of fish in streams that drained agricultural, urban, and mixed-land-use settings. Pesticides were less common in groundwater. More than 80% of urban streams and more than 50% of agricultural streams had concentrations in water of at least one pesticide that exceeded a water quality benchmark for aquatic life, which suggests the need for further control of pesticide releases into the environment. What then are the most hazardous pesticides on ecosystems and their consequences?

6.2.2.1   The Most Hazardous Pesticides Affecting Ecosystems

How do pesticides affect ecosystems? As presented by one source, pesticides can travel great distances through the environment [22]. When sprayed on crops or in gardens, pesticides can be blown by the

> A study reports a growing number of pollinator species worldwide are being driven toward extinction.

wind to other areas. They can also flow with rainwater into nearby streams or can seep through the soil into groundwater. Some pesticides can remain in the environment for many years and pass from one organism to another. In general, insecticides generally are the most toxic pesticides to the environment, followed by fungicides and herbicides.

The most hazardous pesticides include those that can be distinguished on the basis of water solubility or fat solubility. Water-soluble pesticides are easily transported from the target area into groundwater and streams since the pesticides become dissolved in the water. Fat-soluble pesticides are readily absorbed in the tissues of insects, fish, and other animals, often resulting in extended persistence in food chains.

Organochlorine pesticides such as DDT are fat-soluble pesticides. When there is a small amount of pesticide in the environment, it will enter the bodies of the animals that are low in the food chain, e.g., grasshoppers. Even though there is only a small amount of the toxicant in each grasshopper, shrews or other predators will receive a larger amount of the toxicant in its body because the predator will eat many grasshoppers. When the secondary consumer is eaten (e.g., shrews), a higher-level predator (e.g., an owl) will consume all of its toxicants, plus those of all the other prey it eats. This means that the higher the trophic level, the greater the concentration of toxicants. This process is bioamplification.

Therefore, the top carnivore that has the higher trophic level (e.g., owl) will be the most badly affected as it will have obtained the most concentrated amount of toxicants. This will produce a decline in the population of the top predator (e.g., owl), causing an increase in the population of shrews as there would not be as many of their predators, and thereby leading to a decrease in the population of grasshoppers [22]. This biomagnification process is a major challenge to the proper application of pesticides for crop and gardening use.

A pesticide of particular concern in regard to ecological harm is atrazine. Despite being banned in more than 35 countries, including the entire European Union, it remains the second-most used herbicide in the U.S. after glyphosate. In a draft assessment by EPA is part of a legal agreement between the Center and the agency, EPA found that atrazine is likely to harm 1,013 protected species, or 56% of all endangered plants and animals in the U.S. Species harmed include the highly endangered whooping crane, California red-legged frog and San Joaquin kit fox. In September 2020, the Trump EPA announced it would be reapproving atrazine for the next 15 years, eliminating longstanding safeguards for children's health, and allowing 50% more atrazine to end up in U.S. waterways [23].

6.2.2.2 Effects of Pesticides on Pollinators

The effects of pesticides on specific members of an ecosystem become consequential to public and ecosystem health when the effects are broad in impact. An important example is the effect of pesticides on pollinators. A 2-year study conducted

> Between US$ 235 billion and US$ 577 billion worth of annual global food production relies on direct contributions by pollinators.

by the Intergovernmental Science-Policy Platform on Biodiversity and Ecosystem Services (IPBES) was the first investigation of the global status of pollinators [24]. The study reported a growing number of pollinator species worldwide are being driven toward extinction by diverse pressures, many of them anthropogenic, threatening millions of livelihoods and hundreds of billions of dollars of food supplies. Pollinated crops include those that provide fruit, vegetables, seeds, nuts, and oils. Many of these are important dietary sources of vitamins and minerals, without which the risks of malnutrition might be expected to increase. Between US$2 35 and US$ 577 billion worth of annual global food production relies on direct contributions by pollinators.

In addition to food crops, pollinators contribute to crops that provide biofuels (e.g., canola and palm oils), fibers (e.g., cotton), medicines, forage for livestock, and construction materials. Moreover, nearly 90% of all wild flowering plants

> A correlation was noted between honey bee colony losses and a neonicotinoid usage patterns across England and Wales.

depend at least to some extent on animal pollination.

The IPBES assessment found that an estimated 16% of vertebrate pollinators are threatened with global extinction—increasing to 30% for island species—with a trend toward more extinction. Although most insect pollinators have not been assessed at a global level, regional, and national assessments indicate high levels of threat, particularly for bees and butterflies—with often more than 40% of invertebrate species threatened locally. Declines in regional wild pollinators have been confirmed for areas in North-Western Europe and North America. The assessment found that pesticides, including neonicotinoid insecticides, threaten pollinators worldwide, although the long-term effects are still unknown [24].

Several studies of the effects of neonicotinoid pesticides on the mortality of bees have been reported. Neonicotinoids are compounds that are structurally similar to nicotine, the addictive ingredient in tobacco (Chapter 5). In a large-scale field study, Fera Science researchers combined large-scale pesticide usage and yield observations from oilseed rape with those detailing honey bee colony losses over an 11-year period. The findings revealed a correlation between honey bee colony losses and national-scale imidacloprid (a neonicotinoid) usage patterns across England and Wales [25].

In another study, a research team from Bern, Switzerland, and Wolfville, Canada, found that honey bee queens, which are crucial to colony functioning, are severely affected by two neonicotinoid insecticides [26]. These and other investigations led the EU in 2013 to ban most neonicotinoids for use on flowering crops and spring-sown crops but approved sulfoxaflor, a neonicotinoid, in July 2015 on the basis that it would not have any unacceptable effects on the environment.

In May 2018, the Court of Justice of the European Union backed a near-total EU ban of three neonicotinoids because of their harmful effect on the health of both wild bees and honey bees.

In May 2018, the Court of Justice of the European Union backed a near-total EU ban of three neonicotinoids (clothianidin, thiamethoxam, and imidacloprid) because of their harmful effect on the health of both wild bees and honey bees [27]. In the U.S., legislation to protect native bees and honeybees from pesticide exposure was signed into law in New Jersey [28]. Bill A-3398 requires pesticide applicators to notify both honeybee and native beekeepers when they are applying pesticide within three miles of a registered beehive or bee yard.

In 2019 results from a study by South Dakota State University researchers extended toxicological concern about neonicotinoids [29]. Researchers found that white-tailed deer with high levels of neonicotinoid pesticide in their spleens developed defects such as smaller reproductive organs, pronounced overbites, and declined thyroid function. Fawns with elevated levels of the pesticide in their spleens were found to be generally smaller and less healthy than deer with less of the chemical in their organs. The study marks the first time neonicotinoid pesticide consumption has been linked to birth defects in large mammals.

In regard to bees, the EPA took two separate actions in 2019. In May 2019, the agency canceled the registration of 12 neonicotinoid-based products [30]. This cancellation was effected under the agency's FIFRAct authorities. Among the 12 pesticides canceled in the U.S., seven were for seed coating products used by farmers. In a separate decision, in June 2019 the agency made an emergency exception for 11 U.S. states to use sulfoxaflor on cotton and sorghum crops. A study published in 2018 found sulfoxaflor inhibited bumblebee reproduction. An interesting sidelight to EPA's emergency exception declaration is that the EPA's Office of Inspector General reported in 2018 that the agency did not have processes in place to determine how its emergency measures impact human and environmental health [31].

6.2.2.3 Effects of Herbicides on Ecosystems

Herbicides are widely used in agriculture and gardening for purposes of weed control. Weeds are viewed by farmers and gardeners as pests because they compete with crops for water, soil nutrients, and space. Prior to the development of chemical herbicides, weeds were removed by manual means such as plowing and hoeing. While farmers and agribusinesses favor herbicides in part due to their ease of application, consequences to ecosystems are not often considered.

6.2.2.3.1 Effects of Glyphosate on Ecosystems

Turning from insects to plants, the overuse of an herbicide, glyphosate, has produced weeds that are resistant to the herbicide. Glyphosate comprised 57% of all the herbicides used in the U.S. on corn and soybeans in 2013, according to the USDA. The agency has now identified 14 species of glyphosate-resistant weeds in the U.S., and 32 have been documented worldwide, according to a government-industry-university coalition that tracks the issue globally [32]. Of note, glyphosate, the active ingredient in the herbicide Roundup® has become the most heavily used agricultural chemical in the history of herbicides. A study estimated that globally, about 9.4 million tons of the chemical have been sprayed onto fields. Environmental and health authorities are investigating the efficacy of using this herbicide, given that in March 2015 the International Agency for Research on Cancer (IARC) unanimously determined that Roundup® is "probably carcinogenic to humans" [33].

> Glyphosate comprised 57% of all the herbicides used in the U.S. on corn and soybeans in 2013.

6.2.2.3.2 Effects of Dicamba on Ecosystems

Dicamba is an herbicide that has found wide acceptance for application on glyphosate-resistant weeds. But it allegedly has a propensity for drifting away from crops on which it is applied [34]. But in 2017, manufacturers of dicamba released new formulations of the herbicide designed to be sprayed over new genetically modified soybean and cotton crops and marketed to be less volatile, making it less likely to drift onto adjacent areas. However, the manufacturing companies face at least 130 lawsuits in the US. over dicamba. Many of those seek to be class-action lawsuits that would represent the thousands of farmers—largely growers of non-resistant soybeans but also specialty crop farmers—whose fields have allegedly been damaged. In terms of impacts on ecosystems, herbicides that drift can cause loss of vegetation that protects against soil erosion and food for feral animals.

6.2.3 FEDERAL INSECTICIDE, FUNGICIDE, AND RODENTICIDE ACT, 1947

Chemicals designed to kill what humans deem as pests have been part of humankind's experience. For example, both arsenic and hydrogen cyanide were used for pest control but were eventually abandoned as pesticides due to their high toxicity and health hazard to humans. The period of post–World War II saw the development and expanded use of synthetic pesticides, such as DDT [35]. Because pesticides are specifically designed to kill living creatures, concern gradually evolved about potential adverse effects on human and ecosystem health. This section will give a history of pesticide policymaking in the U.S. and elsewhere.

> The FIFRAct, as amended, requires EPA to regulate the sale and use of pesticides in the U.S. through registration and labeling of an estimated 21,000 pesticide products.

Although federal pesticide legislation was first enacted in 1910, its aim was to reduce economic exploitation of farmers by manufacturers and distributors of adulterated or ineffective pesticides. Congress did not address the potential risks to human health posed by pesticide products until it enacted the 1947 version of the FIFRAct. The U.S. Department of Agriculture (USDA) became responsible for administering the pesticide statutes during this period. However, responsibility was shifted to EPA when that agency was created in 1970. Broader congressional concerns about long- and short-term toxic effects of pesticide exposure on pesticide applicators, wildlife, non-target insects and birds, and food consumers subsequently led to a complete revision of the FIFRAct in 1972 (Table 6.1). The 1972 law, as amended, is the basis of current federal policy. Substantial changes were made to the FIFRAct in 1988 in order to accelerate the process of reregistering pesticides, and again in 1996. The 1996 amendments facilitated the registration of pesticides for special (so-called "minor") uses, reauthorization of collection of fees to support reregistration, and a requirement to coordinate regulations between the FIFRAct and the Food, Drug, and Cosmetic Act (FDCAct).

As detailed by Schierow [36], the FIFRAct, as amended, requires EPA to regulate the sale and use of pesticides in the U.S. through registration and labeling of the estimated 21,000 pesticide products currently in use [36]. The act directs EPA to restrict the use of pesticides as necessary in order to prevent unreasonable adverse effects on humans and the environment, taking into account the costs and benefits of various pesticide uses. The FIFRAct prohibits the sale of any pesticide in the U.S. unless it is

TABLE 6.1
FIFRAct Amendments [28]

Year	Act
1947	Federal Insecticide, Fungicide, and Rodenticide Act (FIFRAct)
1964	Amendments
1972	Federal Environmental Pesticide Control Act
1975	FIFRAct Extension
1978	Federal Pesticide Act
1980	Amendments
1988	Amendments
1990	Food, Agriculture, Conservation, and Trade Act
1991	Food, Agriculture, Conservation, and Trade Act Amendments
1996	Food Quality Protection Act

registered and labeled according to approved uses and restrictions. It is a violation of the law to use a pesticide in a manner that is inconsistent with the label's instructions. EPA registers each pesticide for each approved use, e.g., to control boll weevils on cotton. In addition, the FIFRAct requires EPA to reregister older pesticides based on new data that meet current regulatory and scientific standards. Establishments that manufacture or sell pesticide products must be registered by EPA. Facility managers are required to keep certain records and to allow inspections by EPA or state regulatory representatives.

6.3 HEALTH IMPACTS OF TOXIC SUBSTANCES

It is not hyperbole to state that persons in industrialized countries are literally awash with hazardous chemical substances. Potentially hazardous substances, depending on multiple factors, are found in cosmetics, food containers,

> *Pesticides* are broadly defined in the FIFRAct Section 2(u) as chemicals and other products used to kill, repel, or control pests.

personal hygiene products, lawn applications, clothing, furniture, and other items of contemporary daily use. Some substances have been evaluated as to their potential toxicity, if any; but many have not. Described in this section are some of the principal adverse health impacts of environmental toxicants and the U.S. policies that have been implemented to prevent or reduce their impacts.

6.3.1 IMPACTS OF TOXIC SUBSTANCES ON HUMAN HEALTH

Adverse effects on health can be caused by many chemical substances in the environment. The nature and effects depend on such factors as the potency of the substance, the route and extent of exposure, and an individual's personal characteristics such as genetics, age, and health status. As shown in Table 6.2, all of the body's major organs and organ systems can be potentially affected by exposure to chemicals that can be

TABLE 6.2
Toxicity Endpoints and Alphabetized Illustrative Toxic Substances [37]

Endpoint	Example Substances
Cancer	Arsenic, asbestos, beryllium, cadmium, chromium, PAHs
Cardiovascular diseases	Carbon monoxide, lead, ozone
Developmental disorders	Cadmium, endocrine disruptors, lead, mercury
Endocrine disruption	Bisphenol-A (BPA), atrazine, phthalates, perchlorate
Immune dysfunction	Formaldehyde
Liver disease	Ethyl alcohol, carbontetrachloride
Nervous system disorders	Lead, manganese, methyl mercury, organophosphates, PCBs, formaldehyde
Reproductive disorders	Cadmium, endocrine disruptors, DDT, PCBs, phthalates
Respiratory diseases	Nitrogen dioxide, particulate matter, sulfur dioxide
Skin diseases	Dioxins, nickel, pentachlorophenol

toxic under applicable circumstances. Of policy relevance, policies to prevent human and ecological exposure to hazardous substances have increased in scope and importance in concert with increased toxicological knowledge. The public health implications of toxic substances can be especially great when a toxic substance is pervasive or widely spread within an environmental medium, as demonstrated by the following discussions of two toxic substances of topical relevance.

6.3.1.1   Health Effects of Exposure to Lead

Perhaps the most widely pervasive, health-destructive toxicant is lead. As a naturally occurring element, lead has over millennia of time found multiple uses in the hands of humankind. But with lead's utilitarian purposes came the experience that exposure to lead could cause adverse health effects, including death. It is interesting to note that the earliest recorded human exposure to lead occurred at least 250,000 years ago, as measured in the teeth of two young Neanderthal children [38]. The exposure of the children to lead was deemed indicative of short-term exposure from ingestion of contaminated food or water or inhalation from fires containing lead.

> More than three-quarters of global lead consumption is for the manufacture of lead-acid batteries for motor vehicles.

As background, lead is a naturally occurring toxic metal found in the Earth's crust. As noted by the WHO, the widespread use of lead has resulted in extensive environmental contamination, with human exposure and significant public health problems ensuing in many parts of the world. Important sources of environmental contamination include mining, smelting, manufacturing and recycling activities, and, in some countries, the continued use of leaded paint, leaded gasoline, and leaded aviation fuel. More than three-quarters of global lead consumption is for the manufacture of lead-acid batteries for motor vehicles. Lead is, however, also used in many other products, for example, pigments, paints, solder, stained glass, crystal vessels, ammunition, ceramic glazes, jeweler, toys, and in some cosmetics and traditional medicines. As with the Flint, Michigan, episode, drinking water delivered through lead pipes or pipes joined with lead solder might contain lead (Chapter 3). Much of the lead in global commerce is now obtained from recycling [39].

> Until eventually removed, tetraethyl lead was added to gasoline supplies in early model cars to help reduce engine knocking and boost octane ratings.

Exposure to lead has tracked the growth in numbers and the distribution of humans globally. And the adverse health effects have similarly corresponded. As a modern-day environmental health hazard, exposure to lead has been estimated to cause 412,000 premature deaths in the U.S. annually [40]. Researchers at Simon Fraser University analyzed data from the National Health and Nutrition Examination Survey (NHANES). Between 1988 and 1994, U.S. experts gave blood-lead tests to 30,000 randomly selected Americans, from infants to the elderly, and then followed up with people in 2011. The study conducted at Simon Fraser looked at 14,000 of those people who were 20 years or older when tested, and examined relevant death certificates. People with the highest lead levels had a 37% greater risk than normal of a premature death and a 70% greater risk of death from cardiovascular disease, leading the researchers to infer that lead could be responsible for as much as 18% of all mortalities, or 412,000 U.S. deaths. However, other researchers warned outside factors could lead to an "overestimation of the effect of concentrations of lead in blood, particularly from socioeconomic and occupational factors."

As was discussed in Chapter 2, lead is one of the six criteria for air pollutants. Until eventually removed, tetraethyl lead was added to gasoline supplies in early model cars to help reduce engine knocking and boost octane ratings. And as an additive in gasoline, engine combustion of gasoline produced engine exhaust containing lead, which became a significant source of lead exposure to children and adults, raising blood lead levels. Given the known association between prenatal exposure to lead and the adverse effects on children's cognitive development, it was a public health celebratory success when lead was removed from gasoline.

Another pervasive source of lead exposure comes from the legacy of lead-based paint, which was used in the U.S. for decades, until banned as an additive. Lead was added to paint to accelerate drying, increase durability, maintain a fresh appearance, and resist moisture that causes corrosion. Lead-based paint used in older housing became a public health problem when young children ate paint chips, called pica behavior, and were thereby accidentally exposed to lead. Some lead exposures were lethal, depending on the amount of paint ingested. Cities and states found themselves having to respond to an epidemic of childhood lead poisonings.

> Lead was added to paint to accelerate drying, increase durability, maintain a fresh appearance, and resist moisture that causes corrosion.

For some states, removing lead-based paint and conducting health surveillance on children with potential or actual exposure to household lead sources became a pressing financial obligation. In 2006, the state of Rhode Island successfully litigated three paint companies known to have produced lead-based paint in past years [41]. This sent a shock wave throughout the paint industry since costs to them could run in the billions of dollars nationwide as other states pursue their own litigation. Given the public health gravity of these two examples from the U.S. experience with lead, one would expect the potential benefits of the TSCAct would be substantive in regard to preventing the adverse effects of toxic substances.

6.3.1.2 Health Effects of Exposure to PFAS

In addition to lead, a second set of toxicants, PFAS, has drawn the topical attention of environmentalists, public health specialists, and government policymakers, primarily due to the ubiquity of this category of chemicals. Per- and polyfluoroalkyl substances are generally referred to by their plural acronym, PFAS. PFAS are resistant to water, oil, and heat, and their use has expanded rapidly since they were developed by two U.S. chemical companies in the mid-20th century. Their nonstick qualities make them useful in products as diverse as food wrappers, umbrellas, tents, carpets, and firefighting foam. The chemicals are also used in the manufacture of

> PFAS's nonstick qualities make them useful in products as diverse as food wrappers, umbrellas, tents, carpets, and firefighting foam.

plastic and rubber and in insulation for wiring [42]. Their broad presence in many commercial products makes them ubiquitous as a matter of public health concern.

They're also known as "forever chemicals," because the molecular bonds that form them can take thousands of years to degrade, meaning that they accumulate both in the environment and in our bodies. There are thousands of PFAS chemicals, many of them little-understood byproducts. The chemicals were distributed, purposefully and inadvertently, by several chemical companies for generations. A nonprofit group estimated that the chemicals are in the drinking water of about 100 million Americans and further asserted that 297 military installations are contaminated with PFAS, potentially posing a health hazard to nearby communities [42].

The Agency for Toxic Substances and Disease Registry (ATSDR) conducted a review IN 2018 of the data concerning PFAS and drew the following conclusions: "The health effects of PFOS, PFOA, PFHxS, and PFNA have been more widely studied than other per- and polyfluoroalkyl substances (PFAS). Some, but not all, studies in humans with PFAS exposure have shown that certain PFAS may:

- affect growth, learning, and behavior of infants and older children
- increased women's infertility, premature births, and low birthweight [43]
- interfere with the body's natural hormones
- increase cholesterol levels
- affect the immune system
- increase the risk of cancer" [44].

PFOS and PFOA—the compounds most frequently cited by regulators because they have received more scrutiny—have been linked to cancer, conditions affecting the liver, thyroid and pancreas, ulcerative colitis, hormone and immune system interference, high cholesterol, pre-eclampsia in pregnant women, and negative effects on growth, learning, and behavior in infants and children [44]. Based on these adverse health effects, several government agencies have set health advisory levels for PFOA. The EPA's health advisory is 70 ppt in drinking water. The state of Vermont has set a water safety level of 20 ppt in drinking water, and the state of New Jersey has set a 14 ppt drinking water threshold for PFOA in 2019 [44].

> Seventeen rivers, lakes, streams, and ponds throughout Michigan have "do not eat" fish advisories because of PFOS contamination.

Of policy and public health, significance is EPA's decision not to identify PFAS chemicals as hazardous substances under the Superfund law, an action that would trigger remedial actions at uncontrolled waste sites contaminated with PFAS [45]. Potential sites for remediation could include military installations where PFAS contamination has been documented.

Of special import for public health is the finding that PFAS has been measured in breast milk worldwide. Research documenting PFAS levels >100 ppt has been reported in breast milk in Canada, China, Germany, Hungary, Spain, and Malaysia.

> PFAS levels >100 ppt in breast milk have been reported in Canada, China, Germany, Hungary, Spain, and Malaysia.

The highest average level was found in the Ehime region of Japan, where human breast milk contained PFAS at an average level of 232 ppt in 2008 [46].

The state of Michigan has been particularly hard struck by discoveries of PFAS contamination in water and soil [47]. The state may have more than 11,000 sites contaminated with these once-common chemicals. Regulators have identified 46 sites statewide with levels greater than the EPA's health limit in groundwater. Seventeen rivers, lakes, streams, and ponds throughout Michigan have "do not eat" fish advisories, or limitations on consumption of fish, because of PFOS contamination. In 2018, the U.S. Department of Defense (DOD) identified 651 military sites across the U.S. where there are known or suspected releases of PFOS and PFOA through the use of firefighting foam [48]. The 651 figure currently includes only sites where the DOD is known to be the source of PFAS contamination. The military has provided bottled water and filters to the affected areas. The full public health implications and potential impact await further investigation, but it is already evident that PFAS contamination is widespread in the U.S. [49].

In 2018 New Jersey became the first U.S. state to regulate its drinking water for the presence of PFNAs. The state's Department of Environmental Protection rule will cap the amount of compounds in water [50]. The New Jersey rule amends the state's Safe Drinking Water Act to set a maximum contaminant level of 13 parts per trillion of PFNAs starting in 2019. The state took regulatory action in the absence of federal standards for the compounds.

An environmental organization, the Environmental Working Group (EWG) estimated in 2018, using EPA data, that 110 million Americans may be contaminated with PFAS [49]. However, a follow-up study released in 2020 by the EWG alleged that the PFAS contamination is far worse than previously estimated with some of the highest levels found in Miami, Philadelphia, and New Orleans (Chapter 3). Of tap water samples taken by EWG from 44 sites in 31 states and Washington D.C., only one location, Meridian, Mississippi, which relies on 700 ft deep wells, had no detectable PFAS. Only Seattle and Tuscaloosa, Alabama, had levels below 1 part per trillion (ppt), the limit EWG recommends. EPA has recommended that water contain no more than 70 parts ppt of PFAS, although it is not a water quality standard, leading several U.S. states to enact laws requiring lower levels of PFAS for drinking water.

In February 2020, EPA announced it would regulate "forever chemicals" [51]. EPA's decision to regulate PFAS starts a 2-year period for the agency to determine what the new mandatory maximum contamination level (MCL) should be. Once that is formally proposed, the agency has another 18 months to finalize its drinking water requirement. In the interim, U.S. state standards will remain in effect.

In an international setting, in 2019 Denmark became the first country to ban PFAS chemicals [53]. The ban covers the use of PFAS compounds in food contact materials of cardboard and paper.

WISCONSIN PFAS CLEANUP

Milwaukee, WI, August 5, 2019: Johnson Controls International set aside $140 million to address contamination from firefighting foams for use in environmental cleanup in northeastern Wisconsin. The Tyco plant cleanup is the largest known PFAS contamination in Wisconsin, according to the state Department of Natural Resources [52].

The Danish government said it would continue to be possible to use recycled paper and paper for food packaging, but said PFAS compounds must be separated from the food with a barrier which ensures that they don't migrate into the food.

Perspective: The descriptions of the health hazards presented by glyphosate and the PFOS toxicants once again illustrate the gap between the development of a new chemical and its subsequent release into commerce. The gap is the failure to adequately evaluate the nascent chemical for any evidence of toxicity. This gap can result in members of a population at risk becoming the sentinels for any adverse effects of the untested chemical. And that reality presents challenges to the public health community.

6.3.1.3 Toxic Substances and Children's Health

A society is not sustainable without children. This truth has been a common sense from the origins of societal clustering. But prior to the development of vaccines and other medical interventions, many children succumbed to childhood diseases. In the 20th century, public health programs of childhood vaccinations, improved nutrition, and better education of parents all contributed to improved mortality rates for children. Unfortunately, environmental hazards coincident with the Chemical Age of industrialized nations have reintroduced some health problems for children [54]. Some of the chemical hazards in children's environment have prompted medical authorities to issue health warnings.

CLASS ACTION LITIGATION

Melbourne: October 29, 2019—Up to 40,000 people who live and work on land contaminated by PFAS are suing the Australian Government, arguing their property values have plummeted [56].

In one instance, evidence that an entire class of pesticides threatens the health of children and pregnant women is so arresting that the substances should be banned, an expert panel of toxicologists recommended in 2018. The toxicologists noted that exposure to organophosphates (OPs) increases the risk of reduced IQs, memory and attention deficits, and autism for prenatal children [55]. But any ban could only be taken by EPA under its FIFRAct authorities on pesticides. Nevertheless, published recommendations from expert panels or committees can be beneficial to policymakers who have public health goals to pursue.

The interplay between EPA regulations, children's health, and science policy is illustrated by the EPA's struggle with a regulatory rule concerning the pesticide chlorpyrifos, a widely used pesticide [57]. The Obama administration announced in 2015 that it would ban chlorpyrifos after EPA studies had shown the pesticide had the potential to damage brain development in children. However, in July 2019 and later, December 4, 2020, the Trump administration EPA announced that it would not ban chlorpyrifos. This decision was counter to EPA's science staff who had determined that the pesticide was linked to serious health problems in children. In making the chlorpyrifos ruling, the EPA justified its decision by announcing that the data supporting objections to the use of the pesticide was "not sufficiently valid, complete, or reliable." The agency added that it would continue to monitor the safety of chlorpyrifos through 2022 [57]. On February 6, 2020, the sole producer of chlorpyrifos,

Corteva Inc., announced that it will end production of chlorpyrifos by the end of 2020. The company cited shrinking demand and withering sales for the pesticide [58]. The decision comes after California in February 2020 banned sales of chlorpyrifos after state regulators said it had been linked to health issues in children, including brain damage.

6.3.1.3.1 Exposure to Lead and Children's Health

The worst example of a chemical hazard that impacts young children is environmental exposure to lead. As previously noted in this chapter, this historically well-known toxicant was added in the early 20th century as tetraethyllead to gasoline and to paint for commercial purposes, without due regard for any human health consequences. As a result, generations of young children suffered lead intoxication that caused neurological problems, developmental issues, and impaired social functioning. And as the Flint, Michigan, example of a municipal water supply contaminated with lead, as discussed in Chapter 3, illustrates the legacy of lead in children remains a public health challenge.

The public health impacts on children who experience exposure to lead are characterized by the WHO as follows [28]:

> **WHO: Lead exposure is estimated to account for 674,000 deaths per year with the highest burden in low- and middle-income countries.**

- Lead is a cumulative toxicant that affects multiple body systems and is particularly harmful to young children.
- Lead exposure is estimated to account for 674,000 deaths per year with the highest burden in low- and middle-income countries.
- Lead exposure is estimated to account for 9.8% of the global burden of idiopathic intellectual disability, 4% of the global burden of ischemic heart disease, and 5% of the global burden of stroke.
- Lead in the body is distributed to the brain, liver, kidney, and bones. It is stored in the teeth and bones, where it accumulates over time. Human exposure is usually assessed through the measurement of lead in blood.
- There is no known level of lead exposure that is considered safe.
- Lead poisoning is entirely preventable.

The actual number of children in the U.S. with elevated blood lead levels (BLLs) probably exceeds previously reported numbers, according to researchers at the Public Health Institute's California Environmental Health Tracking Program [59]. Elevated BLLS were those that exceeded 10 µg/dL. Investigators reported their analysis, using NHANES data for the years 1999–2010, estimated 1.2 million children had elevated BLLs, twice the number estimated by the CDC. The investigators also reported a wide variability across states in regard to testing children for lead poisoning.

Young children are particularly vulnerable to the toxic effects of lead and can suffer profound and permanent adverse health effects, particularly affecting the development of the brain and nervous system. Lead also causes long-term harm in adults, including an increased risk of high blood pressure and kidney damage. Exposure of pregnant women to high levels of lead can cause miscarriage, stillbirth, premature birth and low birth weight, as well as minor malformations.

The sources of children's exposure to lead include drinking water, air pollution, food, and lead-based paint in older housing. Two sources of children's exposure to lead were recognized by policymakers and public health actions ensued. In 1996, as noted in Chapter 2, lead additives were banned in U.S. gasoline supplies under provisions of the federal Clean Air Act. Similarly, the U.S. Consumer Product Safety Commission banned lead paint in 1977 in residential properties and public buildings, along with toys and furniture containing lead paint. While both actions of removing lead from environmental sources, air and paint, comported with the public health principle of primary prevention, i.e., hazard mitigation, the legacy of lead in housing and some public drinking water supplies remains a major public health challenge.

> WHO: There is no known level of lead exposure that is considered safe.

Concerning lead in drinking water supplies, children exposed to lead in water from school drinking fountains is a national problem. The problem of lead in children's drinking water surfaced during the Flint, Michigan, episode when public water supplies were reconnected to a municipal water supply contaminated with lead. This episode, which is described in Chapter 3, stimulated a national discussion about lead in children's water supplies. Partially in response, the Government Accountability Office (GAO) conducted an analysis of lead in U.S. school drinking water supplies. And quoting them, "An estimated 43% of school districts, serving 35 million students, tested for lead in school drinking water in 2016 or 2017, according to GAO's nationwide survey of school districts. An estimated 41% of school districts, serving 12 million students, had not tested for lead" [60].

GAO's survey showed that, among school districts that did test, an estimated 37% found elevated lead (lead at levels greater than their selected threshold for taking remedial action). All school districts that found elevated lead in drinking water reported taking steps to reduce or eliminate exposure to lead, including replacing water fountains, installing filters or new fixtures, or providing bottled water [55]. The cost to fix a school system's lead in drinking water problem is high. For example, the estimated cost to the Oakland, California, public school system is $38 million to address high lead levels in water taps at its schools [61]. Illustrated in Figure 6.3 are two public school drinking water fountains that were removed from service due to their dispensing lead in drinking water. Due to limited funds, the school began providing bottled water to students upon request.

> GAO: An estimated 41% of school districts, serving 12 million U.S. students, had not tested for lead.

Lead paint in older U.S. housing stock remains a public health challenge to young children. Children who reside in homes bearing lead paint can be exposed to lead via ingestion of paint chips, inhalation of lead dust, and contact with lead in soil adjacent to the housing. A 1995 national survey conducted for the EPA of lead paint in U.S. housing estimated that 64 million homes of the privately owned housing units built before 1980 had lead-based paint somewhere in the building. Twelve million (±1 million) of these homes were occupied by families with children under the age of 7 years old. An estimated 49 million privately owned homes had lead-based paint in

FIGURE 6.3 School water fountains dispensing lead in water and removed from service. (safewater.org, 2020, https://www.safewater.org/news/2018/9/11/yukon-shuts-off-some-school-water-fountains-because-of-lead-levels)

their interiors. Fourteen million homes, 19% of the pre-1980 housing stock, had more than 5 ft² of damaged lead-based paint. Nearly half of them (47%) had excessive dust lead levels [62].

A significant policy issue concerns the presence and removal of lead-based paint in older housing. The question is one of who pays for remediation? As a policy question, it is akin to the issue of who pays for the remediation of uncontrolled hazardous waste sites, as discussed in Volume 2, Chapter 2. The litigation principle has become one of the polluters who pay for remediation of their pollution. So, in the instance of lead-based paint in older housing, litigation against the paint companies that produced the paint has been initiated, for example, in California.

> A significant policy issue concerns the removal of lead-based paint in older housing. The question is one of who pays for remediation?

In the year 2000, Santa Clara County, joined later by six other counties and the cities of Oakland, San Diego, and San Francisco, filed suit against three major paint companies that had produced lead-based paint [63]. In 2014, a superior court ruled in favor of the plaintiffs and his ruling established a state fund to strip lead paint from doors, windows, and floors of homes. The three paint companies were required to pay into the state fund. This ruling was appealed to a state appellate court, which upheld the lower court's ruling. In 2018, the U.S. Supreme Court upheld the original court decision, leaving the three paint companies responsible for funding the California state fund for removal of lead-based paint in pre-1951 homes in California [64]. In 2019, after a 19-year legal struggle, the state of California finally succeeded in getting the companies to fund a remediation program, albeit on a much smaller scale than originally sought. Sherwin-Williams, ConAgra Grocery Products Co., and NL Industries have agreed to a $305 million settlement, according to a filing in Santa Clara County Superior Court [65]. Legal experts observed that the settlement marked a rare success for a public nuisance claim, under which counties and municipalities can sue corporations for past activities—including those conducted decades ago—they say have harmed communities.

6.3.1.3.2 Social Justice Issues

Exposure to environmental sources of lead can be construed as a social justice issue if the exposure, especially to children, occurs due to inequities in social factors such as income, cultural practices, and public policies. As noted by CDC, there is no safe detectable level of

> For African-American Black children nationwide, one in four residing in pre-1950 housing and one in six living in poverty presented with an ≥5 µg/dL.

lead (Pb) in the blood of young children, and in the U.S., predominantly African-American Black children are exposed to more Pb and present with the highest mean blood lead levels (BLLs). A research investigation examined risk factors for early childhood Pb exposure [66]. Overall, Black children had an adjusted +0.83 µg/dL blood Pb and a 2.8 times higher odds of having an Elevated BLL (EBLL) ≥5 µg/dL. For African-American Black children nationwide, one in four residing in pre-1950 housing and one in six living in poverty presented with an EBLL. The researchers concluded that significant nationwide U.S. racial disparity in blood Pb outcomes persist for predominantly African-American Black children even after correcting for risk factors and other variables.

6.3.1.3.3 Exposure to Toxicants on Reproductive and Developmental Health

Several medical groups have taken policy stands against children's exposure to toxic substances in the environment. For instance, the International Federation of Gynecology and Obstetrics (FIGO) was the first global reproductive health organization to take a stand on human exposure to toxic chemicals. Miscarriage and still birth, impaired fetal growth, congenital malformations, impaired or reduced neuro-development and cognitive function, and an increase in cancer, attention problems, ADHD behaviors, and hyperactivity are among the list of adverse health outcomes linked to chemicals such as pesticides, air pollutants, plastics, solvents, and more, according to the FIGO opinion.

The cost of childhood diseases related to environmental toxicants and pollutants in air, food, water, soil, and in homes and neighborhoods was calculated to be $76.6 billion in 2008 in the U.S. FIGO proposes that physicians, midwives, and other reproductive health professionals

> In 2008, an international medical group estimated the cost of childhood diseases related to environmental toxicants to be $76.6 billion in the U.S.

advocate for policies to prevent exposure to toxic environmental chemicals; work to ensure a healthful food system for all; make environmental health part of health care; and champion environmental justice [67].

Other medical groups are also becoming more proactive in expressing concern about the adverse health effects of hazardous chemicals. In 2015, the American Academy of Pediatrics signed a petition to the Consumer Product Safety Commission seeking to ban products that contain organohalogen flame retardants [68]. Similarly, the Endocrine Society, following a review of published scientific literature, concluded there is strong mechanistic, experimental, animal, and epidemiological evidence for endocrine disruption. Obesity and diabetes, female reproduction, male reproduction, hormone-sensitive cancers in females, prostate cancer, thyroid, and neurodevelopment and neuroendocrine systems were cited as being associated with exposure to EDCs [69].

The scientific literature contains many publications that relate various environmental toxicants to adverse health effects in children, fetuses, and pregnant women. Illustrated in Table 6.3 is a sample of such investigations. Especially noteworthy are findings that suggest transgenerational toxic effects can occur when pregnant mothers are exposed to specific hazardous chemicals, signaling that future generations will share in the adverse health effects. While the studies cited in the table are but

TABLE 6.3
Adverse Health Effects of Children Exposed to Selected Hazardous Chemicals

Toxicant	Effect	References
Phthalates	Children exposed to phthalates prenatally and as 3-year olds have decreased motor skills later in their childhood	[70/71]
bisphenol-A (BPA)	Mothers of newborns with lower birthweights had significantly higher BPA levels in their urine	[72/73]
Lead	Elevated blood lead levels in early childhood raise the risk of developing (ADHD) in children, with boys being especially vulnerable	[37/74]
DDT	Elevated levels of DDT in the mother's blood were associated with almost a four-fold increase in her daughter's risk of breast cancer	[71/75]
BPA	80% of teenagers had BPA, an endocrine disruptor, in their blood and urine	[73/76]
Insecticides	Children who had been exposed to insecticides indoors were 47% more likely to have leukemia and 43% more likely to have lymphoma	[74/77]
Pb	Toddlers exposed to lead struggled in school more than those who had not been exposed. As teens, they committed crimes more frequently	[75/78]
Pb	Pregnant women with high levels of lead in their blood not only affect the fetal cells of their unborn children but also their grandchildren	[76/79]
Pb, OP pesticides, methyl mercury	The three environmental exposures together would decrease 1.6 IQ point in each of 25.5 million children	[77/80]
Phthalates	Pregnant women with higher levels of the hormone hCG, which is targeted by phthalates, birthed boys with an abnormality in anogenital distance	[78/81]
Phthalates	Women exposed to high levels of phthalate are more likely to have high blood pressure during pregnancy	[79/82]

a sample of the literature, they still raise health concerns about the potential wide breath of adverse effects on children and pregnant women. Additional science will be required for both clarifications of effects as well as verification of findings.

6.3.1.4 Health Effects of Endocrine Disruptors

Toxicology as a science and an academic discipline has evolved slowly over the 20th century. Early studies were simply mortality investigations. Gradually over the middle- and late-20th century, science began to incorporate studies of putative toxic substances on induction of cancer, mutations, adverse reproduction, and effects on other organ systems, e.g., respiratory and neurologic. In the late 20th century work by Dr. Theo Colborn (1927–2014), an environmental scientist with the World Wildlife Fund identified adverse effects of some environmental toxicants on the endocrine system [83]. The endocrine system consists of the pituitary gland, thyroid gland, parathyroid glands, adrenal glands, pancreas, ovaries (in females), and testicles (in males). Shown in Figure 6.4 is a photo of Dr. Colborn taken in the early years of the 21st century.

As observed by Colborn and colleagues, "The endocrine system is involved in every stage of life, including conception, development in the womb and from birth throughout early life, puberty, adulthood, and senescence. It does this through control of the other vital systems that orchestrate metabolism, immune function, reproduction, intelligence, behavior, etc. The endocrine system acts through signaling molecules, including hormones such as estrogens, androgens, thyroid hormones, and insulin, as well as

> Some environmental toxicants have the capacity to mimic some of the physiological effects of naturally occurring hormones. This mechanism is termed *endocrine disruption.*

FIGURE 6.4 Dr. Theodora (Theo) Colborn, champion of endocrine disruption research, circa early 21st century. (https://en.wikipedia.org/wiki/Theo_Colborn#/media/File:Theo_Colborn_(1927-2014).jpg, 2020.)

brain neurotransmitters and immune cytokines (which are also hormones) and other signaling molecules in the body" [84].

As Colborn and other investigators discovered, some environmental toxicants have the capacity to mimic some of the physiological effects of naturally occurring hormones. This mechanism is termed *endocrine disruption*, and the mimicking substances are called *endocrine disruptors*. *Endocrine-disrupting chemicals* (EDCs) is another term used by investigators. One's hormones literally shape a person's physiological and anatomical character.

A review by the WHO of EDC studies concluded, "[...] endocrine systems are very similar across vertebrate species and [...] endocrine effects manifest themselves independently of species. Effects shown in wildlife or experimental

> The WHO has identified ~800 chemicals that are known or suspected to be endocrine disruptors, yet only a few have been investigated.

animals might also occur in humans if they are exposed to EDCs at a vulnerable time and at concentrations leading to alterations of endocrine regulation. Of special concern are effects on the early development of both humans and wildlife, as these effects are often irreversible and may not become evident until later in life" [85]. The WHO has identified ~800 chemicals that are known or suspected to be endocrine disruptors, yet only a few have been investigated. Included on the list are the following, several of which are rather common in the environmental media: bisphenol A, dioxin, atrazine, phthalates, perchlorate, fire retardants, lead, arsenic, mercury, perfluorinated chemicals, organophosphate pesticides, and glycol ethers [86].

Congruent with the WHO's statement are findings relating to sperm count and reproductive abnormalities. In particular, scientists have reported sperm counts have been dropping; infant boys are developing more genital abnormalities;

> The health costs associated with human exposure to EDCs are estimated to be at least $175 billion (U.S.) per year in Europe alone.

more girls are experiencing early puberty; and adult women appear to be suffering declining egg quality and more miscarriages [87]. Sperm concentration declined significantly between 1973 and 2011 by 0.70 million/mL/year. Further, scientists have reported genital anomalies in a range of species, including unusually small penises in alligators, otters, and minks. In some areas, significant numbers of fish, frogs, and turtles have exhibited both male and female organs. To what extent these adverse reproductive effects are attributable to EDCs remains to be determined.

A study of the health costs associated with human exposure to EDCs estimated an increased risk of serious health problems costing at least $175 billion (U.S.) per year in Europe alone [88]. Reviewers of the study opined that the health care costs in the U.S. would approximate those in Europe. The researchers detailed the costs related to three types of conditions: neurological effects, such as attention deficit disorders; obesity and diabetes; and male reproductive disorders, including infertility. The biggest estimated costs, by far, were associated with chemicals' reported effects on children's developing brains.

The researchers concluded that there is a >99% chance that EDCs are contributing to the diseases. The estimate was limited to a handful of chemicals commonly found in human bodies: bisphenol-A (BPA), used in hard plastics, food-can linings, and paper receipts; two phthalates used as plasticizers in vinyl products; DDE, the breakdown product of the banned insecticide DDT; organophosphate pesticides, including chlorpyrifos used on grain, fruit, and other crops; and brominated flame retardants known as PBDEs that were extensively used in furniture foams until they were banned in Europe and the U.S. BPA, DDE, and the phthalates were examined for their links to obesity and diabetes, phthalates for male reproductive effects, and flame retardants and organophosphate pesticides for neurological effects [88].

To put $175 billion in perspective, it exceeds the combined proposed 2016 budgets for the U.S. Department of Education, Department of Health and Human Services, National Park Service, and EPA combined.

The ecological consequences of EDCs are the subject of a substantial published literature on the consequences of EDCs as pollutants in lakes, rivers, and streams. Of special note, the association between EDCs and feminizing effects in fish is a basis of ecosystem concern. For example, 85% of male smallmouth bass tested in or nearby 19 National Wildlife Refuges in the U.S. Northeast had signs of female reproductive parts, according to a study conducted by the U.S. Geological Survey and U.S. Fish and Wildlife Service. Findings also reported that 27% of male largemouth bass in the testing sites were intersex. Investigators interpreted these findings as evidence of EDC pollution [89]. In a similar report, some male black bass and sunfish in North Carolina rivers were found to have eggs in their testes [90]. In a laboratory study, researchers from the University of Wisconsin-Milwaukee exposed young fathead minnows to water containing levels of metformin, a commonly used diabetes drug, often found in wastewater effluent. Eighty-four percent of 31 metformin-exposed male fish exhibited feminized reproductive organs [91]. On a larger geographic scale, a research geologist with the USGS found hormone-disrupting compounds—called alkylphenols—passing through wastewater treatment plants and contaminating rivers and fish in the Great Lakes and Upper Mississippi River regions [92]. These and other published studies indicate that EDCs that pollute waterbodies are a hazard to ecosystem health.

> Of special note, the association between EDCs and feminizing effects in fish is a basis of ecosystem concern.

6.3.2 IMPACTS OF TOXIC SUBSTANCES ON ECOSYSTEM HEALTH

As was the impact of pesticides on ecosystems, substances covered under the TSCAct also have the potential for deleterious impacts on ecosystem health. Several environmental toxicants were described in prior chapters of this book as pollutants in air, water, and food and the effects on human and ecosystem health. A few more examples will solidify the fact that the Chemical Age has—and continues to—spread chemical substances into various environmental media and the life existing in those media. For example, chemists at the University of Aberdeen found

> The Chemical Age has—and continues to—spread chemical substances into various environmental media and the life existing in those media.

Cd in all the organs, including the brains, of 21 adult long-finned pilot whales that had stranded in 2012. The whales had died in a mass grounding between Anstruther and Pittenweem in Fife, Scotland, in September 2012. The investigators interpreted their findings as clear evidence that whales are absorbing high levels of Cd and toxic heavy metals [93]. Whether or not the Cd in brain tissues was associated with the whales' beaching is unknown.

In a separate kind of investigation, the global fervor for gold has produced severe ecosystem effects in areas where gold mining was conducted without regard for environmental consequences. The majority of the world's gold is extracted from open-pit mines, where huge volumes of earth are scoured away and processed for trace elements. The environmental organization Earthworks estimates "that, to produce enough raw gold to make a single ring, 20 tons of rock and soil are dislodged and discarded. Much of this waste carries with it mercury and cyanide, which are used to extract the gold from the rock. The resulting erosion clogs streams and rivers and can eventually taint marine ecosystems far downstream of the mine site. Exposing the deep earth to air and water also causes chemical reactions that produce sulfuric acid, which can leak into drainage systems. Air quality is also compromised by gold mining, which releases hundreds of tons of airborne elemental mercury every year" [94].

On a more positive note, a review of literature study by the Scripps Institution of Oceanography, La Jolla, California, reported in 2016 that fish in today's oceans contain far lower levels of Hg, DDT, and other toxicants than at any time in the past four decades [95]. The researchers looked at nearly 2,700 studies of pollutants found in fish samples taken globally between 1969 and 2012. They saw steady, significant drops in the concentrations of a wide range of contaminants known to accumulate in fish from about 50% for Hg to more than 90% for PCBs. The investigators attributed these decreases to clean-water regulations, lawsuits, and other forms of public pressure have led to bans or sharp reductions in the use of industrial and agricultural contaminants that migrate to creeks, rivers, and oceans [95]. In a similar theme of regulatory impact, paper companies, recyclers, and water treatment plants agreed to fund another $46 million to restore wildlife and habitat in northeastern Wisconsin as part of a massive PCB cleanup in the Fox River and Green Bay. Federal, state, and the Oneida and Menominee tribes settled on an arrangement with the parties deemed responsible for releasing PCBs into waterbodies. This brought the total Natural Resources Damage Assessment claim to $106 million. The settlements are aimed at remediating damage to wildlife as PCBs are being dredged out of sediments [96].

The settlements are aimed at remediating damage to wildlife as PCBs are being dredged out of sediments [96].

Perspective: Global monitoring data indicate that hazardous chemicals continue to be released into environmental media. As a matter of environmental health, chemical contamination of waterbodies and terrestrial resources must remain a concern for human and ecosystem health. But data also indicate that regulatory and other policies are having an impact in reducing the release of hazardous chemicals into the environment. From a global perspective, environmental health policies can be effective if developed, implement, and monitored.

6.4   U.S. POLICIES TO CONTROL TOXIC SUBSTANCES

The U.S. federal government, commencing in 1960, has become increasingly active in efforts to control toxic substances in the general environment. To be described in this section are the major federal statutes pertaining to toxic substances and the federal agencies involved with implementing the statutes.

6.4.1   FEDERAL HAZARDOUS SUBSTANCES ACT, 1960

One of the early federal statutes on hazardous substances is the FHSAct of 1960. This act requires precautionary labeling on the immediate container of hazardous household products to help consumers

> The FHSAct requires precautionary labeling on the immediate container of hazardous household products.

safely store and use those products and to give them information about immediate first aid steps to take if an accident happens. The act also allows the Consumer Product Safety Commission (CPSC) to ban certain products that are so dangerous or the nature of the hazard is such that the labeling the act requires is not adequate to protect consumers [97].

The FHSAct only covers products that, during reasonably foreseeable purchase, storage, or use, might be brought into or around a place where people live. Products used or stored in a garage, shed, carport, or other building that is part of the household are also covered. The act requires hazardous household products ("hazardous substances") to bear labeling that alerts consumers to the potential hazards that those products present and that tells them what they need to do to protect themselves and their children from those hazards.

Whether a product must bear a warning label depends on its contents and the likelihood that consumers will be exposed to any hazards it presents. To require labeling, a product must first be toxic, corrosive, flammable or combustible, an irritant, or a strong sensitizer, or

> Whether a product must bear a warning label depends on its contents and the likelihood that consumers will be exposed to any hazards it presents.

it must generate pressure through decomposition, heat, or other means. Further, the product must have the potential to cause substantial personal injury or substantial illness during or as a result of any customary or reasonably foreseeable handling or use, including reasonably foreseeable ingestion by children.

Where it is appropriate, regulations issued under the FHSAct specify the tests to perform to evaluate a product for a specific hazard. The definitions specify [97]: (1) A product is toxic if it can produce personal injury or illness to humans when it is inhaled, swallowed, or absorbed through the skin and contain certain tests on animals to determine whether a product can cause immediate injury. In addition, a product is toxic if it can cause long-term chronic effects like cancer, birth defects, or neurotoxicity. (2) A product is corrosive if it destroys living tissue such as skin or eyes by chemical action. (3) A product is an irritant if it is not corrosive and causes a substantial injury to the area of the body that it comes in contact with. Irritation can occur after immediate, prolonged, or repeated contact. (4) A strong sensitizer is a product that the Commission declares by regulation has a significant potential to cause hypersensitivity [...]. (5) The flammability of a product depends on the results of testing [...], (6) Products that generate pressure, through decomposition, heat, or other means include aerosols, fireworks that contain explosive powder, and certain pool chemicals that, when their containers are heated by sunlight, for example, start to react and generate pressure in the containers.

The label on the immediate package of a hazardous product, and any outer wrapping or container that might cover up the label on the package must have the following information in English [97]: (1) The name and business address of the manufacturer, packer, distributor, or seller; (2) The common or usual or chemical name of each hazardous ingredient; (3) The signal word "Danger" for products that are corrosive, extremely flammable, or highly toxic; (4) The signal word "Caution" or "Warning" for all other hazardous products; (5) An affirmative statement of the principal hazard or hazards that the product presents, [...]; (6) Precautionary statements telling users what they must do or what actions they must avoid to protect themselves; (7) Where it is appropriate, instructions for first aid treatment to perform in the event that the product injures someone; (8) The word "Poison" for a product that is highly toxic, in addition to the signal word "Danger"; (9) If a product requires special care in handling or storage, instructions for consumers to follow to protect themselves; and (10) The statement "Keep out of the reach of children" [...].

There are no formal guidelines. However, among the things to consider are: (1) How the contents and form of the product might cause an injury; (2) the product's intended handling, use, and storage; and (3) any accidents that might foreseeably happen during handling, use, or storage that could hurt the purchaser, user or others, including young children who might get into the package of the product. Further details about the FHSAct are available from the CPSC.

6.4.2 TOXIC SUBSTANCES CONTROL ACT, 1976

Health and ecological concerns about hazardous substances in the general environment gradually expanded past just the matter of pesticides, in part, due to concerns expressed by various environmental organizations. Congress responded with the TSCAct, an action with initial public health promise, but subsequently found to be ineffective.

6.4.2.1 History

Federal legislation to control toxic substances was originally proposed in 1971 by the President's Council on Environmental Quality during the Richard Nixon (R-CA) administration. Its report, *Toxic Substances*, defined a need for comprehensive legislation to identify and control chemicals whose manufacture, processing, distribution, use, and/or disposal was potentially dangerous and not adequately regulated under other environmental statutes. The enactment of the TSCAct of 1976 was influenced by episodes of environmental contamination such as the contamination of the Hudson River and other waterways by PCBs, the threat of stratospheric ozone depletion from chlorofluorocarbon (CFC) emissions, and contamination of agricultural produce by polybrominated biphenyls (PBBs) in the state of Michigan. The episodes, together with more exact estimates of the costs of imposing toxic substances controls, opened the way for the final passage of the legislation. President Gerald Ford (R-MI) signed the TSCAct into law on October 11, 1976 [98].

The TSCAct directs EPA to execute the following key actions [98]:
- Require manufacturers and processors to conduct tests for existing chemicals,
- Prevent future risks through premarket screening and regulatory tracking of new chemical products,
- Control unreasonable risks already known or as they are discovered for existing chemicals,
- Gather and disseminate information about chemical production, use, and possible adverse effects on human health and the environment.

At the time of the TSCAct's enactment, the law allowed continued production of the 62,000 chemicals already in commercial use, which were called *existing chemicals*. Another 18,000 chemicals have been introduced into commerce since 1976, known as *new chemicals*. In sum, ~80,000 chemicals potentially fall under the regulatory provisions of the TSCAct. However, the chemical industry asserts that only about 15,000 chemicals are actively made, which would reduce their testing burden [99].

> Approximately 80,000 chemicals potentially fall under the regulatory provisions of the TSCAct, although only 15,000 are in use.

The TSCAct authorizes EPA to screen existing and new chemicals used in manufacturing and commerce in order to identify potentially dangerous products or uses that should be subject to federal control. As enacted, the TSCAct also included a provision requiring EPA to take specific measures to control the risks from polychlorinated biphenyls (PCBs). Subsequently, three titles have been added to address concerns about other specific toxic substances: asbestos in 1986, radon in 1988, and lead in 1992.

EPA may require manufacturers and processors of chemicals to conduct and report the results of tests to determine the effects of potentially dangerous chemicals on living organisms. Based on test results and other information, EPA may regulate the manufacture, importation, processing, distribution, use, and/or disposal of any chemical that presents an unreasonable risk of injury to human health or the environment. A variety of regulatory tools are available to EPA under the TSCAct, ranging in severity from a total ban on production, import, and use to a requirement that a product must bear a warning label at the point of sale.

6.4.2.2   Amendments to the TSCAct, 1986
Starting in 1986, several important amendments to the TSCAct provide important public health authorizations to EPA and other federal agencies in order to undertake

programs on asbestos, radon, and lead. Two amendments are specific to reducing the hazard of asbestos in schools. The Asbestos Hazard Emergency Response Act of 1986 amends the TSCAct to direct the EPA Administrator to promulgate regulations for asbestos hazard abatement in schools and set standards for ambient

> The Asbestos Hazard Emergency Response Act of 1986 amends the TSCAct to direct the EPA Administrator to promulgate regulations for asbestos hazard abatement in schools.

interior concentrations of asbestos after completion of response actions in schools. Other key provisions include: informing and protecting the public during the phases of asbestos abatement, authorizing each state governor to establish administrative procedures for reviewing school asbestos management plans, directing the EPA Administrator to make grants to local educational agencies, and making local educational agencies liable for civil penalties.

The Asbestos School Hazard Abatement Reauthorization Act of 1989 amended the 1986 act by deleting certain reporting requirements of states, directed state governors to maintain records on asbestos in schools, and made accreditation requirements of schools' asbestos removal workers applicable to persons working with asbestos in public or commercial buildings [98].

Further regulatory action by the EPA on the use of asbestos occurred on June 24, 2019. EPA's rule entitled "Restrictions on Discontinued Uses of Asbestos; Significant New Use Rule" took effect on that date [100]. The rule's summary states, "Under the Toxic Substances Control Act (TSCA), EPA is promulgating a rule to ensure that any discontinued uses of asbestos cannot re-enter the marketplace without EPA review, closing a loophole in the regulatory regime for asbestos." While this rule closes the door on discontinued uses of asbestos, requiring any new uses to comply with TSCAct requirements, critics of the rule aver that the rule falls short of an outright ban on asbestos, as has occurred in other countries [101].

The TSCAct has been amended twice for the purpose of reducing the risk of radon gas in the ambient air of residential buildings. The Radon Program Demonstration Act of 1988 established the national goal of making the air within

> The TSCAct has been amended twice for the purpose of reducing the risk of radon gas in the ambient air of residential buildings.

buildings as free of radon as the outside ambient air. The act contains several significant provisions. EPA is directed to make available to the public information about radon's hazards, develop model construction standards for buildings, assist state radon programs, provide technical assistance to states, make grants to states on an annual basis for radon assessment and mitigation, and establish regional radon training centers in at least three institutions of higher learning. The Omnibus Budget Reconciliation Act of 1990 authorized EPA to conduct research on radon and radon progeny measurement methods and mandated an EPA study on the feasibility of establishing a mandatory radon proficiency testing program [98].

Of particular importance to public health, given the toxicity of lead in the environment, Title X of the Housing and Community Development Act of 1992 amended several federal statutes, including the TSCAct, for the purpose of reducing the health hazard of lead in community and workplace environments. The act directs the Department of Housing and Urban Development to assess lead-based paint hazards in federally assisted housing and requires housing agencies to take action on evaluating and reducing lead-based hazards. The act amends the TSCAct by requiring that contractors and laboratories be federally certified. EPA is directed to conduct a comprehensive program to promote safe, effective, and affordable monitoring, detection, and abatement of lead-based paint and other lead exposure hazards. Also, the Secretary of Labor was directed to issue an interim final regulation for workers' exposure to lead in the construction industry.

6.4.2.3 Public Health Effectiveness of the TSCAct

Unfortunately, the potential consequential benefits to the public's health of the TSCAct did materialize. Of the major environmental health laws, the TSCAct stands out as the major disappointment in public health performance. While there have been some positive impacts, particularly due to the act's amendments, the larger promise of the TSCAct has not been realized. At its core, the TSCAct provides EPA with the authority to assess and control chemicals in commerce (i.e., existing chemicals) and new chemicals proposed for manufacture. The intent is to protect the public from "unreasonable risk" to human health and the environment. Given these laudable purposes, why hasn't the TSCAct lived up to its potential as an environmental health force?

One reason why the TSCAct has failed is because of the large number of chemicals, 80,000, that fall under EPA's regulatory coverage. In theory, EPA could require producers of these chemicals to conduct toxicity testing under the

> One reason why the TSCAct failed is because of the large number of chemicals, 80,000, that fall under EPA's regulatory coverage.

TSCAct's authorities. However, under the TSCAct, EPA must find that a chemical presents an "unreasonable risk" before the agency can mandate toxicity testing. Moreover, EPA must determine that any risks are not outweighed by a chemical's economic and societal benefits for each way in which the substance might be used [98]. These risk and benefits determinations pose a significant challenge to EPA, owing to deficiencies in toxicological data for many substances and uncertainties in substances' benefits.

The shortcomings of the TSCAct have been described by former EPA Assistant Administrator Lynn Goldman [102]. She observed, "TSCA has not proven to be a successful tool for managing existing chemicals; indeed, it has created a situation in which new chemicals, which may be more benign, are subject to substantially more risk management activities and reviews than older and possibly more risky ones (which are not managed at all). Likewise, the TSCA procedure of referring chemicals to other EPA programs or agencies for risk management has not been effective." Concerning existing chemicals, only five[3] have been regulated under the TSCAct. As a perspective, more than 60,000 chemicals comprise the EPA inventory of existing chemicals. A major reason for EPA's failure to regulate more existing chemicals is

---

[3] PCBs, chloroflurocarbons, dioxin, asbestos, and hexavalent chromium.

the TSCAct's unreasonable risk provision, which sets a hurdle too high for the routine regulation of chemicals [102].

New chemicals are also regulated under the TSCAct's provisions. Imposition of these provisions is meant to serve as primary prevention measures to keep hazardous substances out of commerce. As Goldman observes, "EPA's process of premanufacture approval is the only safeguard used by the federal government to guard against such risks." "Since 1992, very little progress has been made by EPA in addressing the impacts of new chemicals" [102].

In 2004, the Government Accountability Office (GAO)[4] released a comprehensive study of EPA's TSCAct authorities and programs [103]. The shortcomings of the TSCAct as an effective public health instrument were the salient findings. The GAO stated that they reviewed EPA's TSCAct's

> Of the 32,000 new chemicals submitted to EPA by chemical companies only about 570 were designated for chemical companies to submit premanufacture notices.

efforts "[t]o control the risks of new chemicals not yet in commerce, (2) assess the risks of existing chemicals used in commerce, and (3) publicly disclose information provided by chemical companies under TSCA."

The GAO's primary findings, in order of the study's three purposes, were as follows. Regarding new chemicals, since 1979, when EPA began reviewing chemicals for potential placement on the TSCAct's inventory, the GAO found that, on average, about 700 new chemicals are introduced into commerce each year. Of the 32,000 new chemicals submitted to EPA by chemical companies only about 570 were designated for chemical companies to submit premanufacture notices for any significant new uses of the chemical, thereby providing EPA with the data to assess risks to human health or the environment from new uses of the chemical.

More disturbing, EPA estimated that most premanufacture notices do not include test data of any type, and only about 15% include health or safety test data. EPA reported to the GAO that they had taken actions to reduce the risks of more than 3,500 of the 32,000 new chemicals they had reviewed. Of public health significance, GAO concluded, "EPA's reviews of new chemicals provide limited assurance that health and environmental risks are identified before the chemicals enter commerce" (ibid., p. 2). In regard to existing chemicals, GAO found that while EPA has authority under the TSCAct to require chemical companies to develop test data after an EPA finding of need, this authority has been used for fewer than 200 of the 62,000 chemicals in commerce since 1979 (ibid., p. 7). GAO concluded that "EPA does not routinely assess the risks of all existing chemicals and EPA faces challenges in obtaining the information necessary to do so" (ibid., p. 7). As noted by GAO, in the late 1990s, in cooperation with chemical companies and national environmental groups, EPA implemented its High Production Volume (HPV) Challenge Program [103]. Under this program, chemical companies voluntarily provide test data on about 2,800 chemicals produced or imported in amounts of 1 million pounds or more annually. While this testing program seems quite positive in terms of potential new chemical

---

[4] Previously named the General Accounting Office.

data, there has been no assessment to date of the program's quality and utility for EPA's chemical regulatory purposes.

As to the third part of GAO's study, according to EPA officials, about 95% of premanufacturing notices for new chemicals submitted by chemical companies contain some information that is claimed by companies as being confidential business information (ibid., p. 7). GAO opined that this limits EPA's ability to share health-relevant information with the public, including state environmental and health agencies.

GAO recommended that Congress provide EPA with additional authorities under the TSCAct to improve its assessment of chemical risks. It was also recommended that the EPA Administrator

ENFORCEMENT EXAMPLE—RRP

Washington, D.C.—April 17, 2014, Lowe's Home Centers agreed to implement a comprehensive, corporate-wide compliance program to ensure that the contractors it hires will minimize lead dust from home renovation activities, as required by the federal Lead Renovation, Repair, and Painting (RRP) Rule. The company will also pay a $500,000 civil penalty, which is the largest ever for violations of the RRP Rule [104].

take specific actions to improve EPA's management of its chemicals programs. But given the fact that Congress has failed over almost 30 years to improve the TSCAct, any acceptance of GAO's recommendation will be problematic.

Perspective: If the TSCAct's authorities, as administered by EPA, have led to regulating only five existing chemicals over the life of the statute and regulatory actions taken on only about 10% of new chemicals, one can ask why the TSCAct was not changed for the better. In other words, why hasn't such an important law been fixed? The answer lies in part in the legislative challenges and uncertainties when amending any major federal statute. Bringing any existing statute back before Congress or a state legislature always runs the risk of changes for the worst. As a policy, it is sometimes better to deal with the "devil we know" than with an unknown one!

6.4.3 RESIDENTIAL LEAD-BASED PAINT HAZARD REDUCTION ACT, 1992

In 1992, Congress enacted the Act, which regulates the disclosure of lead-based paint in sales and leases transactions involving residential properties built before 1978 [105]. Congress charged HUD and EPA with developing regulations to enforce the 1992 law. On December 6, 1996, the regulations developed by the two agencies requiring disclosure of lead-based paint and lead-based paint hazards in connection with residential housing sale and lease transactions went into effect. After 1 year of compliance assistance, the agencies began enforcing the disclosure regulation in 1998. Enforcement actions have occurred and substantial fines have been imposed, according to the two agencies.

6.4.4 LAUTENBERG CHEMICAL SAFETY FOR THE 21ST CENTURY ACT, 2016

For persons knowledgeable of environmental health policymaking, the failure of the TSCAct of 1987 was well known and was rather a mystery as to why Congress did not fix the statute. Congress did not act until 2016 when sufficient pressure dictated otherwise. This is described in the ensuing history of the Lautenberg Chemical Safety for the 21st century Act.

Of the body of federal statutes on environmental health and attendant policies, the TSCAct of 1976 stands alone as an abject failure. Under that law, environmental and public health organizations expressed concern that the chemical industry was allowed to put products on the market without safety testing and to keep many of its formulas secret, using "trade secrets" provisions of the TSCAct.

> Of the body of federal statutes on environmental health and attendant policies, the TSCAct of 1976 stands alone as an abject failure.

In particular, EPA regulators were prohibited by the TSCAct provisions from taking action unless they could **prove** (emphasis added) a chemical poses an "unreasonable risk"—a threshold so burdensome that EPA couldn't even ban asbestos, a well-documented carcinogen that is the cause of the deaths of thousands of persons who died from mesothelioma, a lung cancer disease. Although some fledgling discussions on how to fix the TSCAct were held over the years by some members of Congress, no updating of the law occurred until 2016 when the Lautenberg Chemical Safety for the 21st century Act was enacted. This act makes significant changes to the TSCAct and provides EPA with new authorities to regulate toxic substances. President Obama signed the act into law on 22 June 2016. The bill was named by bipartisan political action for Senator Frank R. Lautenberg (D-NJ), whose 30-years tenure in the Senate included support for environmental health policymaking. Shown in Figure 6.5 is an image of Senator Lautenberg taken circa year 2000.

> The Lautenberg Chemical Safety Act will require EPA to restrict the use of any chemical that the agency finds to present an unreasonable risk.

FIGURE 6.5 Senator Frank R. Lautenberg (D-NJ) (1924–2013) circa year 2000. (Congress. gov, 2020, https://www.congress.gov/member/frank-lautenberg/L000123.)

The new TSCA rewrite requires EPA to restrict the use of any chemical that the agency finds to present an unreasonable risk. Certain exemptions are available for substances deemed essential to national defense, for example. EPA receives more authority to order safety tests for chemicals and set deadlines for the agency to determine whether dangerous compounds should be restricted or forced off the market. EPA will also be required to take additional steps to ensure pregnant women, children, and other vulnerable populations are protected [106].

Overall, the bill gives EPA the authority to immediately begin a risk evaluation of any chemical it designates as high priory, such as asbestos. It also requires upfront substantiation of industry's claims that disclosure of confidential data could damage a firm's business and mandates that so-called "confidential business information" protections expire after 10 years unless renewed. Agency officials still will have only 90 days to judge a new chemical before it can enter the market. But EPA will be able to order testing without years of rulemaking and will be required to identify high-priority chemicals for review, with an initial focus on about 90 compounds.

In addition, the measure also authorizes EPA to conduct testing to determine whether a chemical should be a high priority for a safety review. Decisions made by EPA will pre-empt existing and future state laws to restrict chemicals, in order to create uniform national regulations. The agreement also specifies that if EPA fails to follow through with plans to regulate a chemical within a 3.5-year period, then states are free to act [106].

In 2016 EPA selected ten common chemicals for toxicity evaluation under provisions of the Lautenberg Act. Over the next 3 years, the agency will collect information on the uses of the 10 chemicals, extent of human exposure, hazard, persistence in the environment, and other factors. From this information, EPA will decide whether any among the ten poses an "unreasonable risk" to the environment or human health. For those that do, the EPA has 2 years to create regulations that mitigate the risk. The list includes the following chemicals: 1,4-dioxane, 1-bromopropane, asbestos, carbon tetrachloride, cyclic aliphatic bromide cluster, methylene chloride, N-methylpyrrolidone, pigment violet 29, trichloroethylene, and tetrachloroethylene (also known as perchloroethylene) [107].

> In 2016, EPA selected ten common chemicals for toxicity evaluation under provisions of the Lautenberg Act.

Perspective: The politics of this action by the U.S. Congress are the same as other actions by Congress when yielding to pressure exerted by vested interest groups concerned about U.S. states' policy-making. In this example of interdicting hazardous chemicals prior to their introduction into commerce, the chemical and allied industries preferred not to have to deal with individual states, given that chemical regulations would likely differ across states. One can understand the practicality of the chemical industry's political position, but by essentially diminishing individual states' role in regulating toxic substances, conditions specific to an individual state get lost as influences on a state's policy-making. As with clothing, one size might not fit all.

Additionally, environmental groups had long advocated the need for the TSCAct reform but were unpersuasive in garnering Congressional support, given other priorities in Congress, e.g., budget deficits. But the confluence of environmental interests by the chemical industry and environmental organizations gave Congress over a 6-year period of intense negotiations the compromises necessary to enact what became the Lautenberg statute. Whether EPA can effectuate the Lautenberg Act's provisions any more effectively than those of the TSCAct will be a matter for history to report. However, adding further uncertainty as to the effectiveness of the Lautenberg Act is the Trump administration's stated preference for lesser regulatory action by EPA, together with some likely judicial actions by U.S. states and commercial interests litigating for purpose of obtaining legal clarification on the Lautenberg Act's statutory language.

6.4.5 THE FOOD QUALITY PROTECTION ACT, 1996

Policymaking by elected officials is sometimes difficult for the public to fathom for a variety of reasons. One reason is when existing policies seem to conflict or overlap. This can occur when policies are enacted by different policymakers at different times of enactment. On some occasions, a policymaking body, e.g., U.S. Congress, will enact "bridging" legislation whose purpose is to clarify or resolve conflicting authorities between existing policies. An example is the Food Quality Protection Act of 1996.

In 1996, Congress enacted major legislation that changed how pesticides are regulated. The FQPAct revises the FIFRAct and the federal FDCAct. The FQPAct legislation constituted the first major revision in decades in U.S. pesticides laws. This dramatically altered how pesticides are registered, used, and monitored in the food chain. The legislation was passed without a dissenting vote in either the House of Representatives or Senate and signed into law by President Bill Clinton (D-AR).

The overall purpose of the FQPAct is to protect the public from pesticide residues found in the processed and unprocessed foods they eat. Essentially, the FQPAct amended the FIFRAct and the FDCAct so that a single health-based standard would be issued to alleviate problems concerning the inconsistencies between the statutes. The health-based standard would be based on a "reasonable certainty of no harm".

The FQPAct provides a standard for pesticide residues in both raw and processed foods. The standard is "reasonable certainty of no harm." The law requires EPA to review all pesticide tolerances within 10 years, giving particular atten-

> The overall purpose of the FQPAct is to protect the public from pesticide residues found in the processed and unprocessed foods they eat.

tion to exposure of young children to pesticide residues. Furthermore, EPA must consider a substance's potential to disrupt endocrine function when setting tolerances. The statute requires EPA to give consideration to effects of pesticides on the public's health, requiring the Secretary of DHHS to provide information to EPA on pesticides that protect the public's health [108].

6.5 U.S. AGENCIES WITH TOXIC SUBSTANCES AUTHORITIES

In addition to EPA and FDA, there are other U.S. federal government agencies that have responsibilities in regard to hazardous substances in the environment. In particular, the U.S. Department of Agriculture and the U.S. Department of Labor have statutory responsibilities in terms of control of various hazardous substances in the environment. Further, additional resources that bear on the research on the toxicology of select environmental toxicants and investigations of incidents of chemical releases will be described in this section.

6.5.1   U.S. DEPARTMENT OF LABOR

The Occupational Safety and Health Administration (OSHA) of the U.S. Department of Labor has the responsibility to set workplace standards under the provisions of the Occupational Safety and

> Approximately 500 PELs have been established by OSHA. However, many of these limits are outdated.

Health Act of 1970. Specifically, 29 CFR 1910 Subpart Z, 1915 Subpart Z, and 1926 Subparts D and Z of the OSHAct direct OSHA to establish, promulgate, and enforce workplace permissible exposure limits (PELs) to protect workers against the adverse health effects of exposure to hazardous substances and other hazards to workers. This responsibility includes limits on the airborne concentrations of hazardous chemicals in the ambient air of workplaces. Most OSHA PELs are 8-hour time-weighted averages (TWA), although there are also Ceiling and Peak limits, and many chemicals include a skin designation to warn against skin contact. Approximately 500 PELs have been established. However, as acknowledged by OSHA, many of these limits are outdated. Also, there are many substances for which OSHA does not have workplace exposure limits [109].

Given the shortcomings of OSHA's listed PELs, OHSA has provided employers, workers, and other interested parties with a list of alternative occupational exposure limits that may serve to better protect workers, OSHA has chosen to present a side-by-side table with the California/OSHA PELs, the NIOSH Recommended Exposure Limits (RELs), and the American Conference of Governmental Industrial Hygienists (ACGIH)® Threshold Limit Values (TLVs)®. The tables list air concentration limits but do not include notations for skin injury, absorption, or sensitization.

As an illustration of OSHA's challenges in updating its PELs, in May 2016, OSHA promulgated its final rule on a new permissible exposure limit for respirable crystalline silica—50 $\mu g/m^3$ of air averaged during an 8-hour shift. According to OSHA, silica exposure is a serious threat to nearly 2 million U.S. workers, including more than 100,000 whose jobs involve stone cutting, rock drilling, and blasting and foundry work. OSHA estimates that the new safety limits will save nearly 700 lives and prevent 1,600 new cases of silicosis annually. The agency also estimates that when fully implemented, the rule would result in annual financial benefits of $2.8 to $4.7 billion, benefits that far exceed the rule's annual costs [110]. It is noteworthy to realize that this was the first revision of OSHA's silica PEL in 75 years. The updated PEL is half the previous limit for general industry and five times lower than the previous limit for construction work. The rule covers engineering controls, protective clothing, medical surveillance, and other issues. OSHA presents the rule as two standards—one for general industry and maritime and the other for construction [111].

6.5.2 U.S. CHEMICAL SAFETY BOARD

The U.S. Chemical Safety Board was authorized by the Clean Air Act Amendments of 1990 and became operational in January 1998. The Senate legislative history states: "The principal role of the new chemical safety board is to investigate accidents to determine the conditions and circumstances which led up to the event and to identify the cause or causes so that similar events might be prevented" [112]. Congress gave the CSB a unique statutory mission and provided in law that no other agency or executive branch official may direct the activities of the Board. Following the successful model of the National Transportation Safety Board and the Department of Transportation, Congress directed that the CSB's investigative function be completely independent of the rulemaking, inspection, and enforcement authorities of EPA and OSHA. Congress recognized that the Board's investigations would identify chemical hazards that were not addressed by those agencies [112].

The legislative history states: [T]he investigations conducted by agencies with dual responsibilities tend to focus on violations of existing rules as the cause of the accident almost to the exclusion of

> The Trump administration's 2019 budget proposal calls for ending the CSB, as did its 2018 budget request.

other contributing factors for which no enforcement or compliance actions can be taken. The purpose of an accident investigation (as authorized here) is to determine the cause or causes of an accident whether or not those causes were in violation of any current and enforceable requirement [112]. Both accident investigations and hazard investigations lead to new safety recommendations, which are the Board's principal tool for achieving positive change. Recommendations are issued to government agencies, companies, trade associations, labor unions, and other groups. Implementation of each safety recommendation is tracked and monitored by CSB staff. When recommended actions have been completed satisfactorily, the recommendation may be closed by a Board vote. According to the CSB, it has issued 780 recommendations subsequent to its investigations [112].

The CSB recommendations have the potential for preventing similar chemical events in the future, a policy consistent with the principle of public health, although the impact of the CSB's recommendations lacks current analysis by any academic resource. The CSB's operations and fate have been a target of the Trump administration. The Trump administration's 2019 budget proposal calls for ending the CSB, as did its 2018 budget request [113]. However, only Congress can determine budget appropriations. Abolishment of the CSB is a component of the Trump administration's stated political philosophy of reducing the number and impact of regulations by federal government agencies, even though the CSB does not issue regulations.

6.5.3 NATIONAL TOXICOLOGY PROGRAM

The National Environmental Health Sciences (NIEHS) provides the scientific and administrative leadership within the U.S. Department of Health and Human Services (DHHS) for the National Toxicology Program (NTP). The NTP had begun as a program conceived and administered by the National Cancer Institute (NCI), a component of the National Institutes of Health (NIH). NCI was reacting to environmental and Congressional pressures to investigate the carcinogenicity of chemicals found in the general environment. NCI's response was a program largely devoted to testing specific toxicants for carcinogenicity, using laboratory animals under controlled exposure conditions. The testing was conducted by commercial toxicology testing laboratories, using a study protocol designed by NCI. Unfortunately for the NCI, one of the major contractors was found inadequate and their alleged poor-quality work became the subject of critical newsmedia reports and articles in prestigious scientific journals such as *Science*. Weary of the negative publication, the Secretary of DHHS transferred the NTP to the NIH's National Institute of Environmental Health Science (NIEHS) for the program's administration.

In 1981, under NIEHS's administration, the NTP became the federal government's principal program for assessing the toxicity of substances found in the general environment. As a matter of policy, the NTP receives scrutiny and advice from standing extramural committees comprising experts in toxicology and related disciplines.

A major activity of the NTP is to coordinate the preparation of a biennial report for DHHS on substances judged to be carcinogenic by U.S. federal government scientists. As history, a 1978 Congressional mandate to Section 301(b)

> The NTP is the federal government's principal program for assessing the toxicity of substances found in the general environment.

(4) of the Public Health Service Act, as amended, requires that the Secretary of the DHHS publish an annual report that contains a list of all substances that either are known to be human carcinogens or may reasonably be anticipated to be human carcinogens and to which a significant number of persons residing in the U.S. are exposed.

The first Report on Carcinogens (RoC) was published in 1980 and published annually until 1993 when the reporting requirement was changed to biennial. According to the NTP, since the RoC inception in 1978, the NTP has used scientifically rigorous processes and established listing criteria to evaluate substances for the RoC. The listing categories for each substance nominated for listing are two: (1) Known to Be Human Carcinogen or (2) Reasonably Anticipated to Be Human Carcinogen. The Report on Carcinogens is a cumulative report that includes 243 listings since 1980. The NTP provides details on the listing process, and the review process undergone by each RoC [114].

These biennial reports to Congress on carcinogenic substances (singly or as mixtures) draw the attention of both domestic and international audiences. Domestic audiences span the gamut of industry and environmental interests. Sometimes the listing by the NTP of particular substances, for example, formaldehyde and styrene, can bring pressure from elected policymakers. As one example, an attempt was initiated in 2012 by a Member of Congress to remove funds from the NTP's annual federal budget, resulting in cancellation of the RoC [115]. This effort reflected industry dissatisfaction with the RoC that listed these two chemicals as potential carcinogens. Although this effort by the Member failed, this example does illustrate the political scrutiny that some RoCs receive.

6.6 U.S. STATE AND CITIES POLICIES ON TOXIC SUBSTANCES

Some U.S. states and a few cities have implemented legislation on aspects of hazardous substances. But in general, most U.S. states and cities have ceded to EPA the principal responsibilities of protecting the public against adverse effects of exposure to hazardous environmental substances. As such, states will develop policies and devote resources in support of their responsibilities under federal environmental statutes (e.g., CAAct), which is an example of federalism. And there are, of course, some exceptions, which is almost always the situation, given the authorities given to states, territories, and tribes under provisions of the U.S. Constitution. Moreover, some U.S. states and cities choose to act in the absence of federal policies and legislation. In this section are described three U.S. states' programs on controlling adverse effects of contact with hazardous substances. Two examples of U.S. cities' toxic substances regulations are also included. Further described are some states' trends in legislating consumers' right-to-know policies concerning hazardous chemicals.

6.6.1 STATE OF CALIFORNIA

The State of California is rich in resources, social programs, and diverse population. The state has often set the course for environmental health policymaking by the federal government and other U.S. states. An example was

> California's Proposition 65 requires the State to publish a list of chemicals known to cause cancer or birth defects or other reproductive harm.

described in Chapter 2 (Air Quality), wherein California commenced policies on air pollution in advance of other states and the federal government. Commensurate with this history, California voters in 1986 approved a ballot initiative to address their growing concerns about exposure to toxic chemicals. That initiative became the Safe Drinking Water and Toxic Enforcement Act of 1986, better known by its original name of Proposition 65, often called "Prop 65." In California, propositions approved by voters must be implemented by the California Legislature. In law, Prop 65 requires the State to publish a list of chemicals known to cause cancer or birth defects or other reproductive harm. This list, which must be updated at least once a year, has grown to include ~800 chemicals since it was first published in 1987. Prop 65 requires businesses to notify Californians about significant amounts of chemicals in the products they purchase, in their homes or workplaces, or that are released into the environment. California' Office of Environmental Health Hazard Assessment (OEHHA) administers the state's Proposition 65 program [116].

The list contains a wide range of naturally occurring and synthetic chemicals that are known to cause cancer or birth defects or other reproductive harm. These chemicals include additives or ingredients in pesticides, common household products, food, drugs, dyes, and solvents. Listed chemicals might also be used in manufacturing and construction, or they may be byproducts of chemical processes, such as motor vehicle exhaust.

There are four ways for a chemical to be added to the Prop 65 list. A chemical can be listed if either of two independent committees of scientists and health professionals finds that the chemical has been clearly shown to cause cancer or birth defects or other reproductive harm. These two committees the Carcinogen Identification Committee (CIC) and the Developmental and Reproductive Toxicant (DART) Identification Committee are part of OEHHA's Science Advisory Board. The second way for a chemical to be listed is if an organization designated as an "authoritative body" by the CIC or DART Identification Committee has identified it as causing cancer or birth defects or other reproductive harm. The following organizations have been designated by the OEHHA as authoritative bodies: EPA, FDA, NIOSH, NTP, and IARC.

The third way for a chemical to be listed is if an agency of the state or federal government requires that it be labeled or identified as causing cancer or birth defects or other reproductive harm. Most chemicals listed in this manner are prescription drugs that are required by the FDA to contain warnings relating to cancer or birth defects or other reproductive harm. The fourth way requires the listing of chemicals meeting certain scientific criteria and identified in the California Labor Code as causing cancer or birth defects or other reproductive harm. This four-pronged method established the initial chemical list following voter approval of Prop 65 in 1986 and continues to be used as a basis for listing as appropriate.

Under the provision of Prop 65, businesses are required to provide a "clear and reasonable" warning before knowingly and intentionally exposing anyone to a listed chemical. This warning can be given by a variety of means, such as by labeling a consumer product, posting signs at the workplace, distributing notices at a rental housing complex, or publishing notices in a newspaper. Once a chemical is listed, businesses have 12 months to comply with warning requirements.

Prop 65 also prohibits companies that do business within California from knowingly discharging listed chemicals into sources of drinking water. Once a chemical is listed, businesses have 20 months to comply with the discharge prohibition. Businesses with fewer than 10 employees

> California's Prop 65 prohibits companies that do business within California from knowingly discharging listed chemicals into sources of drinking water.

and government agencies are exempt from Proposition 65's warning requirements and prohibition on discharges into drinking water sources. Businesses are also exempt from the warning requirement and discharge prohibition if the exposures they cause are so low as to create no significant risk of cancer or birth defects or other reproductive.

The OEHHA also develops numerical guidance levels, known as "safe harbor numbers" (described in State regulations) for determining whether a warning is necessary or whether discharges of a chemical into drinking water sources are prohibited. OEHHA has developed safe harbor levels. A business has "safe harbor" from Proposition 65 warning requirements or discharge prohibitions if exposure to a chemical occurs at or below these levels with perfluorooctanoic acid (PFOA).

6.6.2 Examples of U.S. Cities' Toxic Substances Policies

In October 2017, the city and county of San Francisco amended its Environment Code to ban the sale in San Francisco of upholstered furniture and juvenile products made with or containing an added flame-retardant chemical [117]. The ordinance expressed concern that organohalogens and some organophosphorous flame retardant chemicals exhibit one or more of the key characteristics of a class of synthetic organic compounds commonly referred to as Persistent Organic Pollutants (POPs), in that they are bioaccumulative, persistent, capable of long-range transport, and/or toxic.

As a second example of a U.S. city enacting its own toxic substances ordinance, the city of Portland, Maine, instituted a ban on synthetic pesticides in 2018. The ordinance mandates that only organic pesticides may be used on public

> Portland, Maine, instituted in 2018 a ban on synthetic pesticides. The ordinance mandates that only organic pesticides may be used.

and private properties [118]. The ordinance was the product of pressure from a grassroots group that lobbies for environmental stewardship.

Perspective: Several U.S. states and some cities have enacted toxic substances policies, as previously described. To some extent, this policymaking occurs because of special needs and the presence of supporters of environmental health policies. But another factor is surely the perceived belief that federal policies on control of toxic substances have been a failure. And in the example of the TSCAct, the perception of inadequacy is reality.

6.7 WHO'S POLICES ON TOXIC SUBSTANCES

The WHO is active in several areas relevant to preventing the public health impacts of hazardous substances. The organization is a partner with the UN Environment Programme (UNEP) in implementing their Health and

> The WHO and the UNEP noted in 2006 that environmental hazards are responsible for an estimated 25% of the total burden of disease globally.

Environment Linkages Initiative (HELI). The initiative is a global effort by WHO and UNEP to assist developing countries' policymakers on issues of environmental threats to health. The two UN organizations noted in 2006 that environmental hazards were responsible for an estimated 25% of the total burden of disease globally, and nearly 35% in regions such as sub-Saharan Africa [119].

The HELI encourages countries to address health and environment linkages as integral to economic development. The two organizations assert that the HELI supports the valuation of ecosystem "services" to human health and well-being—services ranging from climate regulation to provision/replenishment of air, water, food, and energy sources, and generally healthy living and working environments. HELI activities include country-level pilot projects and refinement of assessment tools to support decision-making [120].

6.7.1 INTERNATIONAL AGENCY FOR RESEARCH ON CANCER

The International Agency for Research on Cancer (IARC) is a component organization of WHO. It was created in May 1, 1965, and is based in Lyon, France. IARC's mission "[i]s to coordinate and conduct research on the causes of carcinogenesis, and

TABLE 6.4
IARC's Programs of Work [122]

Program	Illustrative Example
Monitoring global cancer occurrence	Studying cancer incidence, mortality, and survival in many countries
Identifying the causes of cancer	More than 870 agents and exposures have been examined for evidence of carcinogenicity
Elucidation of mechanisms of carcinogenesis	Laboratory research examines the interaction between carcinogens and DNA
Developing scientific strategies for cancer control	Programs are directed to finding ways to prevent human cancer

to develop scientific strategies for cancer control" [121]. IARC is involved in both epidemiological and laboratory research and disseminates scientific information through publications, meetings, courses, and fellowships. IARC's program of work has four main objectives, as listed in Table 6.4. Of the four program areas listed in the table, identifying the causes of cancer has received the greatest public attention, primarily because of the issuance of cancer risk documents on individual chemical and physical agents.

Since 1970, IARC has published assessments of the carcinogenic risks to humans from a variety of agents, mixtures of agents, and exposure circumstances. These assessments, known as the IARC Monographs, are prepared by

> A significant feature of IARC Monographs is the classification of a chemical or physical agent's potential to cause cancer in humans.

international experts, assisted by IARC staff. Each monograph is prepared by an international working group that is specific to the agent under review. More than 870 agents (chemicals, groups of chemicals, complex mixtures, occupational exposures, cultural habits, biological or physical agents) have been evaluated [121]. Each monograph includes basic information about an agent's physical and chemical properties, methods of analysis, production volumes, toxicological data, and epidemiological findings. Sections of the monographs review the evidence for the agent's carcinogenicity. The monographs are available to an international audience of researchers, public health officials, and regulatory authorities. The monographs are particularly relevant to developing countries, where resources to develop similar documents might be lacking.

In the course of developing the IARC Monographs, working groups are asked to categorize each agent or exposure circumstance as to its carcinogenicity. Over time, IARC has developed guidelines for use in the categorization process. Although the guidelines provide considerable direction to a monograph's working group, scientists' professional judgment is still required. For example, different scientists might disagree over the quality and implications of the same toxicological study or epidemiological investigation. These disagreements are usually worked out in the course

TABLE 6.5
Comparison of IARC [124] and EPA [125] Carcinogen Groups

IARC	EPA
Group 1: Carcinogenic to humans	**Group A**: Carcinogenic to humans
Group 2A: Probably carcinogenic to humans	**Group B**: Likely to be carcinogenic to humans
Group 2B: Possibly carcinogenic to humans	
Group 3: Not classifiable as to its carcinogenicity to humans	**Group C**: Suggestive evidence of carcinogenic potential
Group 4: Probably not carcinogenic to humans	**Group D**: Inadequate information to assess carcinogenic potential
	Group E: Not likely to be carcinogenic to humans

of assigning a category (e.g., Group 2A) of carcinogenicity for a particular agent [123]. Shown in Table 6.5 are IARC's current categories of carcinogens.

Also shown in Table 6.5 is a comparison of IARC's grouping of carcinogens and those of EPA. There are obvious similarities and some minor differences in wording. Even though the two sets of carcinogen categories have very similar wording, occasionally IARC and EPA will come to different conclusions as to a compound's carcinogenicity. This is because IARC and EPA workgroups might differ when reviewing the same scientific data as to what is "sufficient" evidence. However, both sets of categories serve their purpose of providing guidance on weight-of-evidence assessment for the carcinogenicity of individual chemical compounds and mixtures.

6.7.2 INTERNATIONAL PROGRAMME ON CHEMICAL SAFETY

The IPCS resulted from the UN Conference on the Human Environment, held in Stockholm in 1972. From the conference came the recommendation that programs, to be guided by the WHO, should be undertaken for the early warning and prevention of harmful effects of chemicals to which human populations were being exposed [126]. The IPCS functions through the cooperation of WHO, UNEP, and the International Labour Organization. These three UN organizations coordinate the development of technical reports, share personnel and other resources, and work together on education programs that address the impacts of chemical hazards on human health. The IPCS also develops recommendations about hazardous toxicants and makes recommendations to mitigate their effects [127].

The two main roles of the IPCS are to establish the scientific health and environmental risk assessment basis for the safe use of chemicals and to strengthen national capabilities for chemical safety. The latter role is particularly important for developing countries, which often lack the technical and economic

The two main roles of the IPCS are to establish the scientific health and environmental risk assessment basis for the safe use of chemicals and to strengthen national capabilities for chemical safety.

resources to develop national programs in chemical safety. The WHO has the overall administrative responsibility for the work of the IPCS, working through a central office in Geneva, Switzerland. IPCS's work is divided into four main areas: risk assessment of specific chemicals, risk assessment of methodologies, risk assessments for food safety, and management of chemical exposures [126]. Much of the IPCS work is conducted in collaboration with regional and national organizations that address chemical safety issues. These organizations include the U.S. EPA, the U.S. National Institute of Environmental Health Sciences, the U.S. Agency for Toxic Substances and Disease Registry, the European Commission, the International Life Sciences Institute, the International Union of Pure and Applied Chemistry, the International Union of Toxicology, and others.

The IPCS develops and coordinates several products and services of considerable importance to global environmental health. In particular, several information resources—some of which overlap each other—on chemical substances are available to environmental and health officials, as well as the general public. These documents include the following [126,127]:

- Environmental Health Criteria (EHC) documents, which are reasonably comprehensive reports of a substance's toxicity, exposure routes, and human health effects. Recommended Exposure Levels (RELs) are usually contained in each document [128]. Approximately 250 chemicals have been subjects of EHC documents. The primary audience for these documents consists of national policymakers, environmental and health officials, and government and private sector risk assessors.
- International Chemical Safety Cards (ICSCs) are cards that summarize essential health and safety information on chemicals. They are intended for use by workers and employers in factories, agriculture, construction, and other workplaces. They provide their users with a quick, credible resource for use in preventing chemical emergencies and responding to them if they occur. ICSCs are similar to Material Safety Data Sheets developed by chemical producers and some national governments.
- Concise International Chemical Assessment Documents (CICADs) are summary documents that provide information on the relevant scientific information pertinent to the adverse effects of a specific substance on human health and the environment. As stated by the IPCS, "The primary objective of CICADs is characterization of hazard and dose-response from exposure to a chemical. CICADs are not a summary of all available data on a particular chemical; rather, they include only that information considered critical for characterization of the risk posed by the chemical" [128]. The primary audience appears to be practicing risk assessors, whether in government or industry.

Methodological publications are part of an effort to improve the methodology of chemical risk assessment, developed by expert panels convened by the IPCS [124]. The documents include such documents as Human Exposure Assessment, Biomarkers in Risk Assessment, Principles for Evaluating Health Risks to Reproduction Associated with Exposure to Chemicals, and Guidelines on Studies in Environmental Epidemiology. The documents are used by national governments, professional organizations, and individual risk assessors. The IPCS also conducts regional and local training sessions in risk assessment, using their methodological publications as teaching materials.

Chemical incidents and emergencies are global problems, irrespective of whether they occur in industrialized or developing countries. Such incidents include spills of oil from tankers, explosions in chemical factories, and mishaps

> INTOX is an IPCS database primarily directed to poison control centers and health-care providers who respond to chemical poisonings.

in overland transportation of chemical products and substances. The primary role of IPCS in such episodes is to interact with public health and medical authorities. More specifically, the IPCS provides guidance and training to member states in their planning on how to respond to chemical incidents and emergencies. The IPCS also serves as a source of technical information, advice, and assistance on the health implications of chemical incidents. In particular, WHO keeps a World Directory of Poisons Centers for access by first responders and health professions responding to chemical incidents and emergencies.

INCHEM is an IPCS database that offers access to "[t]housands of searchable full-text documents from international bodies on chemical risks and chemical risk management" [122,129]. The database can be accessed through the internet and is free of charge. Included in the INCHEM database are the IPCS's EHCs, CICADS, Health and Safety Guides, International Chemical Safety Cards, and documents from non-IPCS sources. This database would seem to have a broad-based audience, ranging from emergency responders to academic researchers.

INTOX [130] is an IPCS database primarily directed to poison control centers and health care providers who respond to chemical poisonings. Poison centers, in particular, need information on the toxicity of toxins and toxicants when caring for victims of exposure to both natural hazards (e.g., snake venom) as well as anthropogenic chemicals (e.g., industrial solvents). INTOX gives health professionals direct access to a database that will assist them in the diagnosis and treatment of poisonings, complemented by data management software. The INTOX system is a primary resource for health professionals in developing countries, where local databases on poisonings might not exist.

6.7.3 GLOBAL ALLIANCE TO ELIMINATE LEAD PAINT, 2006

The Global Alliance to Eliminate Lead Paint is a cooperative initiative begun in 2006 and jointly led by the WHO and the UNEP [131]. Its focus is to catalyze the efforts to achieve international goals to prevent children's exposure to lead from paints containing lead and to minimize

> The UNEP reports that as of September 30, 2018, 71 countries have legally binding controls to limit the production, import, and sale of lead paints.

occupational exposures to lead paint. Its broad objective is to promote a phase-out of the manufacture and sale of paints containing lead and eventually to eliminate the risks that such paints pose. Lead is one of ten chemicals of major public health concern to the IPCS.

The overall aim of the Global Alliance to Eliminate Lead Paint is stated as, "to prevent children's exposure to paint containing lead and to minimize occupational exposure to lead paint. The goal is to phase out the manufacture and sale of lead paint and to eliminate the lead poisoning risks" [131]. The UNEP reports that as of September 30, 2018, 71 countries have legally binding controls to limit the production, import, and sale of lead paints, which is 36.8% of all countries [132].

6.7.4 WORLD HEALTH ASSEMBLY'S RESOLUTION ON CHEMICALS MANAGEMENT, 2016

The World Health Assembly is the decision-making body of WHO. It is attended by delegations from all 192 WHO Member States and focuses on a specific health agenda prepared by its Executive Board. The main functions of the World Health Assembly are to determine the policies of the Organization, appoint the WHO Director-General, supervise financial policies, and review and approve the proposed program budget. The Health Assembly is held annually in Geneva, Switzerland. At the 69th World Health Assembly, May 23–28, 2016, Member Nations urged WHO:

1. "To engage proactively, including by strengthening the role of the health sector, in actions to soundly manage chemicals and waste at the national, regional and international levels in order to minimize the risk of adverse health impacts of chemicals throughout their life cycle;
2. To develop and strengthen, as appropriate, multisectoral cooperation at the national, regional, and international levels in order to minimize and prevent significant adverse impacts of chemicals and waste on health, including within the health sector itself;
3. To take account of the Strategic Approach's overall orientation and guidance towards the 2020 goal, including the health sector priorities, as well as the strategy for strengthening engagement of the health sector, and consider Emerging Policy Issues and Other Issues of Concern, and to take immediate action where possible and where appropriate to accelerate progress towards the 2020 goal;
4. To encourage all relevant stakeholders of the health sector to participate in the Strategic Approach and to ensure appropriate linkages with their national and regional Strategic Approach focal points, and to participate in the reports on progress for the Strategic Approach;
5. To strengthen individual, institutional and networking capacities at the national and regional levels to ensure successful implementation of the Strategic Approach;
6. To encourage health sector participation in the intersessional process established through the fourth session of the International Conference on Chemicals Management to prepare recommendations regarding the Strategic Approach and the sound management of chemicals and waste beyond 2020, including in the third meeting of the Open-Ended Working Group;

7. To continue and, where feasible, increase support, including financial or in-kind scientific and logistic support to WHO's Secretariat's regional and global efforts on chemicals safety and waste management, as appropriate;

8. To pursue additional initiatives aimed at mobilizing national and, as appropriate, international resources, including for the health sector, for the sound management of chemicals and waste;

9. To strengthen international cooperation to address health impacts of chemicals and waste, including through facilitating transfer of expertise, technologies and scientific data to implement the Strategic Approach, as well as exchanging good practices" [133].

In pursuit of these urgings from the World Health Assembly, the Director-General of The WHO was directed: (1) To develop, in consultation with Member States, and other relevant stakeholders, a road map for the health sector at the national, regional, and international levels toward achieving the 2020 goal and contributing to relevant targets of the 2030 Agenda for Sustainable Development, taking into account the overall orientation and guidance of the Strategic Approach to International Chemicals Management, and the intersessional process to prepare recommendations regarding the Strategic Approach and the sound management of chemicals and waste beyond 2020 established through the fourth session of the International Conference on Chemicals Management and building on WHO's existing relevant work. [...] [134].

Perspective: The WHO, as the primary health organization within the structure of the UN, performs the invaluable task of protecting and promoting the planet's human health? The organization has increasingly become active in issues of environmental health, several of which were described in this section. With the urging of the World Health Assembly, WHO will provide additional leadership in environmental health, assuming that resources are commensurate with the organization's responsibilities.

6.8   HAZARD INTERVENTIONS

What is called in this book as the Chemical Age has brought many benefits to humankind, along with substantial problems. One can assert that chemical pesticides have improved food crop production and quality of food products by reducing the presence and impact of food pests. Further, insecticides and fungicides have been useful for combating insects and fungi that can cause serious adverse human health effects, for example, mosquitoes that transmit the Zika virus. One can also assert that chemicals are beneficial for producing components used in commercial products such as vehicles, clothing, appliances, furniture, and such. But the Chemical Age has also brought deleterious health consequences to human and ecosystem health, as described in this chapter. Hazard interventions are required as means to render these consequences to acceptable terms.

1. Hazardous chemical substances should continue to be subject to societal controls, given their potential to harm health.

2. Supporting policymakers who advocate for safe production, distribution, and consumer use of chemicals is necessary if human and ecosystem health is to be protected.

3. Education about the hazards of environmental hazardous substances should be a component of elementary schools.

4. Labels on containers of hazardous substances should be informative, up-to-date, and monitored for accuracy.

5. Alternatives to use of chemical pesticides should be considered when and wherever practicable. As an interesting example, Michigan farmers who add structures or manage their land to attract birds, bats, and other vertebrates have boosted profits, reduced pesticide use, and conserved vulnerable wildlife [135].

6. Organic products should be considered for food and other domestic supplies in order to reduce the intake of pesticides left as residue on non-organic food products. In this regard, a study by the University of California researchers found an organic diet significantly reduced neonicotinoid, OP, pyrethroid, and 2,4-D pesticide exposures in U.S. families [136].

7. Children's exposure to hazardous chemical substances should be monitored by parents and mitigated where possible.

8. Public health departments should incorporate environmental health expertise and programs as essential ingredients of their public service.

9. Individuals can reduce their exposure to hazardous substances by appropriate disposal of containers and other products that contained hazardous substances.

10. Governments and private sector entities should seek ways to replenish elements of ecosystems harmed by hazardous environmental toxicants. An example is the state of Minnesota's program to help bolster the population of rusty patched bumblebees, a one prevalent species in decline due to habitat loss and pesticides applied to crops, gardens, and lawns [137]. Under the plan, the state will pay homeowners populate their lawns with bee-friendly plants.

6.9   SUMMARY

The five major U.S. policies on the control of hazardous chemical substances in the general environment are presented in this chapter. While other chapters have discussed chemical pollutants in air, water, food, and waste, this chapter deals with policies that are specific to hazardous substances found in general commerce. The five U.S. policies specific to control of toxic substances were discussed, along with those of the EU, and the WHO. Impacts of hazardous substances on human and ecosystem health were presented herein. Each of the federal statutes on the control of pesticides and toxic substances described in this chapter has the public health objective of preventing or reducing human contact with chemical substances that could exert toxic effects. Each statute has some interesting policies of relevance to public health. Of the discussed statutes, the FIFRAct is the oldest, dating to 1910 when federal pesticides legislation was first enacted by Congress. The core purpose of the FIFRAct is to control the release into the environment of pesticides and other chemical substances expressly designed to kill specific life forms.

There are several FIFRAct policies of importance to public health. The FIFRAct requires that pesticides must be registered with EPA and used only under prescribed conditions of application. This can be considered as the permit policy, without calling it such in the FIFRAct. It is also a kind of command and control policy, in that manufacturers are commanded to register their products with EPA, which has authority to control how the products are used. Two other policies of relevance to public health practice include: (1) public disclosure of pesticides information, unless it is classified as a trade secret; and (2) holding pesticides imported into the U.S. to the same requirements as domestically produced pesticides. The former policy is a statement of the public's right-to-know; the latter policy closes a potential gap in the distribution and application of pesticides in the U.S. It is an extra measure of prevention that is consistent with hazard elimination practices by public health officials.

The TSCAct was intended by Congress to regulate chemical substances. In particular, substances that have toxic properties are to be banned, or given restricted use, from commerce in the U.S. In a sense, this is a kind of quarantine for toxic substances. The act also adopted the policy of requiring chemical producers, importers, and processors to give premanufacture notification (PMN) to EPA. This information is to be used by EPA for evaluating chemicals' potential adverse effects on human health and the environment. This policy places responsibility on the chemical industry to test their products and furnish the information the EPA as a component of the PMN. This is an example of accountability as public policy in action. Unfortunately, as noted in this chapter, the TSCAct was a failure in terms of regulating hazardous substances, since language in the act made it essentially impossible for EPA to act.

The FQPAct is primarily about updating and strengthening the regulation and control of pesticides. In particular, the act targets the need for extra protection of children potentially exposed to pesticides. The act directs EPA to apply an additional safety factor of ten in risk assessments where children may be at risk of exposure. This policy, extra protection for children, is consistent with the public health practice of special attention given to vulnerable populations.

The Federal Hazardous Substances Act requires companies that produce commercial hazardous substance products to label the products in ways to facilitate consumers' protection. As such, the embedded policy is that of the public's right-to-know. An auxiliary policy is that of manufacturer's responsibility to inform government (i.e., CPSC) of their products' properties.

Also discussed in this chapter were policies on hazardous substances developed and implemented by the EU, the WHO, and three countries. Both international organizations' policies on controlling the release into environmental media and preventing adverse effects on human and ecosystem health are noteworthy. EU directives and regulations that pertain to hazardous substances require chemical manufacturers to provide toxicological data and environmental impact information to the EU for review and registration. WHO programs are primarily informational, with the needs of developing countries paramount in the organization's actions. WHO's IARC work on identifying carcinogens has a global impact on public health, given that the agency's pronouncements are used globally to shape programs of chemical interdictions.

A description of toxic substances policies in China, India, and Brazil identified some common goals for preventing the adverse effects of exposure to toxic substances.

REFERENCES

1. Grandjean, P. and M. Bellanger. 2017. Calculation of the disease burden associated with environmental chemical exposures: Application of toxicological information in health economic estimation. *Environ. Health* 16:123.
2. Kurek, J., P. W. MacKeigan, S. Veinot, et al. 2019. Ecological legacy of DDT archived in lake sediments from Eastern Canada. *Environ. Sci. Technol.* 53(13):7316–25.
3. Brown, T. P. 2006. Pesticides and Parkinson's disease: Is there a link? *Environ. Health Perspect.* 114(2):156–64.
4. Gubellini, P. and P. Kachidian. 2015. Animal models of Parkinson's disease: An updated overview. *Rev. Neurol. (Paris)*. 171(11):750–61. doi: 10.1016/j.neurol.2015.07.011.
5. NCAP (Northwest Coalition for Alternatives to Pesticides). 1999. Are pesticides hazardous to our health? *J. Pesticide Reform* 19(2):4–5.
6. Lee, J. L., A. Blair, J. A. Hoppin, et al. 2004. Cancer incidence among pesticide applicators exposed to chlorpyrifos in the agricultural health study. *J. Nat. Cancer Inst.* 96:1781–9.
7. Whyatt, R. M., V. Rauh, D. B. Barr, et al. 2004. Prenatal insecticide exposures and birth weight and length among an urban minority cohort. *Environ. Health Perspect.* 112(10):1125–32.
8. Bao, W., B. Liu, D. W. Simonsen, et al. 2019. Association between exposure to pyrethroid insecticides and risk of all-cause and cause-specific mortality in the general us adult population. *JAMA Intern. Med.* 180(3):367–374.
9. Kasler, G. 2018. Court bans popular farm pesticide defended by Trump: What it means for farms, workers, kids. *The Sacramento Bee*, August 9.
10. Samayoa, M. 2020. Oregon moves to phase out most uses of a controversial pesticide by 2023. *Oregon Public Broadcasting*, December 15.
11. Chow, L. 2016. Monsanto's glyphosate most heavily used weed killer in history. *Eco Watch*, February 2.
12. Henderson, A. M., J. A. Gervais, B. Luukinen, et al. 2019. Glyphosate general fact sheet. Corvallis: National Pesticide Information Center, Oregon State University Extension Services.
13. IARC. 2015. IARC monographs volume 112: Evaluation of five organophosphate insecticides and herbicides. Lyon: International Agency for Research on Cancer.
14. Baum Hedlund Aristei Goldman. Consumer Attorneys. 2019. Where is glyphosate banned? Los Angeles, CA.
15. Green, M. 2019. Roundup ingredient found in cereals marketed toward kids: Study. *The Hill*, June 12.
16. Polansek, T. 2020. U.S. EPA reaffirms that glyphosate does not cause cancer. *Reuters*, January 30.
17. Cohen, P. 2020. Roundup maker to pay $10 billion to settle cancer suits. *The New York Times*, June 24.
18. Erickson, B. E. 2020. US EPA reapproves atrazine. *C&EN*, September 21.
19. Lerro, C. C., J. N. Hofmann, G. Andreotti, et al. 2020. Dicamba use and cancer incidence in the agricultural health study: An updated analysis. *Int. J. Epidem.* 49(4):1326–37.
20. Frazin, R. 2020. EPA reapproves use of pesticide previously struck down in court. *The Hill*, October 21.
21. USGS (U.S. Geological Survey). 2006. Pesticides in the nation's streams and ground water, 1992–2001. Washington, D.C.: National Water-Quality Assessment Program.

22. Warsi, F. 2004. Pesticides. http://farhanwarsi.tripod.com/id9.html.

23. Donley, N. 2020. EPA: Widely used pesticide atrazine likely harms more than 1,000 endangered species. Press Release: Center for Biological Diversity, November 5.

24. IPBES (Intergovernmental Science-Policy Platform on Biodiversity and Ecosystem Services). 2016. Assessment details, options for safeguarding pollinators. Bonn: IPBES Secretariat, February 26.

25. Budge, G. E., D. Garthwaite, A. Crowe, et al. 2015. Evidence for pollinator cost and farming benefits of neonicotinoid seed coatings on oilseed rape. *Nat. Sci. Rep.* 5. doi: 10.1038/srep12574.

26. Williams, G. R., A. Troxler, G. Retschnig, et al. 2015. Neonicotinoid insecticides severely affect honey bee queens. *Sci. Rep.* 5:14621. doi: 10.1038/srep14621.

27. Van Trigt, E. 2018. Bees at risk: Near-total ban of neonicotinoids backed by ECJ. *Peace Palace Library*, May 31.

28. Brunetti, M. 2018. Bills to protect bees from pesticides become law. PressofAtlanticCity. com. January 16.

29. Lowrey, N. 2019. SDSU study shows world's most common pesticide a danger to deer. *South Dakota News Watch*, October 16.

30. Allington, A. 2019. EPA curbs use of 12 bee-harming pesticides. *Bloomberg Environment News*, May 21.

31. Beitsch, R. 2019. EPA will allow use of pesticide harmful to bees. *The Hill*, June 17.

32. Smith, M. 2015. Four decades of herbicide use is creating zombie weeds that just won't die. *Vice News*, September 24.

33. Main, D. 2016. Glyphosate now the most-used agricultural chemical ever. *Newsweek*, February 2.

34. Hettinger, J. 2020. For dicamba lawsuits, Bader verdict is just the beginning. *Investigate Midwest*, February 20.

35. Muir, P. 2012. A history of pesticide use. Corvallis: Oregon State University, Department of Botany and Plant Pathology.

36. Schierow, L. 1999. Federal Insecticide, Fungicide, and Rodenticide Act. Summaries of environmental laws administered by the EPA. Congressional Research Service, http://www.NLE/CRSreports/BriefingBooks/Laws/l.cfm.

37. Pyle, L. 2015. BPA in mothers' urine linked to low birth weights in China. *Environmental Health News*, September 24.

38. Smith, T. M., C. Austin, D. R. Green, et al. 2018. Wintertime stress, nursing, and lead exposure in Neanderthal children. *Sci. Adv.* 4(10):eaau9483. doi: 10.1126/sciadv. aau9483.

39. WHO (World Health Organization). 2016. Lead poisoning and health. Geneva: Office of Director-General, Media Centre.

40. Hornung, R. W., S. Rauch, P. Auinger, and R. W. Allen. 2018. Low-level lead exposure and mortality in US adults: A population-based cohort study. *Lancet Public Health* 3(4):PE177–E184.

41. Creswell, J. 2006. The nuisance that may cost billions. *The New York Times*, 2 April.

42. Benesh, M. and A. Lothspeich. 2019. Mapping chemical contamination at 206 military sites. Washington, D.C.: Environmental Working Group, Headquarters, March 6.

43. Bjorhus, J. 2020. PFAS-laced water caused infertility, premature births and low birth-weight, study says. *Star Tribune*, September 11.

44. ATSDR. 2018. Per- and polyfluoroalkyl substances (PFAS) and your health. Atlanta: Agency for Toxic Substances and Disease Registry.

45. Faber, S. 2019. EPA on 'forever chemicals': Let them drink polluted water. *The Hill*, October 2.

46. Lerner, S. 2019. High levels of toxic PFAS chemicals pollute breast milk around the world. *The Intercept*, April 30.

47. Matheny, K. 2019. PFAS contamination is Michigan's biggest environmental crisis in 40 years. *Detroit Free Press*, April 26.

48. Beitsch, R. 2020. Military sees surge in sites with 'forever chemical' contamination. *The Hill*, March 16.

49. Reuters. 2020. US drinking water contamination with 'forever chemicals' far worse than scientists thought. *The Guardian*, January 22.

50. Kummer, F. 2018. N.J. is first state to regulate toxic PFNAs in drinking water. *The Philadelphia Inquirer*, September 4.

51. Beitsch, R. 2020. EPA will regulate 'forever chemicals' in drinking water. *The Hill*, February 20.

52. Bergquist, L. 2019. Johnson Controls International takes $140 million charge for cleanup of 'forever' chemicals. *Milwaukee Journal Sentinel*, August 5.

53. Hunt, K. 2019. Denmark just became the first country to ban PFAS 'forever chemicals' from food packaging. *CNN News*, September 4.

54. NRC (National Research Council). 1993. Pesticides in the diets of infants and children. Washington, D.C.: National Academy Press.

55. Neslen, A. 2018. Ban entire pesticide class to protect children's health, experts say. *The Guardian*, October 25.

56. Barraud, A. and A. Kelsey-Sugg. 2019. Landmark class action over PFAS contamination in Australia announced by Erin Brockovich. *Law Report on ABC RN*, October 29.

57. Friedman, L. 2019. E.P.A. won't ban chlorpyrifos, pesticide tied to children's health. *The New York Times*, July 18.

58. Staff. 2020. Manufacturer to stop making pesticide linked to brain damage. *AP News*, February 6.

59. Roberts, E. M. and P. B. English. 2016. Analysis of multiple-variable missing-not-at-random survey data for child lead surveillance using NHANES. *Stat. Med.* 35(29):5417–29. doi: 10.1002/sim.7067.

60. GAO (Government Accountability Office). 2018. Lead testing of school drinking water would benefit from improved federal guidance. GAO-18-382. Washington, D.C.

61. Tadayon, T. 2018. It could cost Oakland schools $38 million to fix lead contamination. *East Bay Times*, February 8.

62. EPA (Environmental Protection Agency). 1995. Report on the national survey of lead-based paint in housing. https://www.epa.gov/sites/production/files/documents/r95-003.pdf.

63. Egelko, B. 2017. California judge says companies must remove pre-1951 lead paint in homes. *San Francisco Chronicle*, November 15.

64. Egelko, B. 2018. Supreme court: Companies on hook for lead paint removal. *San Francisco Chronicle*, October 15.

65. Schneyer, J. 2019. California settles decades-long lawsuit over lead paint, but outcome is bittersweet. *Reuters*, July 17.

66. Yeter, D., E. C. Banks, and M. Aschmer. 2020. Disparity in risk factor severity for early childhood blood lead among predominantly African-American black children: The 1999 to 2010 US NHANES. *Int. J. Environn. Res. Pub. Health* 17(5):1552.

67. Kurtzman, L. 2015. International ob-gyn group urges greater efforts to prevent toxic chemical exposure. University of California, San Francisco. *News Release*, 30 September.

68. Peebles, L. 2015. Major medical groups increasingly warning of toxic chemical risks to unborn babies. *The Huffington Post*, October 21.

69. Gore, A. C., V. A. Chappelll, S. E. Fenton, et al. 2015. Executive summary to EDC-2: The Endocrine Society's second scientific statement on endocrine-disrupting chemicals. *Endocr. Rev.* 36(6):593–602.

70. ATSDR (Agency for Toxic Substances and Disease Registry). 2004. ATSDR Tox profiles. Atlanta: U.S. Department of Health and Human Services, Public Health Service, Division of Toxicology.

71. Hoffman, M. 2018. Early blood lead levels linked to risk of ADHD, especially in boys. *MD Magazine*, June 5.

72. Bienkowski, B. 2019. More bad phthalate news: Early life exposure linked to decreased motor skills. *Environmental Health News*, February 11.

73. Cha, A. E. 2015. Startling link between pregnant mother's exposure to DDT and daughter's risk of breast cancer. *The Washington Post*, June 17.

74. Syson, N. 2018. Gender-bending plastics chemicals 'linked to cancer and infertility' found in four out of every five teenagers. *The Sun*, February 5.

75. Storrs, C. 2015. Report: Pesticide exposure linked to childhood cancer and lower IQ. *CNN*, September 14.

76. Hawthorne, M. 2015. Studies link childhood lead exposure, violent crime. *Chicago Tribune*, June 6.

77. Tecson, K. 2015. Pregnant women's exposure to lead may have lasting effects, can be passed on to grandkids. *International Business Times*, October 6.

78. Bellinger, D. C. 2012. A strategy for comparing the contributions of environmental chemical and other risk factors to children's neurodevelopment. *Environ. Health Perspect.* 120(4):A165.

79. Nutt, A. E. 2015. Phthalates, found in hundreds of household products, may disrupt sex development of male fetus. *The Washington Post*, March 6.

80. Bienkowski, B. 2015. Vinyl flooring chemical linked to high blood pressure during pregnancy. *Environmental Health News*, October 1.

81. EPA (Environmental Protection Agency). 2017. Risk assessment for carcinogenic effects. https://www.epa.gov/fera/risk-assessment-carcinogenic-effects.

82. Werner, E.F., J. M. Braun, K. Yolton, et al. 2015. The association between maternal urinary phthalate concentrations and blood pressure in pregnancy: The HOME study. *Environ. Health* 14:75–80.

83. Johnson, B. L. 2019. Legacies of hope; Theo Colborn. Amazon Kindle Publishing.

84. TEDX (The Endocrine Disruptor Exchange). 2016. Overview. http.www.tedx.org.

85. WHO (World Health Organization). 2012. State of the science of endocrine disrupting chemicals -2012. Geneva: Inter-organization Programme for the Sound Management of Chemicals.

86. EWG (Environmental Working Group). 2013. Dirty dozen endocrine disruptors. Washington, D.C.: Office of Director.

87. Kristof, N. 2021. What are sperm telling us? *The New York Times*, February 20.

88. Grossman, E. 2015. Chemical exposure linked to billions in health care costs. *National Geographic*, March 5.

89. Bienkowski, B. 2015. Smallmouth and largemouth black bass in wildlife refuges across the US Northeast have female parts, bolstering evidence that estrogenic compounds in our water are messing with fish. *Environmental Health News*, December 17.

90. Bienkowski, B. 2015. Male fish in North Carolina rivers found to have female parts. *Environmental Health News*, August 19.

91. Bienkowski, B. 2015. Exposure to widespread diabetes drug feminizes male fish. *Environmental Health News*, April 28.

92. Bienkowski, B. 2015. Hormone-mimickers widespread in Great Lakes region wastewater, waterways and fish. *Environmental Health News*, March 23.

93. Carrell, S. 2016. Toxic chemicals found in beached whales in Fife. *The Guardian*, February 11.

94. Bland, A. 2014. The environmental disaster that is the gold industry. Smithsonian.com, February 14.

95. Smith, J. E. 2016. Fish toxins at lowest levels in decades. *The San Diego Union-Tribune*, January 30.

96. Murray, P. 2016. Polluters commit to pay $46m to restore habitat in northeastern Wisconsin. *Wisconsin Public Radio*, April 11.

97. CPSC (Consumer Products Safety Commission). 2012. Federal Hazardous Substances Act (FHSA) requirements. http://www.cpsc.gov/es/Business--Manufacturing/ Business-Education/Business-Guidance/FHSA-Requirements/.

98. Schierow, L. 1999. Toxic Substances Control Act. Summaries of environmental laws administered by the EPA. National Library for the Environment. http://www.cnie.org/ nl3/leg-8/k.html.

99. Avril, T. 2003. U.S. chemical regulation testing leaves much unknown. *Philadelphia Inquirer*, November 4.

100. EPA (Environmental Protection Agency). 2019. Restrictions on discontinued uses of asbestos; significant new use rule. Regulations.gov. https://www.regulations.gov/ document?D=EPA-HQ-OPPT-2018-0159-5897.

101. Beitsch, R. 2019. Critics say new EPA rule could reintroduce asbestos use. *The Hill*, April 17.

102. Goldman, L. R. 2002. Preventing pollution? U.S. toxic chemicals and pesticides policies and sustainable development. ELR News & Analysis, Environmental Law Institute, 32 ELR 11018. September.

103. GAO (Government Accountability Office). 2005. Chemical regulation: Options exist to improve EPA's ability to assess health risks and manage its chemical review program. Report GAO-05-458. Washington, D.C.: Office of Comptroller General.

104. EPA (Environmental Protection Agency). 2016. Lowe's home centers, LLC, settlement. Washington, D.C.: Office of Enforcement and Compliance.

105. IREM. 2019. Enforcement of the Residential Lead-Based Paint Hazard Reduction Act. https://www.irem.org/file%20library/iremprivate/publicpolicy/enforcement-paint.pdf.

106. Michaeler, R. 2016. Chemistry world: Deal reached on US chemical regulation reform. http://www.rsc.org/chemistryworld/2016/05/us-chemical-regulation-reform-toxic-substances-control-act.

107. Sneed, A. 2016. EPA begins evaluating 10 common chemicals for toxicity. *Scientific American*, December 7.

108. EPA (Environmental Protection Agency). 2003. Summary of FQPA amendments to FIFRA and FFDCA. http://www.epa.gov/oppfead1/fqpa/fqpa-iss.htm.

109. OSHA (U.S. Occupational Safety and Health Administration). 2016. Chemical hazards and toxic substances. Washington, D.C.: Directorate of Standards and Guidance.

110. Grossman, E. 2015. U.S. to increase worker protection from deadly silica dust for first time in more than 40 years. *In These Times*, December 15.

111. Trotto, S. 2016. OSHA's new silica rule generates praise, criticism. *Safety + Health Magazine*, May 22.

112. CSB (U.S. Chemical Safety Board). 2016. About the CSB. http://www.csb.gov/about-the-csb/.

113. Natter, A. 2019. Trump to again propose eliminating chemical safety board, official says. *Bloomberg News*, March 8.

114. NTP (National Toxicology Program). 2017. History of the report on carcinogens. Research Triangle Park: Office of Director, National Institute of Environmental Health Sciences.

115. Sheppard, K. 2012. Republicans attempt to ax program monitoring carcinogens. *The New York Times*, August 24.

116. OEHHA (California Office of Health Hazard Assessment). 2013. Proposition 65 in plain language. http://oehha.ca.gov/proposition-65/proposition-65-list.

117. City of San Francisco. 2017. Ordinance No. 211-17. San Francisco: Department of Public Health.
118. Billings, R. 2018. Portland's tough new ban on synthetic pesticides allows few exceptions. *Press Herald*, January 4.
119. Prüss-Üstün, A. and C. Corvalán. 2006. Preventing disease through healthy environments. Towards an estimate of the environmental burden of disease. Geneva: Office of Director-General, Media Centre.
120. WHO (World Health Organization). 2016. Health and environment linkages initiative. Geneva: Office of Director-General, Media Centre.
121. IARC (International Agency for Research on Cancer). 2002. Evaluation. http://www.cie.iarc.fr/monoeval/eval/html.
122. IPCS (International Programme on Chemical Safety). 2002. INCHEM. http://www.inchem.org.
123. IARC (International Agency for Research on Cancer). 2016. Agents classified by the IARC monographs, volumes 1–116. Lyon: Office of Director, Communications Group.
124. IPCS (International Programme on Chemical Safety). 2002. Methodological publications. http://www.who.int/pcs/pubs.
125. EPA (Environmental Protection Agency). 2016. Risk assessment for carcinogens. Office of Research and Development. http://www.epa.gov/fera/risk assessment- carcinogens.
126. IPCS (International Programme on Chemical Safety). 2002. About IPCS. http://www.who.int/pcs/html.
127. IPCA (International Program on Chemical Safety). 2010. Action is needed on chemicals of major public health concern. Geneva: World Health Organization.
128. WHO (World Health Organization). 1994. Assessing human health risks of chemicals: Derivation of guidance values for health-based exposure limits. Environmental Health Criteria No. 170. Geneva: International Programme on Chemical Safety.
129. ILO (Internaitonal Labour Organization. 2021. International chemistry safety cards. https://www.ilo.org/safework/info/publications/WCMS_113134/lang--en/index.htm.
130. IPCS (International Programme on Chemical Safety). 2002. INTOX. http://www.intox.org.
131. IPCS (International Programme on Chemical Safety). 2002. Global alliance to eliminate lead paint. https://www.unenvironment.org/explore-topics/chemicals-waste/what-we-do/emerging-issues/global-alliance-eliminate-lead-paint.
132. UNEP (United Nations Environment Programme). 2018. Update on the global status of legal limits on lead in paint. Nairobi: Office of Director General.
133. WHO (World Health Organization). 2016. World Health Assembly. Geneva: Office of Director-General, Media Centre.
134. WHA (Sixty-Ninth World Health Assembly). 2016. Agenda item 13.6. The role of the health sector in the strategic approach to international chemicals management towards the 2020 goal and beyond, May 28. Geneva: Office of Director-General, Media Centre.
135. McGlashen, A. 2018. Protecting crops with predators instead of poisons. *Ensia*, May 31.
136. Fagan, J., L. Bohlen, S. Patton, and K. Klein. 2020. Organic diet intervention significantly reduces urinary pesticide levels in U.S. children and adults. *Environ. Res.* 171:568–75.
137. Katz, B. 2019. Minnesota will pay residents to grow bee-friendly lawns. Smithsonian.com, June 17.

Lessons Learned and Authors' Reflections

The six chapters in this volume are considered by the authors as addressing the principal environmental hazards affecting human and ecological health. This assertion is based on the public health data and bodies of science that have accumulated over the decades in which humans have become more acutely aware of environmental hazards. Covered in this volume were discussions of climate change, air quality, water quality and security, food safety and security, tobacco and other smoked products, and toxic substances in the environment. Each of these six topics was described in terms of adverse effects on human and ecological health and policies intended for mitigation of adverse health effects. Reflections on these six environmental hazards gave rise to the following observations.

- The six environmental hazards are global in their adverse effects. We humans and other creatures with which we share the planet will all experience adverse health effects, depending on the nature, quality, and extent of exposure.
- Climate change is the environmental hazard that has the greatest potential for adverse global health impacts.
- Mitigation policies such as government statutes and accompanying regulations, while generally well purposed to protect public health, are only as effective as their enforcement.
- In regard to environmental hazards, the health of children should be a principal component of mitigation strategies and actions.
- Two principal environmental hazards, water and food, portend significant issues of availability and security as impacted by climate change.
- Government agencies that have statutory and ethical responsibilities for the mitigation of environmental hazards should not become subjects of political interference lacking substantive sociopolitical debate.
- Individuals and other private sector entities have responsibilities for the prevention of adverse consequences of environmental hazards.
- Youth education about environmental hazards and their health effects should be included in school curricula.

Environmental Policy and Public Health

Principal Health Hazards and Mitigation, Volume 1 Workbook

Barry L. Johnson and Maureen Y. Lichtveld

Barry L. Johnson, MSc, PhD, FCR	**Maureen Y. Lichtveld, MD, MPH**
RADM (Ret.), U.S. Public Health Service	Dean, Graduate School of Public Health
Adjunct Professor	Professor, Environmental and
Rollins School of Public Health	Occupational Health
Emory University	Jonas Salk Chair in Population Health
Atlanta, Georgia, 30322	University of Pittsburgh
	Pittsburgh, Pennsylvania 15261

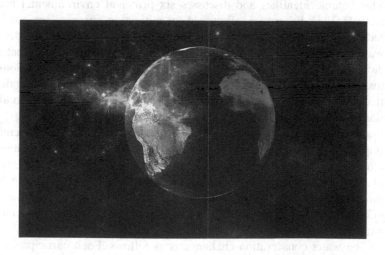

2022

RESEARCH PROJECTS

1. Described in this volume are various policies that are intended to mitigate the effects of principal environmental hazards such as air pollution and food contamination. U.S. federal government environmental statutes were discussed specifically. Research how federal laws are developed and promulgated. Be specific by identifying by name the various committees in Congress that will be involved in enacting an environmental statute.

2. The World Health Organization (WHO) was mentioned in this volume as the global leader on issues of epidemic and pandemic infections due to such environmental hazards as novel viruses and air pollution. Research the origins of the WHO and discuss and compare the sociopolitical aspects of the WHO, for instance, funding mechanism and governance. In your analysis of the work of the WHO, cite in some detail how the WHO coordinated the global eradication of smallpox.

3. The U.S. Centers for Disease Control and Prevention (CDC) was mentioned in this volume as the U.S. national on issues of epidemic and pandemic infections due to such environmental hazards as novel viruses and air pollution. Research the origins of the CDC and discuss and compare the sociopolitical aspects of the CDC, for instance, funding mechanism and governance. In your analysis of the work of the CDC, cite in some detail the agency's involvement, both positively and negatively, in responding to the COVID-19 pandemic.

4. This volume identifies and discusses six principal environmental health hazards. Each hazard was discussed for mitigation purposes various U.S. federal government policies such as statutes. At the core of such statutes was the authority given to regulatory agencies for their development and enforcement of environmental regulations such as control of emissions of toxic materials from electric power plants. Research and describe in detail the legal policies that must be followed in the development and promulgation of federal government environmental regulations.

5. Water Conservation Challenge: Water is essential for life. It is therefore important that this precious resource be used wisely. Its conservation is already important in arid parts of the world, including parts of the U.S. As global population continues to increase and industrialization and agriculture enlarge to meet the needs of a larger global population, along with the effects of climate changes, water conservation will move from important to vital.

   The water conservation challenge is as follows. Each participant must maintain a DAILY log of the estimated amount of water they saved during a 30-day period. The log must specify the date, the reasonable estimated amount of water conserved, the method of conservation, and the estimated amount of water used each day. Water logs can be compared based on policies established by participants in the water conservation challenge.

6. 1.) Discuss the practical importance of the National Environmental Policy Act (NEPA). 2) Review NEPA's policy statement that begins, "The Congress,

recognizing the profound impact...", and rewrite the statement in terms of sustainable development.

7. Several federal environmental statutes include the embedded policy of using permits, a kind of command and control policy. Select one such statute and discuss two alternatives to a permit policy. Using critical thinking, discuss the likely effectiveness of each alternative.

8. Discuss in 50 words or fewer the primary benefit of embedding federalism in federal environmental statutes.

9. Select one of these laws (CAA, CWA, SDWA, FDCA), as amended, and (1) describe how its purpose and provisions have affected you personally. (2) Then assume the law you selected does not exist. Discuss the ramifications of no law (i.e., federal, state, local), and discuss what actions you would take to protect your personal health.

10. Concerning the Food Quality Protection Act, (1) discuss the Act's impact on children's health. (2) Should the Act have repealed the Delaney Clause, formerly a component (and embedded policy) of the Food, Drug, and Cosmetic Act, as amended? Why?

11. Congratulations! You have successfully completed the assigned work. To celebrate, assume that you visit a nearby fast food restaurant and purchase a cheeseburger and a cup of coffee. The cheeseburger consists of six items (bun, beef patty, cheese, tomato, lettuce, and paper wrapping). The cup of coffee consists of four items (coffee, paper cup, plastic lid, artificial sweetener). You also receive four other items (paper napkins, plastic stirrer for the coffee, paper bag, and restaurant operations). Construct a table that contains four columns and 14 rows. Each *row* is labeled for an item that you received (e.g., bun, beef patty, etc.). The four *columns* are labeled according to which entity has involvement in food safety or food operations. More specifically, the columns are to be labeled as follows: Federal, State, Local, Food Service. Fill in the table according to which law might be operative or where the food service would be accountable. For example, the row labeled beef pattie could be filled in as follows:

Food	Federal	State	Local	Food Service
Beef Patty	FMIA, FDCA	State Dept Agri., State Div Pub. Health	County Dept. Health	Food safety operations

12. Discuss the significance to you of the two world summits on environment and development, including a discussion of what you think the U.S. government's role should be in these kinds of summits.

13. Do you accept the premise of the Ghia Hypothesis? (1) State your reasons for or against your acceptance of the hypothesis. (2) Assume that you accept the Ghia Hypothesis, how would that affect how you practice public health, both (1) personally and (2) professionally as a public health practitioner.

14. Of the public health pioneers introduced to you as part of each session's review questions, which one would have made the best dinner companion? Why?
15. (1) What is the most important policy you learned during the EHP part of EOH 570? Why? (2) Which of the public's policy expectations is the most important to you? Why?
16. How would you use the 3 Rs of waste management in managing your household waste? Be specific.
17. Discuss whether Congress should reinstate the Superfund Trust Fund tax on industry. Use critical thinking, citing the Pros and Cons of such an action.
18. Develop a PACM model for reducing noise on a college campus.
19. Should FDA have the authority to regulate tobacco products? Discuss the pros and cons.
20. Where do the funds come from that are used by EPA to remediate NPL sites?
21. How would you use the 3 Rs of waste management in managing your household waste? Be specific.
22. What is the IARC? Where is it headquartered? What controversy has surrounded the development of some of its monographs? What does its work impact global health?

VOLUME 1

CHAPTER 1: CLIMATE CHANGE

1. This chapter presents climate change as a global hazard to humankind. Do you agree? If not, why? Discuss in detail, including any limitations in your knowledge of the position you have taken in responding to this question.
2. The relationship between technology and climate change was discussed in this chapter. Discuss what current technology should be abandoned as a contribution to mitigating climate change. Be specific and describe in detail the pros and cons of your recommendation.
3. In your opinion are elected representatives of the people moving too slowly or outright ignoring the urgency of action for mitigating climate change? If so, what can you personally do about your concern?
4. Taxing a product or process can have both positive and negative effects. For example, taxing tobacco products has been shown to reduce tobacco use among young people. Should the U.S. place a tax on the release of carbon dioxide into the atmosphere? Discuss the political implications of a carbon tax?
5. When discussing climate change with your grandmother, she replies, "Nothing new about hot weather! Whew! I remember all the hot summers on the farm. This climate thing is just what I remember as a young girl working in the corn fields. Tell all your friends not to worry about that climate whatsathing." What do you say and do in reply?
6. Oops, we failed to mention that your grandmother is a Member of the U.S. Congress. What do you say and do in reply to her comment in Question 5?

7. The International Committee on Climate Change is an organization created by the United Nations (UN). Do you agree that the UN is the appropriate body to coordinate global actions to mitigate climate change? Do you have trust in the UN? Is there a more appropriate body to lead the efforts to mitigate climate change? If so, why? If not, why and what body would be better? Detail your reasons for your answers.

8. The Clean Power Plan is summarized in this chapter. As discussed the plan engendered both considerable political support and determined opposition. Discuss your personal assessment of the core goal of the Clean Power Plan and state why you support or oppose the plan.

9. If the dire predictions of food shortages become reality due to climate change, what is your plan to prevent human famine in those geographic areas of persistent or permanent drought? Be specific about the elements of your plan.

10. It is proffered in this chapter that reducing the impact of climate change will require global effort. In this sense, describe three actions that you can execute in a global context that will contribute to mitigation of climate change. Be specific and be detailed.

11. Assume that due to reduction in the use of coal as a source of energy, coal mining has ended in the U.S. Does the federal government have an obligation to subsidize coal companies and coal miners due to government's greenhouse policies that ended the use of coal? Present your answer with an argument for both pros and cons in regard to government subsidies.

12. List, detail, and discuss three actions that you can personally perform as your responsibility for mitigating climate change. Be specific and be real.

13. In the previous question, the word *responsibility* was used. Would the word *duty* have been a better choice? Explant the difference in meaning between the two words. Which word was best for the purpose of the previous question? Discuss your choice of word and relate your choice to your course of academic study.

14. Congratulations! Your research team has discovered and patented a novel method for manufacturing fuel cells in ways that produce no waste in manufacture and yields a sustainable fuel cell that will revolutionize transportation and home energy supply. Your research team has two offers to commercialize your invention: (1) sell the patent to a large energy corporation or (2) donate the patent to an international NGO for their global distribution of fuel cells at cost of manufacture. Which choice do you make and why? Discuss the ethics of your choice.

15. The year is 2050. Sadly, global efforts to mitigate climate change failed. The planet has fallen victim to the predictions of global disaster and uncompromising effects of disasters on many forms of life on the planet. Explain to your descendants why the dire predictions of what went wrong in efforts to mitigate climate change. Be as specific as possible. Your response to this question should be composed as an apology to your descendants.

16. On Earth Day, 2016, the presidents of China and the U.S. met and jointly signed their nations' responsibilities under the Paris Agreement. Was this a

significant event? And why? Was there a common denominator that brought together the world's two largest emitters of carbon into the environment? Discuss the sociopolitical implications of this event.

17. Much of the U.S. government's policymaking on climate change was ultimately shaped by judicial decisions, particularly those of the U.S. federal courts. As a newly appointed member of the U.S. Supreme Court, but a person who has a background both in law and environmental science, do you believe the latter credential might be a hindrance to your consideration of environmental cases brought before the court? Be specific in your reply.

18. Well, after several years of reliable, but polluting service, the family car needs to be replaced. You have been active in protecting your neighborhood's nature preserve and parks. Recently announced was the marketing of a fuel cell vehicle and also a solar-powered electric car. And still on the market are highly fuel-efficient hybrid vehicles, powered by internal combustion engines and batteries. Assuming fuel cell and solar vehicles are each 1.5 times the cost of a hybrid vehicle, which vehicle do you purchase? And why? List the pros and cons of your decision.

19. Decades have passed since the global community of the first half of the 21st century attempted to mitigate climate change. Some success was achieved, and global ambient air temperatures cooled, but insufficiently to fully mitigate the effects of climate change. Your planet's food supply has dramatically changed, in part due to climate change, and partly due to the increase in the global population. Speculate in some detail about what your daily food consumption might comprise. How is your diet different from that of your 20th-century ancestors?

CHAPTER 2: AIR QUALITY

1. Assume you are a senior member of the local health department. Review three journal articles on asthma in children. Using what you consider to be the key public health issues gleaned from the articles, prepare a summary of findings that can be presented to a local underserved community.

2. Under Section 202, Title II, of the CAAct, EPA must establish emission standards for new motor vehicles and engines. (1) Do you agree that EPA should have this authority? Why or why not? Be specific. (2) Using internet and EPA resources, determine the emissions standards for your personal vehicle.

3. Ethanol is used as a gasoline additive for the purpose of lowering vehicle emissions. Using internet resources, discuss the benefits and disadvantages of using ethanol as a gasoline additive.

4. Assume that you are in charge of communications for the local health department. Develop a one-page health alert that can be released to the public and medical community during periods of high ambient air levels of ozone.

5. Summarize for your 10-year-old nephew the primary effects of the criteria air pollutants on human health? List the three most important ways for him to prevent these health effects.

6. For pollution control purposes, several federal environmental statutes include the embedded policy of granting permits to those who pollute a kind of command and control policy. Select one such statute and discuss two alternatives to a permit policy. Using critical thinking, discuss the likely effectiveness of each alternative.

7. EPA's assertion that the CAAct applies to regulation of greenhouse gases was challenged in federal courts, eventually being a matter decided by the U.S. Supreme Court in Massachusetts v. EPA. Discuss in a two-page analysis the Supreme Court's decision and the key themes of the Court's justices.

8. Prepare an essay of appropriate depth that compares the benefits of industrial development versus the costs to public health due to increased air pollution emissions. Is there a way to balance costs vs. benefits? Be specific.

9. EPA data show that air quality has generally improved in the U.S. during recent years. Discuss why this desirable trend has occurred and in your analysis the one factor most responsible for the trend.

10. The state in which you reside requires annual vehicle emission inspections. Do you consider this a regulatory overreach? If so, why? If not, why? Be specific.

11. Your company has recently decided to expand its overseas markets. You have been offered a plum position as director of marketing for a country with a nascent economy and with impressive growth potential. However, the country's air pollution is near WHO's list of most polluted countries. (1) Assume you have no dependents, do you accept the job? If so, why? If not, why? (2) Assume you have a family with young children. Do you accept the job? If so, why? If not, why?

12. How clever you are! As an amateur inventor, you've invented and patented a heating element that uses solar power and can store the energy until needed. Your friend suggests that the element could replace primitive cooking stoves in developing countries, thereby eliminating indoor air pollution. A UN representative suggests that your patent be donated to the UN for their distribution to developing countries. What is your decision? Provide details.

13. Assume that you reside in a large city that has historically struggled to meet federal air quality standards for pollutants caused by dense automobile traffic. Residents of the city are organizing protests against being subjected to polluted air. As a newly elected member of the city council, you are responsible for preparing a list of potential policies on how to improve the city's air quality. In an essay of appropriate depth, outline your top three policy suggestions. Be specific on how each suggestion would improve air quality.

14. In regard to the previous question, which specific air pollutants would be of the greatest concern to you as a member of city council? Describe the basis for your selection and the degree of concern.

15. Your sister and her family are relocating to a new city for residence. She has two young children, ages 4 and 6 years old. She asks for your advice about how to choose between two possible residences in the new city. One residence abuts a major road, but offers good real estate value; the second possible residence is on a cul-de-sac but is 10% greater in purchase cost and

annual taxes. Knowing that your sister has limited income, what advice do you offer? And why?

16. Discuss the relationship between human population increase and any implications for global air pollution. Be specific and provide supporting data.

17. Fire events were mentioned as one source of outdoor air pollution. Using internet resources, are there currently any uncontrolled forest fires in your state? If so, what are the implications for air pollution problems? And who would be the population(s) most at health risk? If your state is currently without any forest fires, select a state that does.

18. Your parents reside in a rural area. The area is site to a large forest, two lakes, and a small river that abuts your parent's property. There are a variety of plants and feral animals. Across the state is a large electric power plant, whose air pollution emissions have been frequently investigated by the state's department of environmental protection. The county's health department has expressed concerns about air pollutants. What concerns would you have in regard to the ecological consequences of the plant's air pollution?

19. It is asserted in this chapter that air pollution is the planet's greatest environmental hazard. Do you agree with this assertion? If so, why? If not, what in your opinion is a greater hazard? Be specific in your answer in respect to "hazard."

CHAPTER 3: WATER QUALITY AND SECURITY

1. The CWAct requires states and tribes to establish ambient water quality standards for bodies of water. Select a lake within your state's borders and determine the applicable water quality standards. Discuss the public health implications of the lake's water quality standards.

2. In the context of public health, discuss the significant differences between a primary drinking water regulation and a secondary drinking water regulation, as found in the SDWAct.

3. If climate-change models are correct, changes in global rainfall patterns are likely, making water conservation a necessary environmental policy. Discuss ten ways that you personally can conserve water, now and in the future. List the ways in descending order of effectiveness in terms of water conservation.

4. For pollution control purposes, several federal environmental statutes include the embedded policy of granting permits to those who pollute, which is a kind of command and control policy. Select one such statute and discuss two alternatives to a permit policy. Using critical thinking, discuss the likely effectiveness of each alternative.

5. Do you purchase bottled water? If so, explain why. If not, explain why. If you purchase bottled water, what do you do with empty plastic bottles?

6. Using internet resources, locate your state's TDML report. What are the key features of the plan? In your opinion, does the plan adequately respond to the provisions of the CWAct?

7. Your community has many residences that have private wells for household use. As a senior member of the local health department, you have been asked to advise the county commission on whether mandatory water quality inspection should occur. Describe your response.

8. The Flint, Michigan, water crisis resulted in the revelation that many schools in the U.S. have drinking water contaminated with lead. Using the EPA SWDIS, conduct a survey of your community's schools as to the quality of school drinking water. Describe the findings in a two-page report.

9. Using internet resources, research the quality of surface and groundwater water supplies in your state. Prepare a two-page report of your findings.

10. What department or agency in your state has responsibility for water quality? Describe the agency's principal responsibilities for protecting water quality.

11. Waterborne diseases can be common in geographic areas that lack potable drinking water and/or where water pollution is inadequately controlled. List three waterborne diseases and describe the consequences to public health, the prevalence of each disease, and actions that can be taken to prevent each disease.

12. Rate the adequacy of your local public water supply, using a scale of 0–5, where 0 represents unacceptable quality and 5 represents water of impeccable quality. Using this scale, ask three persons who reside in your community about their perception of local water quality. Using the rating from all three persons, discuss the implications of your survey.

13. Identify the source of your community's drinking water. Identify any threats to the source as to pollution sources. Describe each threat and suggest methods to interdict them.

14. Research using internet resources, the geographic global areas that are experiencing drought. Select two areas and describe the effects of drought and policies being implementing in response to drought.

15. A primary feature of the CWAct, as amended, is to provide grants to states for the construction or upgrade of wastewater treatment plants. Research the amount of grant money that has been supplied by EPA to your state for wastewater treatment purposes. In your opinion, what should the federal government be financing what is a state's responsibility.

16. Algal contamination of water sources has become a significant problem globally. Prepare a two-page paper that details the extent of the problem in the U.S. and discuss recommendations as how to counteract the contamination.

17. Assume your community is suffering through a prolonged summer drought and that local water restrictions have been announced. Your neighbor continues to water his/her lawn, using sprinklers that thoroughly soak the lawn. Do you feel compelled to take action? If so, what would you do? If not, why not. Be specific and elaborative.

18. An international NGO has announced a humanitarian program to provide low-cost water pumps to an African population that lacks water security. As a charitable person, you are impressed with the purpose of the NGO's proposal. Describe what research you would conduct prior to providing financial assistance to the NGO.

19. Discuss the procedures and steps that EPA uses for ranking priority chemical pollutants under provisions of the Safe Drinking Water Act, as amended.

CHAPTER 4: FOOD SAFETY AND SECURITY

1. Let's consider the matter of food safety. Should food safety be a concern of local health departments through inspections of restaurants and other places of commercial food service? If so, why? If not, why?
2. Assume you were recently hired by an urban municipal health department. Your first assignment is to design a public health program to improve food safety in public establishments. (1) Discuss the nature and impact of foodborne illness that would be of concern to your health department. (2) Using this material, design a public health program to prevent foodborne illness, choosing any four elements of the eight elements shown in Figure 1.1. Use critical thinking, as described in Chapter 1, to the extent possible.
3. Summarize the public health benefits of the FMIAct. Discuss the ethical implications, if any, of the Act.
4. The FDCAct, as amended, gives the FDA the authority to approve drugs to be placed into commerce. Assume that the act did not exist, leaving the manufacturer solely responsible for the safety of their products. Discuss the public health implications of this kind of market-driven arrangement.
5. Visit a local restaurant and look for a posted food inspection report. Describe the impact, if any, on your patronage of the selected restaurant. (1) What aspects of the food inspection report were of greatest importance to your decision? In your opinion, should food inspection reports be available to the public? If so, how? (2) Discuss the pros and cons of making restaurant inspection scores available to the public. (3) Some county health departments post restaurant scores on the internet. Using such a website, select a restaurant known to you and access its restaurant score and other background information. Critique the adequacy of the restaurant inspection information made available to you.
6. As discussed in this chapter, states have a major responsibility for protecting the public against foodborne illnesses. Discuss your state's responsibilities for food safety. Be specific in regard to which state agencies have specific responsibilities.
7. As discussed in the chapter, foodborne illnesses will affect annually about one in six Americans. A substantial but unknown number of illnesses occur because of poor food preparation practices in the home. Discuss some practical means of preventing foodborne illnesses caused by home food preparation.
8. Discuss the pros and cons of giving meat industry inspectors the authority to supplant government meat inspectors.
9. Consider the elements of your most recent meal. Discuss the origins of each major food item. In your discussion, include the geographic location from which each item originated.
10. Discuss three ways that you personally can help reduce the global food shortage. Be specific and elaborate on how your help would benefit those persons facing food insecurity.

11. One food security source has predicted that global food shortages will inevitably lead to vegetarianism as a lifestyle. Assume that this forecast is accurate. Could you accept a vegetarian diet as your choice? Discuss the ramifications of your choice.

12. In 2016 the U.S. Food and Drug Administration promulgated regulations that curtailed the administration of antibiotics in food sources. In your opinion, was this necessary? Discuss the public health benefits of the FDA regulation. Also discuss the economic impact of the FDA decision.

13. Compare the food safety policies of China and India. List three elements in common. Discuss any element of significant difference based on your knowledge of principles of public health.

14. Your grandparents still reside on a farm that has been your ancestors' home for five generations. The farm is located in Vermont, which recently enacted a new law that mandates specific farming methods and practices. These requirements will be an economic burden on your grandparents. In an essay of appropriate depth discuss whether a state should enact this kind of mandate.

15. A food expert has suggested that alternate forms of non-animal protein will be required if global food insecurities are to be avoided. One suggestion is to incorporate insects into the human diet. Discuss how you would react to being served a beetle burger. Discuss any ethical issues that would be a component of your reaction.

16. Assume that you have volunteered to serve a charitable organization that raises funds and delivers food to people residing in areas of food insecurity. Prepare a one-page circular that could be used in your fundraising efforts.

17. Assume that you attended a school that provided students with free breakfast and lunch meals funded by a government program. Using internet resources, prepare a two-page analysis of the purpose, function, benefits, and decrements of a school meals program.

18. Discuss the matter of global food security in the context of sustainable development. Present your findings in an essay of appropriate depth. Begin your essay by defining sustainable development and its tenets.

19. This chapter includes nine potential hazard interventions. Present two additional interventions that should be added to this list. Justify your two contributed interventions with evidence of critical thinking.

CHAPTER 5: TOBACCO PRODUCTS, VAPING DEVICES, MARIJUANA SMOKING

1. Various forms of tobacco are still used by many young people. Using your own experience as a current, former, or non-tobacco user, discuss: (1) the reasons why tobacco use is acceptable to some youth, and (2) the best methods to use to prevent smoking or other tobacco use by youth.

2. Do you as an individual have a personal policy about tobacco use? If so, describe your policy. If not, why not?

3. Does your state or province have any laws or other kinds of policies in regard to control of tobacco use? Describe in summary the purposes of the laws or

policies that you identified? In your opinion, are these laws effective? If so, how? If not, why not?

4. Using internet resources, determine which methods, in your opinion, are the most effective way to quit tobacco use? For example, how effective are nicotine patches for smoking cessation?

5. One policy for reducing tobacco use, particularly for cigarette smoking, is to tax tobacco products. Conduct a local survey on the amount of tax placed on typical tobacco products, e.g., what is the tax on cigarettes, pipe tobacco, smokeless tobacco, and cigars? What is the disposition of the taxes collected by vendors of tobacco products? On an annual basis, how much tobacco tax money does your state collect?

6. In some countries, e.g., China, the government owns the tobacco manufacturing system. As a result, such countries reap all the revenue from domestic and export sales of tobacco products, thereby yielding income for the state. Discuss the ethics, in your opinion, of state-operated manufacturing and sales of tobacco products, knowing the public health impacts of tobacco use.

7. Water pipes have increased in popularity in some locales, attracting younger smokers in particular. Using the material in this chapter, together with material available on the internet, describe three or more hazards presented by smoking water pipes. For each hazard, provide a public health method to reduce or mitigate the hazard.

8. Using the material in this chapter, along with any additional materials, provide an analysis of the value or harm of e-cigarettes when used as a cigarette smoking cessation device. Be specific and provide supportive data for your analysis.

9. In 2020 the Trump administration proposed creating a Center for Tobacco Products, a new federal agency that would replace the FDA's authorities regarding tobacco products. In an analysis of appropriate length, discuss the pros and cons of creating the new agency.

10. In the U.S. there are federal laws prohibiting advertisements of cigarettes on television and similar media. However, there are no legal bans on showing cigarette-smoking characters in motion pictures and television productions. Using the PACM model for policymaking (Chapter 2), outline how you would lead a campaign targeted at the production of smoking-free movies and TV productions.

11. Some business enterprises in the U.S and elsewhere have forged a policy of not hiring cigarette smokers, with the justification that non-smokers are an economic resource to businesses. Describe your support or objection to this kind of policy. Include in your analysis the rights of individuals (i.e., personal freedoms) versus the rights of business entities.

12. Do you or someone you know regularly use some kind of vaping device? If so, state the purpose of vaping. Were you aware of the nicotine content of the inhaled vapor? Express in a two-page essay the reasons for or against vaping.

13. Express why in your opinion marijuana should be permitted as a recreational drug. Select a state where recreational marijuana is permitted and

evaluate any data on the social consequences of the state's decision. In your opinion, should a state control marijuana as an illegal recreation drug?

14. Should tobacco smoking be forbidden in homes in which young children reside? Discuss the pros and cons of action by child protective services (CPS). As a public health specialist, would you support ordnance or other form of health policy that would authorize CPS to remove children from the homes of parents or guardians who smoke?

15. Assume you are a parent of a child who is of age to be impressionable by peer pressure. Describe in detail what you as a parent would do and say to your child in order to discourage social pressures to smoke or otherwise use a tobacco product?

16. Federal laws require warning labels to be placed on cigarette packages and some other tobacco products. Focusing on cigarette packages, prepare three different warning labels, with images, that you would use for discouraging adolescents from smoking cigarettes.

17. Congratulations! Love has arrived. Your love interest has proposed marriage. To celebrate the forthcoming nuptials, a family dinner has been arranged at a local restaurant. Your state of residence permits smoking in closed areas of restaurants. After seating, you take note that most of your beloved's family are cigarette smokers. Following dinner, you learn from a friend that your fiancée has only recently quit smoking. What would you do? Specify a plan of action; including a plan of no action should that be your choice.

18. Using internet resources conduct an analysis of the degree of success associated with the WHO Framework Convention on Tobacco Control. Be specific and provide details on the program's accomplishments as well as missed opportunities to reduce the global use of tobacco products.

19. The U.S. has never ratified WHO's Framework Convention on Tobacco Control, an international treaty that requires approval by the U.S. Senate. Conduct an analysis of why this treaty has never been submitted to the Senate for consideration. Further, opine on whether the treaty merits approval by the U.S. and give reasons for your decision.

CHAPTER 6: TOXIC SUBSTANCES IN THE ENVIRONMENT

1. Concerning the practical significance of the Food Quality Protection Act: (1) Using EPA resources, ascertain the Act's impact on that agency's children's health program; (2) In your opinion, should the act have repealed the Delaney Clause, formerly a component (and embedded policy) of the FDCAct, as amended? Why?

2. The FIFRAct requires EPA to regulate the sale and use of pesticides in the U.S. of products known as "restricted-use pesticides," which are those assessed by EPA to be dangerous to the applicator or to the environment. Using EPA resources identify such a pesticide and discuss why it was classified for restricted use. What special precautions were developed for the pesticide's use and application?

3. Using internet resources of the responsible federal agencies develop a summary of the programs, policies, and progress that comply with Title X of the Housing and Community Development Act of 1992.

4. The TSCAct, as amended, divides chemicals into two broad categories: existing and new. Discuss EPA's regulatory responsibilities and regulatory policies for each category.

5. The TSCAct, Section 7, provides EPA with the authority to take emergency action through the federal district courts in order to control a chemical substance or mixture that presents an imminent and unreasonable risk of serious widespread injury to human health or the environment. Discuss why EPA must work through a court in order to interdict an "imminent" hazard.

6. Assume that you work in a county health department. The county has become infested with mosquitoes, raising anxiety in the public that mosquito-borne diseases could result. Your department decides to use Malathion, a pesticide, to periodically spray those areas known to have high concentrations of mosquitoes. You are assigned the task of informing the public of the department's plans. What do you say to the public?

7. Why should the government require that pesticides be registered?

8. What are trade secrets and how do they relate to the FIFRAct?

9. Examine the warning label on a commercially available pesticide. Discuss its content in the context of personal and public health.

10. Section 26 of the TSCAct permits EPA to impose regulatory controls on categories of chemicals, not just individual chemicals. Discuss the advantages to public health of regulating categories of chemicals.

11. The National Toxicology Program coordinates the preparation of biennial reports to Congress on the subject of chemical carcinogens (RoC). (1) Using internet resources access the most current RoC and identify five listed carcinogens of interest to you. Discuss the RoC's characterization of each of the five. (2) As a public health specialist, what do you consider the public value of the RoC.

12. The Lautenberg Chemical Safety for the 21st century Act is touted by chemical industry representatives and some environmental organizations as an improvement to the TSCAct, as amended. Do you agree? If so, why? If not, why not?

13. Access the TEDX website (*www.TEDX.org*) and prepare an essay of appropriate depth on the public health significance of endocrine disruptors. Be specific and provide references to any material you have cited in your essay. Further, assume that your essay is to be orally presented to a community grassroots environmental group.

14. Using internet and other resources, discuss the EU's REACH policy and program. List the most significant provisions of the REACH program that in your opinion are protective of public health.

15. As the senior public health officer in your local health department, a community group of concerned parents of young children has asked to meet with you in regard to a new issue of concern to them. They have heard via social media about something called "obesogens" and are seeking your advice

on whether the local school district's food program is serving obesogens to their children. Will you meet with them? If so, how will you prepare for the meeting? What will you tell them? Will you involve the school district's food administrators?

16. Congratulations! You have been accepted as a summer intern at a local food distribution company. On your first day of work, you observe a commercial pesticide company is spraying the company's food storage warehouse. As a person well versed in the requirements of the FIFRAct, do you take any action? If so, what? If not, why? Be specific.

17. What is your personal opinion about organic food? Do you purchase organics? If so, why? If not, why not? Be specific.

18. The local school district has outreached to your local public health department for advice on assessing any health risk to children who drink water from school water fountains. Knowing of your expertise in environmental health policies, the department's director assigns you to respond to the school district. What will you do? Be specific.

19. Your cousin's young children are under the care of a senior pediatrician at a local medical practice. A routine blood assay indicated the presence of a minute amount of blood lead in both children. The pediatrician says not to worry because it's only a trace indicator of lead exposure. As a public health specialist, what advice would you provide your cousin? Be specific and cite any references to material that would support your advice.

Index

Printed in the United States
by Baker & Taylor Publisher Services